Lecture Notes in Artificial Intelligence 5998

Edited by R. Goebel, J. Siekmann, and W. Wahlster

Subseries of Lecture Notes in Computer Science

T0223405

Friedhelm Schwenker   Neamat El Gayar (Eds.)

Artificial Neural Networks in Pattern Recognition

4th IAPR TC3 Workshop, ANNPR 2010
Cairo, Egypt, April 11-13, 2010
Proceedings

 Springer

Series Editors

Randy Goebel, University of Alberta, Edmonton, Canada
Jörg Siekmann, University of Saarland, Saarbrücken, Germany
Wolfgang Wahlster, DFKI and University of Saarland, Saarbrücken, Germany

Volume Editors

Friedhelm Schwenker
University of Ulm, Department of Neural Information Processing
Oberer Eselsberg, 89069, Ulm, Germany
E-mail: friedhelm.schwenker@uni-ulm.de

Neamat El Gayar
Nile University, Department of Information Technology
Smart Village - B115, Alex Dessert Road, 12677 Giza (Cairo), Egypt
E-mail: nelgayar@nileuniversity.edu.eg

Library of Congress Control Number: 2010922360

CR Subject Classification (1998): I.2, I.4, H.3, I.5, J.3, F.1

LNCS Sublibrary: SL 7 – Artificial Intelligence

ISSN	0302-9743
ISBN-10	3-642-12158-6 Springer Berlin Heidelberg New York
ISBN-13	978-3-642-12158-6 Springer Berlin Heidelberg New York

This work is subject to copyright. All rights are reserved, whether the whole or part of the material is concerned, specifically the rights of translation, reprinting, re-use of illustrations, recitation, broadcasting, reproduction on microfilms or in any other way, and storage in data banks. Duplication of this publication or parts thereof is permitted only under the provisions of the German Copyright Law of September 9, 1965, in its current version, and permission for use must always be obtained from Springer. Violations are liable to prosecution under the German Copyright Law.

springer.com

© Springer-Verlag Berlin Heidelberg 2010
Printed in Germany

Typesetting: Camera-ready by author, data conversion by Scientific Publishing Services, Chennai, India
Printed on acid-free paper   06/3180

Preface

The 4th IAPR TC3 Workshop on Artificial Neural Networks in Pattern Recognition, ANNPR 2010, was held at Nile University (Egypt), April 11–13, 2010. The workshop was organized by the Technical Committee on Neural Networks and Computational Intelligence (TC3) that is one of the 20 technical committees (TC) of the International Association for Pattern Recognition (IAPR). The scope of TC3 includes computational intelligence approaches, such as fuzzy systems, evolutionary computing and artificial neural networks and their use in various pattern recognition applications.

The major topics of ANNPR are supervised and unsupervised learning, feature selection, pattern recognition in signal and image processing, and applications in data mining or bioinformatics. High quality across such a diverse field of research is achieved through a rigorous and selective review process. For this workshop, 42 papers were submitted and 23 of them were selected for inclusion in the proceedings. The workshop was enriched by three invited talks given by Barbara Hammer, University of Bielefeld, Germany, Amir F. Atiya, Cairo University, Egypt, and Mohamed Kamel, University of Waterloo, Canada.

We would like to thank all authors for the effort they put into their submissions, and the Scientific Committee for taking the time to provide high-quality reviews and selecting the best contributions for the final workshop program. Special thanks are due to the members of the Nile University Organizing Committee, Ahmed Salah, Amira El Baroudy, Esraa Aly, Heba Ezzat, Nesrine Sameh, Rana Salah and Mohamed Zahhar for their indispensable contributions to the registration management and local organization.

ANNPR 2010 was supported by the International Association for Pattern Recognition, the Center for Informatic Sciences at the Nile University, Egypt, and the Institute of Neural Information Processing at the University of Ulm. We wish to express our appreciation to our two financial sponsors: The Information Technology Industry Development Agency (ITIDA), which is the main sponsor of the event, and the Microsoft Innovation Laboratory in Cairo (CMIC). We are grateful to Springer for publishing the ANNPR 2010 proceedings in their LNCS/LNAI series.

January 2010

Neamat El Gayar
Friedhelm Schwenker

Organization

Organizing Committee

Friedhelm Schwenker   University of Ulm, Germany
Neamat El Gayar   Nile University, Cairo, Egypt

Program Committee

Shigeo Abe	José Manuel Inesta	Lionel Prevost
Amir Atiya	Rudolf Kruse	Fabio Roli
Monica Bianchini	Miroslav Kubat	Edmondo Trentin
Horst Bunke	Cheng-Lin Liu	Michel Verleysen
Hisham El Shishiny	Marco Maggini	Terry Windeatt
Markus Hagenbuchner	Simone Marinai	
Barbara Hammer	Erkki Oja	
Tom Heskes	Günther Palm	

Sponsoring Institutions

Nile University, Egypt
The Information Technology Industry Development Agency (ITIDA)
Microsoft Innovation Laboratory in Cairo (CMIC)
University of Ulm, Germany
International Association for Pattern Recognition (IAPR)

Table of Contents

Supervised Learning

Unsupervised Learning

Visual Pattern Recognition

Applications

Invited Talk

Pattern Classification Using a Penalized Likelihood Method

Ahmed Al-Ani[1] and Amir F. Atiya[2]

[1] Faculty of Engineering and Information Technology, Univesity of Technology,
Sydney, Australia
ahmed@eng.uts.edu.au
[2] Department of Computer Engineering, Cairo University, Giza, Egypt
amir@alumni.caltech.edu

Abstract. Penalized likelihood is a well-known theoretically justified approach that has recently attracted attention by the machine learning society. The objective function of the Penalized likelihood consists of the log likelihood of the data minus some term penalizing non-smooth solutions. Subsequently, maximizing this objective function would lead to some sort of trade-off between the faithfulness and the smoothness of the fit. There has been a lot of research to utilize penalized likelihood in regression, however, it is still to be thoroughly investigated in the pattern classification domain. We propose to use a penalty term based on the K-nearest neighbors and an iterative approach to estimate the posterior probabilities. In addition, instead of fixing the value of K for all pattern, we developed a variable K approach, where the number of neighbors can vary from one sample to another. The chosen value of K for a given testing sample is influenced by the K values of its surrounding training samples as well as the most successful K value of all training samples. Comparison with a number of well-known classification methods proved the potential of the proposed method.

1   Introduction

The basic concept behind penalized likelihood is that a good model should possesses two indispensable properties: the goodness of fit and the smoothness of the fit [1], [2]. However, as these two are primarily conflicting goals, a trade-off that suits the given application is pursued. The penalized likelihood approach seeks to achieve that trade-off by defining an overall objective function consisting of the log-likelihood of the data minus a roughness measure, and subsequently maximizing this objective function. The likelihood function is a measure of the faithfulness of the fit, while the roughness function is a penalty term that penalizes non-smooth solutions.

An example of the roughness function is the integral of the square of the second derivative of the function, leading to the following objective function (see [3]):

$$T = \log \text{ likelihood } - \lambda \int f''^2(x)dx \tag{1}$$

F. Schwenker and N. El Gayar (Eds.): ANNPR 2010, LNAI 5998, pp. 1–12, 2010.
© Springer-Verlag Berlin Heidelberg 2010

One example of a penalized likelihood regression is the well-known regression spline model [4]. Most of the penalized regression work focused on finding a complete functional formulation and the optimization is performed mostly in the Hilbert space [5].

For the classification problem the underlying function would be the class posterior probabilities. These are the functions which we attempt to estimate and for which we impose smoothness. Among the works considering penalized likelihood classification is the work of O'Sullivan et al [6], which was subsequently analyzed and extended in many other studies [7], [8], [5], [9], [10]. The basic idea of these approaches is to assume that the class posterior probability (considering a two-class case with classes C_1 and C_2) is modeled as a logit function applied to some (unrestricted) function. This is a mean to enforce the $[0, 1]$ bound on the posterior probability. In some of these works thin-plate spline is used as smoothness penalty, and in some others general smoothness penalties are used with the help of the theory of reproducing kernel hilbert spaces. The problem could be solved through a parametric representation, whose parameters are obtained through Newton-Raphson iteration. A related approach is to consider the logistic regression problem (which is essentially a two-class classification problem) in the framework of penalized likelihood regression (see [11] and see also the generalization to the multinomial logistic regression case in [12]), or the generalized additive model [13] (which also tackles in some way the penalized logistic regression problem).

A different methodology based on a Bayesian paradigm is the Gaussian process classification (GPC) approach [19]. While it does not have a penalized likelihood element in it, it enforces smoothness by defining a Bayesian prior that assigns a higher probability to smooth solutions. Again, imposing a logit function lead to intractable integrals that can only be approximated. Another related approach [15] uses the K-nearest neighbor class memberships in some way to describe the priors. It is a Bayesian approach, with the key parameters being attached some priors and these are then integrated out. Again, the integral is intractable and MCMC is proposed as a way to evaluate it.

In this paper we propose a new penalized likelihood classification method for the two-class case. Rather than insisting on evaluating the posterior probability as a functional form (which makes it generally quite difficult), we evaluate it only for the points we need, that is for the training and the testing points. We use as a measure of roughness the sum of square difference between the posterior of a point and that of its K nearest neighbors. We therefore managed to avoid the use of the logit function, which in all above works was an obstacle to obtaining straightforward analytic solutions. We propose an iterative algorithm that converges to the maximum of the penalized likelihood function in few iterations. While we make use of some kind of pattern distance matrix like in the case of Gaussian process classification, the philosophy and the approach is quite different.

We tested the proposed method on a number of UCI benchmark datasets. As it turns out, it produces a classification performance beating many of the well-known methods (such as SVM and several other methods) and comparable to

GPC (it is generally believed that SVM and GPC are among the best two classification approaches [16]). On the other hand the computation time was much less than that of GPC. Another advantage of the method is that it is entirely based on distances between the training patterns (like the K nearest neighbor classifier and the GPC). So it can handle also non-numeric inputs, for example text inputs whereby some distance function can be defined. The proposed method is also very simple, consisting of only a simple iteration, and requiring little development time to implement it and no sophisticated optimization routines.

The paper is organized as follows. The proposed method is presented in the next section. The following section details the classification algorithm. In Section 5 we present the simulations results, followed by the conclusions section.

2 The Proposed Method

Let $x_m \in \mathcal{R}^L$ denote the feature vectors, with $x_1, \ldots, x_M$ denoting the training patterns, and $x_{M+1}, \ldots, x_{M+N}$ denoting the test patterns. In this work we consider only the two-class case. The class membership y_{gm} for class label g and training pattern x_m is defined as follows: it equals 1 if $x_m \in C_g$ and equals 0 otherwise, where $g \in \{1, 2\}$.

Let $P_{gm} \equiv P(C_g | x_m)$ denote the posterior probability for class C_g, and $\sum_{g=1}^{2} P_{gm} = 1$. The purpose of the proposed method is to estimate the posterior probabilities P_{gm}, both for the training set and the test set. Knowing the posterior probabilities will automatically determine the classification of the patterns. As we will shortly see, the posterior probabilities are obtained by defining the penalized likelihood function and subsequently maximizing it, leading to an iterative algorithm.

The likelihood of the data is given by

$$L = \prod_{m=1}^{M} \prod_{g=1}^{2} P_{gm}^{y_{gm}} \tag{2}$$

Denote by $\mathcal{K}(x_m)$ as the set of K_m-nearest neighbors of point x_m (their indexes), and K_m the size of $\mathcal{K}(x_m)$, which can vary from one pattern to another. We define a roughness function based on the square differences of the posteriors of neighboring data points. Specifically, it is given by

$$R = \sum_{m=1}^{M} \frac{1}{K_m} \sum_{m' \in \mathcal{K}(x_m)} (P_{hm} - P_{hm'})^2 \tag{3}$$

where h is the class that x_m belongs to (either 1 or 2), hence, $y_{hm} = 1$. Note that Eq. 2 can be written as $L = \prod_{m=1}^{M} P_{hm}$. We define our overall objective function as a combination of the log-likelihood function and the roughness function:

$$J = \log(L) - \lambda R \tag{4}$$

$$= \sum_{m=1}^{M} \log(P_{hm}) - \lambda \sum_{m=1}^{M} \frac{1}{K_m} \sum_{m' \in \mathcal{K}(x_m)} (P_{hm} - P_{hm'})^2 \tag{5}$$

The first term in the penalized log-likelihood J focuses on the goodness of fit aspect. It gauges how well that the considered P_{hm}'s fit the observed data (i.e. the given class memberships). The second term serves to penalize the roughness of the underlying posterior function. A posterior surface where its values for neighboring points are close (i.e. having low R) will generally be smooth, and conversely a high R is indicative of a rough or wiggly surface. The goal is to find the posterior probabilities that maximize the penalized log-likelihood J. We will therefore achieve a compromise between faithfully respecting the class memberships of the training data and the smoothness property of the posterior surface, with λ being the parameter that controls the degree of smoothness.

3   The Proposed Algorithm

The goal is to solve the following maximization problem:

Maximize J (given by (5)) w.r.t. the variables: P_{gm}, s.t. $0 \leq P_{gm} \leq 1, m = 1, \ldots, M, g = \{1, 2\}$.

It is easy to see that J is a convex function w.r.t. the P_{gm}'s. Hence the problem has a unique maximum. The algorithm proposed below is based on cycling through all variables, each time optimizing w.r.t. only one of the variables (through a line search). In each step, the optimum w.r.t. one variable can be obtained analytically, as show below.

$$\frac{\partial J}{\partial P_{hm}} = \frac{1}{P_{hm}} - \frac{2\lambda}{K_m} \left[\sum_{m' \in \mathcal{K}(x_m)} (P_{hm} - P_{hm'}) + \sum_{m' \in \mathcal{S}(x_m)} (P_{hm} - P_{hm'}) \right] \quad (6)$$

$$= \frac{1}{P_{hm}} - \frac{2\lambda(K_m + S_m)}{K_m} P_{hm} + \frac{2\lambda}{K_m} \left[\sum_{m' \in \mathcal{K}(x_m)} P_{hm'} + \sum_{m' \in \mathcal{S}(x_m)} P_{hm'} \right] \quad (7)$$

where $\mathcal{S}(x_m)$ is the set of patterns for which x_m is one of the neighbors, S_m is the length of $\mathcal{S}(x_m)$. To find the value of P_{hm} that maximizes J, we need to make the right hand side of Eq. 7 equals 0, which would lead to

$$1 - \left(\frac{2\lambda(K_m + S_m)}{K_m} \right) P_{hm}^2 + \left(\frac{2\lambda}{K_m} \left[\sum_{m' \in \mathcal{K}(x_m)} P_{hm'} + \sum_{m' \in \mathcal{S}(x_m)} P_{hm'} \right] \right) P_{hm} = 0 \quad (8)$$

$$P_{hm}^2 - \left(\frac{1}{K_m + S_m} \left[\sum_{m' \in \mathcal{K}(x_m)} P_{hm'} + \sum_{m' \in \mathcal{S}(x_m)} P_{hm'} \right] \right) P_{hm} - \frac{K_m}{2\lambda(K_m + S_m)} = 0 \quad (9)$$

Eq. 9 is a quadratic equation that can easily be solved. The algorithm of the proposed method is given below.

1. Start with any initial choice e.g. $P_{gm} = 0.5$, $m = 1, \ldots, M$, $g =\in \{1, 2\}$.
2. While the change in the posteriors between the current and previous iteration is greater than a certain threshold $(Thresh)$, execute step 3.
3. For each training pattern, $m = 1$ to M:
   (a) Set:

$$P_{hm} \equiv \frac{1}{2} \bar{P}_{hm} + \frac{1}{2} \sqrt{(\bar{P}_{hm})^2 + \frac{2K_m}{\lambda(K_m + S_m)}} \tag{10}$$

   where

$$\bar{P}_{hm} = \frac{1}{K_m + S_m} \left[\sum_{m' \in \mathcal{K}(x_m)} P_{hm'} + \sum_{m' \in \mathcal{S}(x_m)} P_{hm'} \right] \tag{11}$$

   where K_m is the number of nearest neighbors, $\mathcal{S}(x_m)$ is the set of data points for which x_m is one of the nearest neighbors, and S_m is the size of set $\mathcal{S}(x_m)$. Thus $\bar{P}_{hm}$ is the mean of the values of $P_{hm'}$ for some sort of neighborhood of points around x_m.
   (b) Truncate if P_{hm} goes out of the constraint box, set:

$$P_{hm} = 1 \quad \text{if } P_{hm} > 1, \quad \text{or} \quad P_{hm} = 0 \quad \text{if } P_{hm} < 0 \tag{12}$$

   (c) Let $f \neq h, f \in \{1, 2\}$, set:

$$P_{fm} = 1 - P_{hm} \tag{13}$$

   (d) Let the set of possible K-nearest neighbor values $\mathcal{KK} = \{k_1, k_2, \ldots, k_n\}$, calculate the error associated with using k_j-nearest neighbors, $j = 1 : n$, and set K_m to the value that minimizes the error, E_{mj}, as follows

$$E_{mj} = 1 - \text{mean}(P_{h\mathcal{KK}_j(x_m)})$$
$$K_m = \arg \min_j (E_{mj}) \tag{14}$$

   where $\mathcal{KK}_j(x_m)$ is the set of k_j-nearest neighbors of x_m. One possible choice is to use $\mathcal{KK} = \{3, 5, 7, \ldots, 25\}$.
4. For each test pattern, $m = M + 1$ to $M + N$:
   (a) Find the value of K_m:

$$K_m = \begin{cases} \arg \min_j (\text{mean}(E_{\mathcal{Q}_m j})) & \text{if } \min (\text{mean}(E_{\mathcal{Q}_m j})) < T_E \\ \arg \min_j (\text{mean}(E_{\mathcal{M}j})) & \text{otherwise} \end{cases} \tag{15}$$

   where $\mathcal{Q}_m$ represents the set of training patterns that surrounds x_m (local neighborhood), while $\mathcal{M}$ represents the set of all training patterns. T_E is a threshold, which can be the mean of $E_{\mathcal{M}j}$.
   (b) Calculate the posterior probabilities $P_{gm}, g \in \{1, 2\}$:

$$P_{gm} = \text{mean}(P_{g\mathcal{K}(x_m)}) \tag{16}$$

This algorithm performs an iterated estimation of the posteriors through Eq. 10, which is basically the closed-form outcome of the one-variable search that is performed by cycling through all variables. Eq. 14 shows how the value of K_m can vary from one pattern to another. This can lead to a higher value of J when compared to using a fixed K for all patterns. The iterations should carry on until the change in the posteriors from one cycle till the next is small. Once the algorithm converges, we use the obtained final values of the P_{gm}'s as the estimated posteriors of training data points, which will be used to estimate the posteriors of the test data points. The rationale behind Eq. 15 is to first check the error of the local neighborhood. If the error is small, then the neighborhood would influence the choice of K_m. Otherwise, use the value of K that, on average, is most reliable over all training samples. Recalling that $P_{gm} \equiv P(C_g|x_m)$, then the final classification of a test data point is estimated as class C_1 if $P_{1m} > P_{2m}$, otherwise it is class C_2.

4   Simulation Results

A number of benchmark datasets were used to test the performance of the proposed method. We have compared the performance of the proposed method to that of the following well-known classification methods:

– Bayes classifier ([17], p. 168) with the class-conditional densities estimated according to the Parzen window density estimator (PARZEN) [18]. A key parameter for the Parzen estimator is the width of kernels h. We used the value derived in [18] (Silverman's rule):

$$ h = \hat{\sigma} \left[\frac{4}{(2L + 1)I} \right]^{\frac{1}{L+4}} \tag{17} $$

where $\hat{\sigma}^2 \equiv \sum_{i=1}^{L} S_{ii}/L$ denotes the mean of the diagonal of the sample covariance matrix S, L is the dimension of the space, and I is the number of data points (we used Gaussian kernels).

– Gaussian process classification using the expectation propagation approximation [19]. We used the non-optimized (GPC) and optimized (GPCo) versions. The latter attempts to approximate the integrals in the Gaussian process classification formula. We used the software available in [14].

– Support vector machines (SVM) (Scholkopf and Smola [20]). We used a radial basis function SVM implemented using the OSUsvm toolbox[1]. The values of C and γ for the latter are set using a K-fold validation procedure (we used five-fold validation and allowed C and γ to range between [0.5, 1.5]).

– K-nearest neighbor classifier. The value of K was set using a five-fold validation process (only odd numbers that range between 3 and 25 were considered).

– Evidential K-nearest neighbor (KNNds). This algorithm is based on the Dempster-Shafer theory of evidence taking into account the distance and class label information of the neighbors for generating soft decision vectors [21][2].

---

[1] Obtained from http://downloads.sourceforge.net/svm/osu-svm-3.0.zip

[2] The KNNds software is available at http://www.hds.utc.fr/ tdenoeux/software.htm

We tested all these competing methods on real-world pattern classification problems, mostly from the UCI repository [22]. We also tested those algorithms on the well-known two-spiral classification problem. This dataset consists of points on two inter-wined spirals that cannot be linearly separated, as shown in Fig. 1. Table 1 summarizes the characteristics of the datasets used in this paper.

Patterns that consist of missing values were removed from the datasets. In certain cases, attributes that consist of many missing values were excluded to minimize the number of removed patterns. Categorical attributes were changed to attributes with integer values to enable the chosen algorithms to handle them. For all considered problems the input attributes are first scaled so that they lie in a suitable range. We used 80% of the data as a training set, and the remaining 20% as a test set. We performed 20 runs for each method, each run with a different random train/test partition. Then we average the classification accuracies on the test sets of the 20 runs.

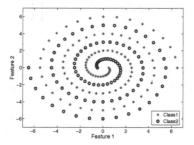

Fig. 1. Two-spiral dataset

Table 1. Datasets used to evaluate the performance of classifiers

Dataset	# Attributes	# Patterns	Class distribution
Two Spiral	2	194	0.50/0.50
Ger. Credit	24	1000	0.70/0.30
Cylinder bands	30	350	0.62/0.38
Blood transfusion	4	748	0.24/0.76
cancer	9	683	0.65/0.35
Haberman's survival	3	306	0.73/0.27
heart	22	267	0.79/0.21
heart SPECT	13	270	0.55/0.45
hill-valley	100	606	0.51/0.49
ionosphere	33	351	0.64/0.36
mammographic	5	814	0.48/0.52
monk	6	432	0.50/0.50
Parkinson	22	195	0.75/0.25
pima	8	768	0.35/0.65
sonar	60	208	0.53/0.47
Tic-tac	9	958	0.65/0.35
wdbc	30	569	0.63/0.37

In order to compare the performance of the various algorithms mentioned above, we used the following measures:

- **Mean classification accuracy (Acc).** This measure gives a general indication about the performance of each classifier.
- **Estimated standard deviation of the accuracy.** It is calculated by dividing the standard deviation of Acc by the square root of the number of runs.
- **Significance test.** A two-tailed paired t-test is performed with significance level of $\alpha = 0.05$. This indicates if there is a significant difference in the performance of two classifiers.
- **Geometric mean error ratio.** For the two classifiers that have errors $a_1, a_2 \ldots, a_n$ and $b_1, b_2 \ldots, b_n$ respectively (n represents the number of runs), the geometric error ratio is:

$$\exp \frac{\sum_{i=1}^{n} \log(ai/bi)}{n} = \sqrt[n]{\prod_{i=1}^{n} ai/bi} \tag{18}$$

This measure reflects the relative performance of one classifier with respect to another. If the outcome is less than 1, then it is an indication that the first classifier outperforms the second classifier in terms of error reduction.
- **Win-Tie-Loss.** This is an important measure, where the three values are the number of datasets for which classifier a obtained better, equal, or worse performance outcomes than classifier b.
- **Sign test.** The p-values of a two-tailed sign test based on the win-tie-loss record. if p is significantly low, then one can conclude that it is unlikely that the outcome was obtained by chance, i.e., the difference between the two classifiers is significant. On the other hand, a higher p value indicates that the two classifiers are not significantly different.

For detailed description of these measures the reader is referred to [23], [24].

The PLC algorithm described in section 3 only needed few iterations to converge for all of the 17 datasets when we set $Thresh$ to 0.01. Table 2 shows the average classification accuracy of the competing methods with the estimated standard deviation. It also shows if (PLC) is significantly different from other classifiers from a statistical viewpoint. For a given dataset, if PLC is significantly better than a certain classifier, then a bullet is displayed next to that classifier's result. On the other hand, an open circle indicates that the classifier is significantly better than PLC. A quick glance at the table would show that there are more bullets than open circles. PLC is found to be particularly better than Parzen, GMM, KNNds and KNN. However, the results indicate that PLC is not significantly better than the Gaussian process and SVM, particularly GPCo. As mentioned earlier, these two classifiers are considered in the literature to be among the best classification approaches.

In order to present a more detailed analysis of the classification results, Table 3 presents other comparison measures. The first row of the table represents the

Table 2. Classification accuracy and estimated standard deviation for the considered classifiers

	Parzen	GPC	GPCo	SVM
2Spiral	33.16 ± 1.31•	47.63 ± 1.51•	50.39 ± 0.29•	47.63 ± 1.61 •
GerCr	68.75 ± 0.65•	76.78 ± 0.57∘	77.50 ± 0.58∘	75.90 ± 0.46∘
bands	68.57 ± 0.84	74.00 ± 1.41∘	75.00 ± 1.29∘	73.14 ± 1.22∘
Btrans	76.97 ± 0.39	76.73 ± 0.36	77.7 ± 0.41	75.97 ± 0.15•
canc	95.84 ± 0.36•	96.82 ± 0.23	96.57 ± 0.32	96.57 ± 0.23
haber	71.48 ± 0.59•	72.87 ± 0.70	73.28 ± 0.77	72.54 ± 0.50
heart	80.19 ± 1.31•	83.96 ± 0.86∘	84.15 ± 1.00	83.68 ± 0.95
heartS	80.19 ± 0.87	83.24 ± 0.83∘	83.70 ± 1.00∘	83.06 ± 0.89∘
hill	53.51 ± 0.74•	50.45 ± 0.58•	51.78 ± 0.51•	49.96 ± 0.57•
ion	88.21 ± 0.8∘	89.57 ± 0.65∘	96.07 ± 0.52∘	94.29 ± 0.55∘
mamm	79.20 ± 0.67	80.68 ± 0.58∘	82.01 ± 0.56∘	80.19 ± 0.62
monk	76.86 ± 0.95•	80.99 ± 1.00	94.65 ± 0.82∘	91.10 ± 0.49∘
parkin	95.00 ± 0.71	83.72 ± 1.10•	93.08 ± 1.10	87.82 ± 1.08•
pima	73.21 ± 0.50	76.33 ± 0.61∘	76.79 ± 0.61∘	76.66 ± 0.57∘
sonar	87.32 ± 1.14	84.63 ± 1.30•	83.66 ± 1.41•	86.34 ± 1.33
tic-tac	87.49 ± 0.38	83.48 ± 0.70•	96.18 ± 0.36∘	95.71 ± 0.29∘
wdbc	97.52 ± 0.31	97.65 ± 0.31	97.26 ± 0.28	97.92 ± 0.27∘

	KNN	KNNds	PLC
2Spiral	75.13 ± 2.10•	75.13 ± 2.10•	79.61 ± 1.09
GerCr	72.48 ± 0.34•	72.15 ± 0.74•	73.88 ± 0.40
bands	68.36 ± 1.30	67.57 ± 0.99	68.93 ± 1.14
Btrans	77.90 ± 0.45	73.27 ± 0.72•	77.5 ± 0.58
canc	96.64 ± 0.31	96.61 ± 0.27	96.68 ± 0.29
haber	73.11 ± 0.68	70.33 ± 1.01•	72.87 ± 0.69
heart	81.79 ± 1.17	80.28 ± 1.43	82.36 ± 1.13
heartS	81.76 ± 1.03	77.78 ± 1.07•	81.48 ± 0.92
hill	54.38 ± 0.77•	55.95 ± 1.12•	59.09 ± 1.00
ion	84.71 ± 0.69•	89.86 ± 0.74∘	86.36 ± 0.68
mamm	79.26 ± 0.70	77.10 ± 0.47•	79.66 ± 0.62
monk	83.72 ± 0.87	83.95 ± 0.80∘	82.50 ± 0.87
parkin	92.56 ± 1.03	92.69 ± 1.04	93.85 ± 1.08
pima	73.64 ± 0.67	73.99 ± 0.47	74.09 ± 0.58
sonar	83.66 ± 1.05•	83.78 ± 1.01•	87.56 ± 1.20
tic-tac	88.04 ± 0.48∘	82.64 ± 0.48•	86.86 ± 0.57
wdbc	96.81 ± 0.31	96.99 ± 0.30	97.17 ± 0.26

mean accuracy across all the datasets. According to this measure PLC is found to be the second best classifier, after GPCo, outperforming all remaining classifiers, including SVM. The table also presents pair-wise comparisons between the classifiers according to their geometric error ratio ($\dot{r}$), and the win-tie-loss (s). Also shown is the p-value of the sign test for the win-tie-loss (p). According to these measures, PLC outperformed Parzen, KNN, and KNNds. In fact the geometric error ratio indicates that PLC is slightly better than GPC and not too different from SVM. On the other hand, the win-tie-loss favors the Gaussian

Table 3. Comparison of averaged classification accuracy, geometric error, win-tie-loss, and p-value of the sign test across all the used datasets

	Parzen	GPC	GPCo	SVM	KNN	KNNds	PLC
Mean Acc.	77.26	78.80	81.75	80.50	80.23	79.42	81.20
Parzen							
$\dot{r}$		1.016	1.333	1.212	1.052	1.020	1.112
s		5-0-12	4-0-13	4-0-13	5-0-12	9-0-8	4-0-13
p		0.1435	0.049	0.049	0.1435	1.000	0.049
GPC							
$\dot{r}$			1.311	1.193	1.035	1.003	1.094
s			3-0-14	9-1-7	10-0-7	12-0-5	9-1-7
p			0.0127	0.8036	0.6291	0.1435	0.8036
GPCo							
$\dot{r}$				0.910	0.789	0.765	0.834
s				14-0-3	12-1-4	13-0-4	12-0-5
p				0.0127	0.0768	0.049	0.1435
SVM							
$\dot{r}$					0.868	0.841	0.917
s					11-0-6	13-0-4	10-0-7
p					0.3323	0.049	0.6291
KNN							
$\dot{r}$						0.969	1.057
s						9-1-7	5-0-12
p						0.8036	0.1435
KNNds							
$\dot{r}$							1.090
s							2-0-15
p							0.0023

Table 4. Execution Time for GPCo and PLC, measured in CPU time (sec). This time includes training time and testing time.

	2Spiral	GerCred	bands	Btrans	cancer	haber	heart	heartS	hill
GPCo	1.63	882.85	22.77	360.82	314.43	11.75	9.58	7.63	82.67
PLC	0.19	2.73	0.56	1.40	1.43	0.36	0.35	0.31	6.06

	ion	mamm	monk	parkin	pima	sonar	tic-tac	wdbc
GPCo	59.81	530.17	106.57	9.31	469.99	10.39	1229.33	238.80
PLC	1.09	3.73	1.25	0.38	3.67	0.62	5.55	2.80

process and support vector machine over PLC. However, as seen from the p-value measure, only GPCo is significantly better than PLC.

The above results indicate that GPCo is the only classifier that is significantly better from a statistical point of view. So, it would be important to compare these two classifiers in terms of computational complexity. Table 4 shows the computation time of both GPCo and PLC for all considered datasets. The table indicates that PLC is considerably faster than GPCo, which represents a great advantage for the proposed algorithm.

5 Conclusion

A new classification method based on penalized likelihood concept is presented in this paper. The method is based on defining a roughness term based on the K-nearest neighbors. We have developed an algorithm that converges to the global optimum in only few iterations. We have also proposed to allow the value of K to vary from one pattern to another, which proved to be useful in maximizing the objective function. When compared with several well-known classification methods, the proposed classifier achieved a performance competitive with the top models, but with less computational time. As such, the proposed approach can be ranked among the top binary classification algorithms.

References

1. Green, P.: Penalized likelihood. In: Encyclopedia of Statistical Sciences, Update vol. 3 (1999)
2. Gu, C., Kim, Y.-J.: Penalized likelihood regression: general formulation and efficient approximation. Canadian Journal of Statistics 29 (2002)
3. Green, P.J., Silverman, B.W.: Nonparametric Regression and Generalized Linear Models: a Roughness Penalty Approach. Chapman and Hall, London (1994)
4. Berry, S.M., Carroll, R.J., Ruppert, D.: Bayesian smoothing and regression splines for measurement error problems. J. Amer. Statist. Assoc. 97, 160–169 (2002)
5. Wahba, G.: Spline Models for Observational Data. SIAM, Philadelphia (1990)
6. OSullivan, F., Yandell, B., Raynor, W.: Automatic smoothing of regression functions in generalized linear models. J. Amer. Statist, Assoc. 81, 96–103 (1986)
7. Gu, C.: Cross-validating non-gaussian data. J. Comput. Graph. Statist. 1, 169–179 (1992)
8. Lu, F., Hill, G.C., Wahba, G., Desiati, P.: Signal probability estimation with penalized likelihood method on weighted data, Technical Report, No. 1106. Department of Statistics, University of Wisconsin (2005)
9. Wahba, G.: Soft and hard classification by reproducing kernel hilbert space methods. Proc. Nat. Acad. Sciences 99, 16524–16530 (2002)
10. Wahba, G., Gu, C., Wang, Y., Chappell, R.: Soft classification, a.k.a. risk estimation, via penalized log likelihood and smoothing spline analysis of variance, Technical Report, No. 899. Department of Statistics, University of Wisconsin (1993)
11. Loader, C.: Local Regression and Likelihood. Springer, Heidelberg (1999)
12. Cawley, G., Talbot, N.L., Girolami, M.: Sparse multinomial logistic regression via bayesian l1 regularisation. In: Proceedings NIPS, pp. 209–216 (2007)
13. Hastie, T., Tibshirani, R.: Generalized Additive Models. Chapman and Hall, Boca Raton (1990)
14. Rasmussen, C.E. (2007), http://www.GaussianProcess.org/gpml/code/index.html
15. Holmes, C.C., Adams, N.M.: A probabilistic nearest neighbour method for statistical pattern recognition. Journal Royal Statistical Society B 64, 295–306 (2002)

16. Jensen, R., Erdogmus, D., Principe, J.C., Eltoft, T.: The laplacian classifier. IEEE Trans. Signal Processing 55, 3262–3271 (2007)
17. Duda, R.O., Hart, P.E., Stork, D.G.: Pattern Classification, 2nd edn. Wiley Interscience, Hoboken (2000)
18. Silverman, B.W.: Density Estimation for Statistics and Data Analysis. Chapman and Hall, Boca Raton (1986)
19. Rasmussen, C.E., Williams, C.K.I.: Gaussian Processes for Machine Learning (Adaptive Computation and Machine Learning). The MIT Press, Cambridge (2005)
20. Scholkopf, B., Smola, A.J.: Learning with Kernels: Support Vector Machines, Regularization, Optimization, and Beyond. MIT Press, Cambridge (2001)
21. Zouhal, L., Denoeux, T.: An evidence-theoretic k-nn rule with parameter optimization. IEEE Trans. Syst. Man Cyber. 28, 263–271 (1998)
22. Asuncion, D.J.: UCI Machine Learning Repository (2007),
    http://www.ics.uci.edu/~mlearn/MLRepository.html
23. Zhang, C.-X., Zhang, J.-S.: Rotboost: a technique for combining roataion forest and adaboost. Pattern Recognition Letters 29, 1524–1536 (2008)
24. Webb, G.: Multiboosting: a technique for combining boosting and wagging. Machine Learning 40, 159–196 (2000)

Evaluation of Feature Selection by Multiclass Kernel Discriminant Analysis

Tsuneyoshi Ishii and Shigeo Abe

Graduate School of Engineering
Kobe University
Rokkodai, Nada, Kobe, Japan
abe@kobe-u.ac.jp
http://www2.kobe-u.ac.jp/~abe/

Abstract. In this paper, we propose and evaluate the feature selection criterion based on kernel discriminant analysis (KDA) for multiclass problems, which finds the number of classes minus one eigenvectors. The selection criterion is the sum of the objective function of KDA, namely the sum of eigenvalues associated with the eigenvectors. In addition to the KDA criterion, we propose a new selection criterion that replaces the between-class scatter in KDA with the sum of square distances between all pairs of classes. To speed up backward feature selection, we introduce block deletion, which deletes many features at a timeC and to enhance generalization ability of the selected features we use cross-validation as a stopping condition.

By computer experiments using benchmark datasets, we show that the KDA criterion has performance comparable with that of the selection criterion based on the SVM-based recognition rate with cross-validation and can reduce computational cost. We also show that the KDA criterion can terminate feature selection stably using cross-validation as a stopping condition.

1 Introduction

Feature selection is to select from the original set of features the minimum subset of features that realizes the maximum generalization ability. To realize this, during the process of feature selection, the generalization ability of a subset of features needs to be estimated. This type of feature selection is called a wrapper method. Instead of estimating the generalization ability, some selection criterion, which is considered to well reflect the generalization ability, is used. This method is called a filter method.

The forward or backward selection method using a selection criterion is widely used. In backward selection, we start from all the features and delete one feature at a time, which deteriorates the selection criterion the least. We delete features until the selection criterion reaches a specified value. In forward selection, we start from an empty set of features and add one feature at a time, which improves the selection criterion the most. We iterate this procedure until the selection

F. Schwenker and N. El Gayar (Eds.): ANNPR 2010, LNAI 5998, pp. 13–24, 2010.
© Springer-Verlag Berlin Heidelberg 2010

criterion reaches a specified value. Because forward or backward selection is slow, we may add or delete more than one feature at a time based on feature ranking, or we may combine backward and forward selection [1].

Because these selection methods are local optimization techniques, global optimality of feature selection is not guaranteed. Usually, backward selection is slower but is more stable in selecting optimal features than forward selection [2]. If a selection criterion is monotonic for deletion or addition of a feature, we can terminate feature selection when the selection criterion violates a predefined value [3].

By the introduction of support vector machines (SVMs), various selection methods suitable for support vector machines have been developed. The selection criterion for filter methods used in the literature is, except for some cases [4,5,6,7,8], the margin [9,10,11,12]. In addition, in most cases, a linear support vector machine is used.

In [4,6], the objective function of kernel discriminant analysis called the KDA criterion, namely the ratio of the between-class scatter and within-class scatter, is proved to be monotonic for the deletion of features for two-class problems, and feature selection based on the KDA criterion was shown to be robust for benchmark data sets.

As a wrapper method, in [13,14], block deletion of features in backward feature selection is proposed using the generalization ability by cross-validation as the selection criterion.

In addition to filter and wrapper methods, the embedded methods combine training and feature selection; because training of support vector machines results in solving a quadratic optimization problem, feature selection can be done by modifying the objective function [15,16,17].

In this paper we discuss backward feature selection based on KDA proposed in [18]. For an n-class problem, KDA gives $n-1$ projection axes. We use as the selection criterion the sum of the objective function values associated with the eigenvectors, which is equivalent to the sum of eigenvalues. To speedup feature selection we use block deletion of features used in [13,14], which deletes features at the same time that give the larger KDA criterion than the threshold value if each is deleted. To stabilize stopping of feature selection, we use cross-validation for the selected sequence by block deletion. Further, to improve the separability measure of KDA, as the between-class scatter, we propose using the scatter between all the class pairs.

We compare the proposed KDA criterion with the SVM-based criterion with cross-validation and the between-class and within-class ratio and demonstrate usefulness of the proposed criterion from the standpoint of selected features and the computation time.

In Sections 2, we summarize KDA and in Section 3, we discuss selection criteria. In Section 4, we explain feature selection methods. In Section 5 we demonstrate the validity of the proposed methods by computer experiments.

2   Kernel Discriminant Analysis for Multiclass Problems

In this section, we explain kernel discriminant analysis for multiclass problems based on [19].

We assume that the center of mapped training data in the feature space is zero. Then the total scatter matrix Q_T and the between-class scatter matrix Q_B are given, respectively, by

$$Q_T = \frac{1}{M} \sum_{k=1}^{n} \sum_{j=1}^{M_k} \phi(\mathbf{x}_{kj}) \phi^\top(\mathbf{x}_{kj}), \tag{1}$$

$$Q_B = \frac{1}{M} \sum_{k=1}^{n} M_k \mathbf{c}_k \mathbf{c}_k^\top, \tag{2}$$

where $\mathbf{x}_{kj}$ is the jth training data for class k, n is the number of classes, M_k is the number of training data for class k, $M = M_1 + \cdots + M_n$, $\mathbf{c}_k$ is the center of class k, and $\phi(\mathbf{x})$ is the mapping function that maps the input space to the high-dimensional feature space.

For n class problems, we obtain $n - 1$ projection axes. Let them be $\mathbf{w}_i$ ($i = 1, \ldots, n-1$). Then the total scatter and the between-class scatter on this axis are given, respectively, by

$$\frac{1}{M} \sum_{k=1}^{n} \sum_{j=1}^{M_k} (\mathbf{w}_i^\top \phi(\mathbf{x}_{kj}))^2 = \mathbf{w}_i^\top Q_T \mathbf{w}_i, \tag{3}$$

$$\frac{1}{M} \sum_{k=1}^{n} M_k (\mathbf{w}_i^\top \mathbf{c}_k)^2 = \mathbf{w}_i^\top Q_B \mathbf{w}_i. \tag{4}$$

We seek the projection axis $\mathbf{w}_i$ that maximizes the between-class scatter and minimizes the total scatter. Namely,

$$\text{maximize} \quad J(\mathbf{w}_i) = \frac{\mathbf{w}_i^\top Q_B \mathbf{w}_i}{\mathbf{w}_i^\top Q_T \mathbf{w}_i}. \tag{5}$$

Here, $\mathbf{w}_i$ can be expressed by the linear combination of the mapped training data:

$$\mathbf{w}_i = \sum_{k=1}^{n} \sum_{j=1}^{M_k} a_i^{kj} \phi(\mathbf{x}_{kj}), \tag{6}$$

where a_i^{kj} are constants.

Substituting (6) into (5), we obtain

$$J(\mathbf{a}_i) = \frac{\mathbf{a}_i^\top K W K \mathbf{a}_i}{\mathbf{a}_i^\top K K \mathbf{a}_i}, \tag{7}$$

where $\mathbf{a}_i = \{a_i^{kj}\}$ $(i = 1, \ldots, n-1, k = 1, \ldots, n, j = 1, \ldots, M_k)$, K is the kernel matrix, and $W = \{W_{ij}\}$ is the block diagonal matrix given by

$$W_{ij} = \begin{cases} \dfrac{1}{M_k} & \mathbf{x}_i, \mathbf{x}_j \in \text{class } k, \\ 0 & \text{otherwise.} \end{cases} \tag{8}$$

Taking the partial derivative of (7) with respect to $\mathbf{w}_i$, and the resulting equation to 0, we obtain the following generalized eigenvalue problem:

$$KWK\mathbf{a}_i = \lambda_i KK\mathbf{a}_i, \tag{9}$$

where λ_i are eigenvalues.

Let singular value decomposition of K be $K = P\Gamma P^\top$, where Γ is the diagonal matrix with nonzero eigenvalues and $P^\top P = I$. Substituting $K = P\Gamma P^\top$ into (7) and replacing $\Gamma P^\top \mathbf{a}_i$ with $\boldsymbol{\beta}_i$, we obtain

$$J(\boldsymbol{\beta}_i) = \frac{\boldsymbol{\beta}_i^\top P^\top W P \boldsymbol{\beta}_i}{\boldsymbol{\beta}_i^\top P^\top P \boldsymbol{\beta}_i} = \frac{\boldsymbol{\beta}_i^\top P^\top W P \boldsymbol{\beta}_i}{\boldsymbol{\beta}_i^\top \boldsymbol{\beta}_i}. \tag{10}$$

Therefore, the resulting eigenvalue problem is

$$P^\top W P \boldsymbol{\beta}_i = \lambda_i \boldsymbol{\beta}_i. \tag{11}$$

Solving (11) for $\boldsymbol{\beta}_i$ we obtain $\mathbf{a}_i$ from $\mathbf{a}_i = P\Gamma^{-1}\boldsymbol{\beta}_i$.

3   Selection Criteria

3.1   KDA Criterion for Multiclass Problems

The feature selection method based on KDA for two-class problems [6] can be extended to multiclass problems but will be architecture dependent. Therefore, we extend the method to multiclass problems using the multiclass KDA. In multiclass KDA, $n-1$ projection axes are obtained for an n-class problem. We propose using the sum of the objective function values associated with the $n-1$ projection axes. We can easily show that

$$J(\mathbf{w}_i) = \lambda_i. \tag{12}$$

Therefore the selection criterion is

$$\sum_{i=1}^{n-1} J(\mathbf{w}_i) = \sum_{i=1}^{n-1} \lambda_i, \tag{13}$$

where λ_i are given by (11). Because the sum of eigenvalues is the trace of the associated matrix, (13) becomes

$$\sum_{i=1}^{n-1} J(\mathbf{w}_i) = \text{trace}\{P^\top W P\}, \tag{14}$$

which leads to speeding up the calculation of the selection criterion. We call this the KDA criterion.

3.2   New Between-Class Scatter

The between-class scatter for multiclass problems is calculated by the square sum of distances between the center of the mapped training data and class centers $\mathbf{c}_k$. Namely, the between-class scatter does not consider the overlap between classes. Suppose for a multiclass problem in which data of different classes do not overlap, we rotate all the data of some classes around the center of the mapped training data until different classes overlap under the constraint that the center of the mapped data does not move. Then, the between-class scatters of the initial and the rotated problems are the same. But this is unfavorable from the standpoint of class separability.

This problem can be avoided if we use the following between-class scatter:

$$Q_{\mathrm{B}} = \sum_{i=1}^{n-1} \sum_{j=i+1}^{n} (\mathbf{c}_i - \mathbf{c}_j)(\mathbf{c}_i - \mathbf{c}_j)^{\top}. \tag{15}$$

The eigenvalue problem becomes

$$KAK\mathbf{a} = \lambda KK\mathbf{a}, \tag{16}$$

where

$$A_{ij} = \begin{cases} \dfrac{n-1}{M_k^2} & \mathbf{x}_i, \mathbf{x}_j \in \text{class } k, \\[2ex] -\dfrac{1}{M_p M_q} & \mathbf{x}_i \in \text{class } p, \mathbf{x}_j \in \text{class } q, p \neq q. \end{cases} \tag{17}$$

The difference with the previous method is that in (9), W is replaced with A. Thus the calculation time will not change very much.

4   Feature Selection Methods

4.1   Backward Feature Selection

We select features by backward feature selection. In sequential backward feature selection, first we calculate the value of the selection criterion using all the features. Then starting from the initial set of features we temporarily delete each feature, calculate the value of the selection criterion, and delete the feature with the largest value of the selection criterion from the set. We iterate feature deletion so long as the value of the selection criterion is larger than the prescribed threshold.

The KDA criterion for two-class problems is nonincreasing for the deletion of features. We assume that this hold for the KDA criterion for multiclass problems. Then to determine the threshold we normalize the selection criterion by that evaluated using all the features. Then we set the threshold smaller than 1. It is difficult to set a proper value but in the following study based on some preliminary experiment we set $\delta = 0.95$ for multiclass problems.

Let the initial number of features be m and F^k and F^k_j denote the set of k features and the set of k features with the jth element temporarily deleted from the set, respectively. And let the selection criterion for F^k_j be $T(F^k_j)$. Then the normalized selection criterion $c(F^k_j)$ is

$$c(F^k_j) = \frac{T(F^k_j)}{T(F^m)}. \tag{18}$$

The procedure of backward feature selection is as follows:

1. Set the initial set of features as $F^m = \{1, \ldots, m\}$, and evaluate the selection criterion $T(F^m)$. Set $k = m$ and go to Step 2.
2. Delete the ith $(i = 1, \ldots, k)$ feature temporarily from F^k and calculate the normalized selection criterion $c(F^k_i)$. For the KDA criterion, if

$$c(F^k_j) > \delta \quad \text{for} \quad j = \arg\max_{i \in F^k} c(F^k_i), \tag{19}$$

   where δ is the threshold for the KDA criterion, go to Step 3. Otherwise stop feature selection.
3. Permanently delete j from F^k:

$$F^{k-1} = F^k - \{j\}. \tag{20}$$

   Then $k \leftarrow k - 1$ and go to Step 2.

4.2   Block Deletion

To speed up sequential backward selection, backward selection with block deletion is proposed [13,14]. To speed up variable selection, we use this method.

In block deletion, we reorder the candidate features with $c(F^k_j) > \delta$ in the descending order of $c(F^k_j)$ and delete the features simultaneously. If the selection criterion after deletion is larger than or equal to the threshold value we continue backward deletion. If not, we delete the lower half of the candidate features and repeat the above procedure until the deletion succeeds. Because one feature can be deleted, the block deletion does not fail. The algorithm is as follows.

1. Calculate $T(F^m)$ and set $k = m$.
2. Calculate $c(F^k_i)$ $(i \in F^k)$ and if

$$c(F^k_i) > \delta \qquad \text{for } i \in F^k,$$

   include i in the candidate set V^k, which is ordered in descending order of $c(F^k_i)$ and go to Step 3. If there is no i, terminate the algorithm.
3. If $|V^k| = 1$, where $|V^k|$ is the number of elements in V^k, delete that element from F^k, $k \leftarrow k - 1$, and go to Step 2. If $|V^k| > 1$ and $c(F^k - V^k) > \delta$, set $F^k \leftarrow F^k - V^k$, $k \leftarrow k - |V^k|$, and go to Step 2. Otherwise, go to Step 4.
4. Delete the lower half of V^k, $k \leftarrow |V^k|$, and go to Step 3.

4.3 Cross-Validation as a Stopping Condition

Usually it is difficult to set a proper value to the threshold δ. To solve this problem, we use the recognition rate of the validation set by cross-validation of the SVM as a stopping condition. Namely, at Step 1 of block deletion in Section 4.2 we calculate the recognition rate, r, of the validation data set in cross-validation of the SVM. And at the end of Step 3, by sequentially deleting features according to the order of V^k, we repeat evaluating the recognition rate of the validation data set so long as it is equal to, or higher than, r. And we stop feature selection if it is lower.

In the following we show the selection algorithm for the sequence of deleted features obtained by block deletion in Section 4.2.

1. Generate the sequence of features $[f_1, \ldots, f_k]$, where f_i is a feature deleted by KDA and k is the number of deleted features.
2. Calculate the recognition rate of the validation data set with all the features, r.
3. Calculate r_i, where r_i is the recognition rate of the validation data set with i features $f_1, \ldots, f_i$ deleted.
4. Find maximum r_i that satisfies $r_i \geq r$ and delete $[f_1, \ldots, f_i]$.

5 Performance Evaluation

5.1 Data Sets and Evaluation Conditions

We performed feature selection normalizing the input range into $[0, 1]$ and using polynomial kernels: $(\mathbf{x}^\top \mathbf{x}' + 1)^d$ or RBF kernels: $\exp(-\gamma \|\mathbf{x} - \mathbf{x}'\|^2)$, where d is the polynomial degree and γ is the width of the radius. We selected the parameter value from $d = [1, 2, 3, 4]$ for polynomial kernels and $\gamma = [0.1, 1, 10]$ for RBF kernels by fivefold cross-validation using the SVM.

In evaluating the classification performance we used the SVM with the same kernel parameter values used for feature selection and the margin parameter value selected from $C = [1, 10, 50, 100, 500, 1000, 2000, 3000, 5000, 8000, 10000, 50000, 100000]$ by fivefold cross-validation.

Table 1 lists the data sets used in the study. It also includes the kernel parameter value determined by cross-validation. In eigenvalue analysis we used the QR algorithm with the error limit for the off-diagonal elements being 10^{-6} and with the maximum iteration number of 100. We used Athlon 64×2 4800+ personal computer running on Linux.

The threshold value for the proposed method was set to $\delta = 0.95$ and we compared the following four selection methods:

1. Sequential backward selection using the proposed criterion with $\delta = 0.95$ (abbreviated as KDA),
2. Block deletion using the proposed criterion with $\delta = 0.95$ (KDA+B),
3. Block deletion using the proposed criterion with cross-validation (KDA+BC),

Table 1. Data sets

Data	Inputs	Classes	Train. Data	Test Data	kernel
Iris	4	3	75	75	$\gamma = 0.1$
Numeral	12	10	810	820	$d = 3$
Thyroid	21	3	3772	3428	$d = 1$
Blood cell	13	12	3097	3100	$\gamma = 10$
Hiragana-13	13	38	8375	8375	$\gamma = 10$
Hiragana-50	50	39	4610	4610	$\gamma = 10$
Satimage [20]	36	6	4435	2000	$\gamma = 10$

4. Block deletion using the SVM with cross-validation (SVM+BC),
5. Kernel class separability (KCS).

SVM+BC and KCS were used for comparing the proposed methods and SVM+BC used the same selection procedure as that of KDA+BC. The only difference is the selection criterion.

Kernel class separability, which is a simplified version of KDA, is a well-used measure and is defined by [21,22]

$$\frac{\sum_{i=1}^{n-1} \sum_{j=i+1}^{n} ||\mathbf{c}_i - \mathbf{c}_j||^2}{\sum_{i=1}^{n} \frac{1}{M_i} \sum_{j=1}^{M_i} ||\phi(\mathbf{x}_{ij}) - \mathbf{c}_i||^2}. \tag{21}$$

We selected the same number of features as that by KDA+BC.

5.2   Experimental Results

Table 2 shows the results. In the table, the "Deleted (Remaining) Features" column lists the deleted features in the order of deletion and the sequence of features in parentheses is that of remaining features. The first row of each data set shows the results using all the features and for each data set, the best recognition rate of the test data is shown in boldface. The "C" column lists the value of C selected by cross-validation of the SVM for the selected features. And "Train." and "Test" columns show the recognition rates of training data and test data, respectively.

First we compare the results for the proposed methods. For hiragana-50 and satimage data sets we could not obtain the selected features by KDA because of slow sequential backward selection. Now compare KDA and KDA+B. If different features were selected, they are shown in boldface. From the table both methods selected the same or similar sets of features. And except for the thyroid data set, both methods gave the similar recognition rates for the test data. Comparing the recognition rates of the test data with those without deleting variables, for all the data sets except for the iris data set, the recognition rates of the both methods were inferior. This means that too many features were deleted because of the improper selection of the threshold value. For example, for the hiragana-50 data by KDA+B, only five features remained and the recognition rate was

Table 2. Performance comparison of feature selection methods

Data	Method	Deleted (Remaining) Features	C	Train.	Test
Iris	—	None	100	100	**97.33**
	KDA	2	3000	97.33	**97.33**
	KDA+B	2	3000	97.33	**97.33**
	KDA+BC	2	3000	97.33	**97.33**
	SVM+BC	1	50	100	**97.33**
	KCS	1	—	100	**97.33**
Numeral	—	None	10^5	100	**99.76**
	KDA	$3, 12, 7, 4, 10, 2, 5$	10^5	100	99.51
	KDA+B	$3, 10, 12, 5, 4, 7, 2$	10^5	100	99.51
	KDA+BC	$3, 10, 12$	500	100	**99.76**
	SVM+BC	$3, 7, 12, 10$	10^5	100	99.51
	KCS	$4, 5, 9$	—	99.26	98.29
Thyroid	—	None	10^5	98.83	97.64
	KDA	$(10, 17, 19, 20)$	10^5	95.20	95.01
	KDA+B	$(2, 18, 3, 10, 17, 19, 20)$	10^5	97.64	96.79
	KDA+BC	$(8, 2, 18, 3, 10, 17, 19, 20)$	10^5	98.73	**97.90**
	SVM+BC	$(3, 8, 17, 19, 20)$	10^5	98.59	97.81
	KCS	$(7, 12, 13, 15, 18, 19, 20, 21)$	—	94.03	93.47
Blood	—	None	50	97.22	**93.55**
	KDA	$1, 8, 13, 10, 11, 6$	500	96.03	92.32
	KDA+B	$1, 8, 13, 10, 11, \mathbf{4}$	50	96.03	92.32
	KDA+BC	$1, 8$	50	97.13	93.16
	SVM+BC	$9, 8, 1, 6$	50	96.96	92.41
	KCS	$5, 6$	—	96.19	92.00
H-13	—	None	500	100	**99.76**
	KDA	$13, 11, 10$	10^5	100	99.62
	KDA+B	$13, \mathbf{3}, \mathbf{12}, 11$	50	99.64	99.53
	KDA+BC	13	1000	100	**99.76**
	SVM+BC	13	1000	100	99.72
	KCS	1	—	99.99	99.55
H-50	—	None	50	100	99.07
	KDA+B	$(14, 18, 28, 30, 33)$	500	99.74	90.65
	KDA+BC	$43, 15, 37$	50	100	99.05
	SVM+BC	$6, 8, 9, 12, 13, 18, 20, 23, 25, 26, 32, 35, 37, 38,$ $42, 43, 44, 47, 49$	100	100	98.52
	KCS	$43, 46, 5$	—	100	**99.08**
Satimage	—	None	1000	97.34	89.20
	KDA+B	$(1, 3, 5, 7, 9, 11, 13, 19, 21, 23, 25, 27, 30, 31,$ $33, 35)$	1000	95.78	87.70
	KDA+BC	$24, 20, 16, 4, 32, 8$	1000	97.47	88.95
	SVM+BC	$(1, 2, 3, 5, 10, 11, 18, 20, 23, 25, 26, 30, 36)$	1000	95.65	89.15
	KCS	$3, 27, 26, 19, 35, 10$	—	96.82	**89.60**

90.65%, which was much lower than 99.07% with all the features. This happened as follows. At the initial stage of feature deletion, deletion of any feature did not decrease the selection criterion. Therefore, we needed to delete features randomly until deletion of a feature led to a decrease of the selection criterion. In such a situation, we needed to use an alternative selection criterion.

By replacing the stopping condition of the threshold value in KDA+B with cross-validation, the selection became much more stable. Out of seven data sets, KDA+BC performed best four times and for the remaining data sets: the blood cell, hiragana-50, and satimage data sets, the differences from the best values were small. From the standpoint of recognition rate of the test data, KDA+BC was better than SVM+BC except for the satimage data set.

For the iris, hiragana-50 and satimage data sets, KCS showed better recognition rates than KDA+BC but for other four data sets, KDA+BC showed better recognition rates. Because KCS is not monotonic for the deletion of features selection was not stable.

In the above evaluation, we used (2) as the between-class scatter. Instead of (2), we used (15) and compared the difference of the selected features for KDA+BC, but there were not much difference between the two. For the thyroid data set, the obtained sequences were different, but the selected features for $\delta = 0.95$ were the same.

Table 3 shows the feature selection time for the four methods. In each problem the shortest time is shown in boldface. For the thyroid data set, we measured the feature selection time confining the value of C in $C = [10000, 50000, 100000]$. By introducing block deletion two to five times speedup was realized. Except for the blood cell and hiragana-13 data sets, KDA+BC was faster than SVM+BC. But for the hiragana-13 data set, SVM+BC was faster because only one feature was deleted.

For two-class problems the KDA criterion is proved to be monotonic for the deletion of features. But for the KDA criterion for multiclass problems, it is an open problem whether the KDA criterion is monotonic. Figure 1 shows the change of KDA criteria for the deletion of features for five data sets. We set $\gamma = 10$, which gave the maximum class separability. Except for the thyroid data set, the KDA criterion monotonically decreased as the features were deleted. For the thyroid data set, until six features were deleted, the KDA criterion monotonically increased. And afterwards, it decreased monotonically. For the KCS criterion, this sort of monotonicity was not observed.

Table 3. Comparison of feature selection time in seconds

Data	KDA	KDA+B	KDA+BC	SVM+BC
Iris	1	1	1	1
Numeral	313	**122**	391	773
Thyroid	87, 303	**25, 686**	31, 822	116, 026
Blood cell	20, 430	**11, 240**	12, 371	5, 432
Hiragana-13	835, 648	165, 408	174, 703	**51, 357**
Hiragana-50	–	**165, 408**	271, 001	404, 445
Satimage	–	82, 737	**12, 664**	166, 182

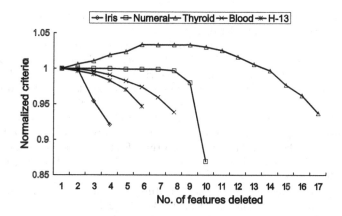

Fig. 1. Monotonicity of KDA Criterion

6   Conclusions

We proposed using the sum of objective function values associated with the eigenvectors of KDA as the selection criterion. This criterion reduces to the sum of eigenvalues of the KDA. To speed up feature selection by backward selection, we proposed to use block deletion of features at a time, and to improve the generalization ability of the selected features we proposed to use cross-validation. We also proposed calculating the between-class scatter using all the class pair distances.

By the computer experiment, we showed that the KDA criterion with block deletion performed better than the recognition rate of the SVM calculated by cross-validation if cross-validation is used to stop feature selection. But feature selection by the proposed between-class scatter did not give much difference from the conventional between-class scatter calculated based on the distances between the class centers and the center of the mapped training data.

References

1. Somol, P., Pudil, P., Novovičová, J., Paclík, P.: Adaptive floating search method in feature selection. Pattern Recognition Letters 20(11-13), 1157–1163 (1999)
2. Abe, S.: Pattern Classification: Neuro-Fuzzy Methods and Their Comparison. Springer, London (2001)
3. Thawonmas, R., Abe, S.: A novel approach to feature selection based on analysis of class regions. IEEE Transactions on Systems, Man, and Cybernetics—Part B 27(2), 196–207 (1997)
4. Ashihara, M., Abe, S.: Feature selection based on kernel discriminant analysis. In: Kollias, S.D., Stafylopatis, A., Duch, W., Oja, E. (eds.) ICANN 2006. LNCS, vol. 4132, pp. 282–291. Springer, Heidelberg (2006)
5. Louw, N., Steel, S.J.: Variable selection in kernel Fisher discriminant analysis by means of recursive feature elimination. Computational Statistics & Data Analysis 51(3), 2043–2055 (2006)

6. Ishii, T., Ashihara, M., Abe, S.: Kernel discriminant analysis based feature selection. Neurocomputing 71(13-15), 2544–2552 (2008)
7. Evgeniou, T., Pontil, M., Papageorgiou, C., Poggio, T.: Image representations for object detection using kernel classifiers. In: Proc. ACCV 2000, pp. 687–692 (2000)
8. Mukherjee, S., Tamayo, P., Slonim, D., Verri, A., Golub, T., Mesirov, J.P., Poggio, T.: Support vector machine classification of microarray data, Technical Report AI Memo 1677, Massachusetts Institute of Technology (1999)
9. Guyon, I., Weston, J., Barnhill, S., Vapnik, V.: Gene selection for cancer classification using support vector machines. Machine Learning 46(1-3), 389–422 (2002)
10. Perkins, S., Lacker, K., Theiler, J.: Grafting: Fast, incremental feature selection by gradient descent in function space. Journal of Machine Learning Research 3, 1333–1356 (2003)
11. Liu, Y., Zheng, Y.F.: FS_SFS: A novel feature selection method for support vector machines. Pattern Recognition 39(7), 1333–1345 (2006)
12. Wang, L.: Feature selection with kernel class separability. Pattern Analysis and Machine Intelligence 30(9), 1534–1546 (2008)
13. Abe, S.: Modified backward feature selection by cross validation. In: Proc. ESANN 2005, pp. 163–168 (2005)
14. Nagatani, T., Abe, S.: Backward variable selection of support vector regressors by block deletion. In: Proc. IJCNN 2007, pp. 2117–2122 (2007)
15. Bradley, P.S., Mangasarian, O.L.: Feature selection via concave minimization and support vector machines. In: Proc. ICML 1998, pp. 82–90 (1998)
16. Brown, M.: Exploring the set of sparse, optimal classifiers. In: Proc. ANNPR 2003, pp. 178–184 (2003)
17. Bo, L., Wang, L., Jiao, L.: Sparse Gaussian processes using backward elimination. In: Wang, J., Yi, Z., Żurada, J.M., Lu, B.-L., Yin, H. (eds.) ISNN 2006, Part 1. LNCS, vol. 3971, pp. 1083–1088. Springer, Heidelberg (2006)
18. Ishii, T., Abe, S.: Feature selection based on kernel discriminant analysis for multiclass problems. In: Proc. IJCNN 2008, pp. 2456–2461 (2008)
19. Baudat, G., Anouar, F.: Generalized discriminant analysis using a kernel approach. Neural Computation 12(10), 2385–2404 (2000)
20. Asuncion, A., Newman, D.J.: UCI machine learning repository (2007), http://www.ics.uci.edu/~mlearn/MLRepository.html
21. Cantú-Paz, E.: Feature subset selection, class separability, and genetic algorithms. In: Deb, K., et al. (eds.) GECCO 2004. LNCS, vol. 3102, pp. 959–970. Springer, Heidelberg (2004)
22. Wang, L., Chan, K.L.: Learning kernel parameters by using class separability measure. In: Sixth Kernel Machines Workshop, In conjunction with Neural Information Processing Systems, NIPS (2002)

Correlation-Based and Causal Feature Selection Analysis for Ensemble Classifiers

Rakkrit Duangsoithong and Terry Windeatt

Center for Vision, Speech and Signal Processing
University of Surrey
Guildford, United Kingdom GU2 7XH
{r.duangsoithong,t.windeatt}@surrey.ac.uk

Abstract. High dimensional feature spaces with relatively few samples usually leads to poor classifier performance for machine learning, neural networks and data mining systems. This paper presents a comparison analysis between correlation-based and causal feature selection for ensemble classifiers. MLP and SVM are used as base classifier and compared with Naive Bayes and Decision Tree. According to the results, correlation-based feature selection algorithm can eliminate more redundant and irrelevant features, provides slightly better accuracy and less complexity than causal feature selection. Ensemble using Bagging algorithm can improve accuracy in both correlation-based and causal feature selection.

Keywords: Correlation-based feature selection, causal feature selection, ensemble classification.

1 Introduction

With improvements in information and technology, many information databases have been created. However, in some applications especially in biomedical area, dataset usually contains hundreds to thousands of features with relatively small sample size and leads to degradation in accuracy and efficiency of system by curse of dimensionality and over-fitting. The resulting classifier works very well with training data but very poorly on testing data.

To overcome this high dimensional feature spaces degradation problem, number of features should be reduced. Basically, there are two methods to reduce the dimension: feature extraction and feature selection. Feature extraction transforms or projects original features to fewer dimensions without using prior knowledge. Nevertheless, it lacks comprehensibility and uses all original features which may be impractical in large feature spaces. On the other hand, feature selection selects optimal feature subsets from original features by removing irrelevant and redundant features. It has the ability to reduce over-fitting, increase classification accuracy, reduce complexity, speed of computation and improve comprehensibility by preserving original semantic of datasets. Normally, clinicians prefer feature selection because of its understandbility and user acceptance.

F. Schwenker and N. El Gayar (Eds.): ANNPR 2010, LNAI 5998, pp. 25–36, 2010.
© Springer-Verlag Berlin Heidelberg 2010

Feature selection is an important pre-processing step whether the classifier is Multilayer Perceptron (MLP), Support Vector Machines (SVM) or any other classifier. Generally, feature selection can be divided into four categories: Filter, Wrapper, Hybrid and Embedded methods [1],[2],[3]. Filter method is independent from learning method used in the classification process and uses measurement techniques such as correlation, distance and consistency measurement to find a good subset from entire set of features. Nevertheless, the selected subset may or may not be appropriate with the learning method. Wrapper method uses pre-determined learning algorithm to evaluate selected feature subsets that are optimum for the learning process. This method has high accuracy but is computationally expensive. Hybrid method combines advantage of both Filter and Wrapper method together. It evaluates features by using an independent measure to find the best subset and then uses a learning algorithm to find the final best subset. Finally, Embedded method interacts with learning algorithm but it is more efficient than Wrapper method because the filter algorithm has been built with the classifier.

Basically, feature selection does not take causal discovery or casuality into account [4]. Nevertheless, in some cases such as when training and testing dataset do not conform to i.i.d. assumption, testing distribution is shifted from manipulation by external agent, causal discovery can provide some benefits for feature selection under these uncertainty conditions. Causality also can learn underlying data structure, provide better understanding of the data generation process and better accuracy and robustness under uncertainty conditions [4].

Normally, causal relationships are uncovered by Bayesian Networks (BNs) which consists of a direct acyclic graph (DAG) that represents dependencies and independencies between variable and joint probability distribution among a set of variables [5].

An ensemble classifier or multiple classifier system (MCS) is another well-known technique to improve system accuracy [6]. Ensemble combines multiple base classifiers to learn a target function and gathers their prediction together. It has ability to increase accuracy of system by combining output of multiple experts to reduce bias and variance, improve efficiency by decomposing complex problem into multiple sub problems and improve reliability by reducing uncertainty. To increase accuracy, each classifier in the ensemble should be diverse or unique in order to reduce total error such as starting with different input, initial weight, random features or random classes [7].

In this paper, we present a comparison analysis between correlation-based and causal feature selection for ensemble classifiers in terms of number of eliminated features, complexity of algorithms and average percent accuracy.

1.1   Related Research

Feature selection and ensemble classification have received attention from many researchers in statistics, machine learning, neural networks and data mining areas for many years. At the beginning of feature selection history, most researchers focused only on removing irrelevant features such as ReliefF [8], FOCUS [9] and

Correlation-based Feature Selection(CFS) [10]. Recently, in Yu and Liu (2004) [11], Fast Correlation-Based Filter (FCBF) algorithm was proposed to remove both irrelevant and redundant features by using Symmetrical Uncertainty (SU) measurement and was successful for reducing high dimensional features while maintaining high accuracy.

In the past few years, learning BNs from observation data has received increasing attention from researchers for many applications such as decision support system, information retrieval, natural language processing, feature selection and gene expression data analysis [12],[13].

The category of BNs can be divided into three approaches: Search-and-Score, Constraint-Based and Hybrid approaches [12],[13]. In Search-and-Score approach, BNs search all possible structures to find the one that provides the maximum score. The second approach, Constraint-Based, uses test of conditional dependencies and independencies from the data by estimation using G^2 statistic test or mutual information, etc. This approach defines structure and orientation from results of the tests based on some assumptions that these tests are accurate. Finally, Hybrid approach uses Constraint-Based approach for conditional independence test (CI test) and then identifies the network that maximizes a scoring function by using Search-and-Score approach [13].

Constraint-Based algorithms are computationally effective and suitable for high dimensional feature spaces. PC algorithm [14], is a pioneer, prototype and well-known global algorithm of Constraint-Based approach for causal discovery. Three Phase Dependency Analysis (TPDA or PowerConstructor) [15] is another global Constraint-Based algorithm that uses mutual information to search and test for CI test instead of using G^2 Statistics test as in PC algorithm. However, both PC and TPDA algorithm use global search to learn from the complete network that can not scale up to more than few hundred features (they can deal with 100 and 255 features for PC and TPDA, respectively) [16]. Sparse Candidate algorithm (SC) [17] is one of the prototype BNs algorithm that can deal with several hundreds of features by using locally candidate set. Nevertheless, SC algorithm has some disadvantages, it may not identify true set of parents and users have to find appropriate k parameter of SC algorithm by themselves [12].

Recently, many Markov Blanket-based algorithms for causal discovery have been studied extensively and they have ability to deal with high dimensional feature spaces such as MMMB, IAMB [16] and HITON [5] algorithms. HITON is a state-of-the-art algorithm that has ability to deal with thousands of features and can be used as an effective feature selection in high dimensional spaces. However, HITON and all other MB-based algorithms may not specify features in Markov Blanket for desired classes or target (MB(T)) when the data is not faithful [20].

2 Theoretical Approach

In our research, two correlation-based feature selection methods, Fast Correlation-Based Filter (FCBF) [11] and Correlation-based Feature Selection with Sequential Forward Floating Search (CFS+SFFS) [10],[18] are compared with causal feature selection algorithms (PC, TPDA, SC and HITON) for Bagging [21] ensemble classifiers (described in Section 2.2) and experimentally compared with different learning algorithms.

2.1 Feature Selection

Fast Correlation-Based Filter (FCBF). FCBF [11] algorithm has two stages: relevance analysis and redundancy analysis.

Relevance Analysis. Normally, correlation is widely used to analyze relevance. In linear systems, correlation can be measured by linear correlation coefficient.

$$r = \frac{\sum_i (x_i - \overline{x_i})(y_i - \overline{y_i})}{\sqrt{\sum_i (x_i - \overline{x_i})^2} \sqrt{\sum_i (y_i - \overline{y_i})^2}} \tag{1}$$

However, most systems in real world applications are non-linear. Correlation in non-linear systems can be measured by using Symmetrical Uncertainty (SU).

$$SU = 2 \left[\frac{IG(X|Y)}{H(X)H(Y)} \right] \tag{2}$$

$$IG(X, Y) = H(X) - H(X|Y) \tag{3}$$

$$H(X) = -\sum_i P(x_i) log_2 P(x_i) \tag{4}$$

where $IG(X|Y)$ is the Information Gain of X after observing variable Y. $H(X)$ and $H(Y)$ are the entropy of variable X and Y, respectively. $P(x_i)$ is the probability of variable x.

SU is the modified version of Information Gain that has range between 0 and 1. FCBF removes irrelevant features by ranking correlation (SU) between feature and class. If SU between feature and class equal to 1, it means that this feature is completely related to that class. On the other hand, if SU is equal to 0, the features are irrelevant to this class.

Redundancy Analysis. After ranking relevant features, FCBF eliminates redundant features from selected features based on SU between feature and class and between feature and feature. Redundant features can be defined from meaning of predominant feature and approximate Markov Blanket. In Yu and Liu (2004) [11], a feature is predominant (both relevant and non redundant feature) if it does not have any approximate Markov blanket in the current set.

Approximate Markov blanket: For two relevant features F_i and F_j ($i \neq j$), F_j forms an approximate Markov blanket for F_i if

$$SU_{j,c} \geq SU_{i,c} \text{ and } SU_{i,j} \geq SU_{i,c} \tag{5}$$

where $SU_{i,c}$ is a correlation between any feature and the class. $SU_{i,j}$ is a correlation between any pair of feature F_i and F_j ($i \neq j$).

Correlation-based Feature Selection (CFS). CFS [10] is one of well-known techniques to rank the relevance of features by measuring correlation between features and classes and between features and other features.

Given number of features k and classes C, CFS defined relevance of features subset by using Pearson's correlation equation

$$Merit_s = \frac{kr_{kc}}{\sqrt{k + (k - 1)r_{kk}}} \tag{6}$$

where $Merit_s$ is relevance of feature subset, r_{kc} is the average linear correlation coefficient between these features and classes and r_{kk} is the average linear correlation coefficient between different features.

Normally, CFS adds (forward selection) or deletes (backward selection) one feature at a time, however, in this research, we used Sequential Forward Floating Search (SFFS) as the search direction.

Sequential Forward Floating Search (SFFS). SFFS [18] is one of a classic heuristic searching method. It is a variation of bidirectional search and sequential forward search (SFS) that has dominant direction on forward search. SFFS removes features (backward elimination) after adding features (forward selection). The number of forward and backward step is not fixed but dynamically controlled depending on the criterion of the selected subset and therefore, no parameter setting is required.

Causal Discovery Algorithm. In this paper, three standard (PC, TPDA, SC) and one state-of-the-art causal discovery algorithms (HITON) are used as causal feature selection methods. In the final output of the causal graph from each algorithm, the unconnected features to classes will be considered as eliminated features.

1. PC Algorithm
PC algorithm [14],[4] is the prototype of constraint-based algorithm. It consists of two phases: Edge Detection and Edge Orientation.

Edge Detection: the algorithm determines directed edge by using conditionally independent condition. The algorithm starts with:

i) Undirected edge with fully connected graph.

ii) Remove a share direct edge between A and B $(A - B)$ iff there is a subset F of features that can present conditional independence $(A, B|F)$.

Edge Orientation: The algorithm discovers V-Structure $A - B - C$ in which $A - C$ is missing.

i) If there are direct edges between $A - B$ and $B - C$ but not $A - C$, then orient edge $A \rightarrow B \leftarrow C$ until no more possible orientation.

ii) If there is a path $A \rightarrow B - C$ and $A - C$ is missing, then $A \rightarrow B \rightarrow C$.

iii) If there is orientation $A \rightarrow B \rightarrow ... \rightarrow C$ and $A - C$ then orient $A \rightarrow C$.

2. Three Phase Dependency Analysis Algorithm (TPDA)

TPDA or PowerConstructor algorithm [15] has three phases: *drafting, thickening and thinning.* In *drafting phase*, mutual information of each pair of nodes is calculated and used to create a graph without loop. After that, in *thickening phase*, edge will be added when that pair of nodes can not be *d-separated.* (node A and B are *d-separated* by node C iff node C blocks every path from node A to node B [12].) The output of this phase is called an independence map (*I-map*). The edge of *I-map* will be removed in *thinning phase* if two nodes of the edge can be *d-separated* and the final output is defined as a *perfect map* [15].

3. Sparse Candidate Algorithm (SC)

SC algorithm has two phases: *restrict* and *maximize steps* [17]. In *restrict step*, candidate sets are chosen by heuristic estimates of size k (define by user) and then a hill climbing search is performed in *maximize step*. In this second step, a network is started with empty graph and one of the operators: *add, delete* or *reverse* that provides the highest score will be chosen and applied to the current network. Finally, the algorithm will be repeated until there is no change in the candidate set [17],[12],[19].

4. HITON Algorithm

HITON [5] is one of state-of-the-art causal discovery algorithms that can be used as feature selection and can scale up to deal with thousands of features. HITON identifies Markov Blanket of the classes (or target) and then removes by backward greedy wrapper search of the features from the Markov Blanket that do not affect the classifier performance [5],[20].

2.2   Ensemble Classifier

Bagging. Bagging [21] or Bootstrap aggregating is one of the earliest, simplest and most popular for ensemble based classifiers. Bagging uses Bootstrap that randomly samples with replacement and combines with majority vote. Bootstrap is the most well-known strategy for injecting randomness to improve generalization performance in multiple classifier systems and provides out-of-bootstrap estimate for selecting classifier parameters [6]. Randomness is desirable since it increases diversity among the base classifiers, which is known to be a necessary condition for improved performance. However, there is an inevitable trade off between accuracy and diversity known as the accuracy/diversity dilemma [6].

3   Experimental Setup

3.1   Dataset

The medical datasets used in this experiment were taken from UCI machine learning repository [22]: heart disease, hepatitis, diabetes and Parkinson dataset and from Causality Challange [23]: lucas and lucap datasets. The details of

Table 1. Datasets

Dataset	Sample	Features	Classes	Missing Values	Data type
Heart Disease	303	13	5	Yes	Numeric (cont. and discrete)
Diabetes	768	8	2	No	Numeric (continuous)
Hepatitis	155	19	2	Yes	Numeric (cont. and discrete)
Parkinson's	195	22	2	No	Numeric (continuous)
Lucas	2000	11	2	No	Numeric (binary)
Lucap	2000	143	2	No	Numeric (binary)

datasets are shown in Table 1. The missing data are replaced by mean and mode of that dataset.

3.2 Evaluation

To evaluate feature selection process we use four widely used classifiers: Naive-Bayes(NB), Multilayer Perceptron (MLP), Support Vector Machines (SVM) and Decision Trees (DT). The parameters of each classifier were chosen based on the highest average accuracy of the experiment datasets from base classifier. MLP has one hidden layer with 16 hidden nodes, learning rate 0.2, momentum 0.3, 500 iterations and uses backpropagation algorithm with sigmoid transfer function. SVMs uses linear kernel and set the regularization value to 0.7 and Decision Trees use pruned C4.5 algorithm. The number of classifiers in Bagging is varied from 1, 5, 10, 25 to 50 classifiers. The threshold value of FCBF algorithm in our research is set at zero for heart disease, diabetes, parkinson and lucas and 1.4 and 0.15 for hepatitis and lucap dataset, respectively.

The classifier results were validated by 10 fold cross validation with 10 repetitions for each experiment and evaluated by average percent of test set accuracy of algorithm. For causal feature selection, PC algorithm using mutual information as statistic test with threshold 0.01 and maximum cadinality equals to 2. In TPDA algorithm, mutual information are used as statistic test with threshold 0.01 and assumed that data is monotone faithful. SC algorithm uses BDeu score function, k = 5 and using Bayesian scoring metric for statistic test. Finally, HI-TON use G^2 statistic test with threshold 0.05, maximum size of the conditional set is set to 3 and provides output as Markov Blanket of the classes.

4   Experimental Result

Table 2 and table 3 show the number of selected features in each analysis and the complexity of the algorithm, respectively.

Figure 1 and 2 present example of the average accuracy of heart disease and lucas dataset. Y-axis presents the average percent accuracy of the classifier and X-axis shows the number of ensemble from 1 to 50 classifiers. Figure 3 and 4 show the average accuracy of six datasets for each classifier and average of all classifiers for all six datasets, respectively. Finally, figure 5 presents the examples of causal graph of lucas data set from PC algorithm.

Table 2. Number of selected features

Dataset	Original Feature	Correlation-Based		Causal			
		FCBF	CFS+SFFS	PC	TPDA	SC	HITON
Heart Disease	13	6	9	13	13	11	4
Diabetes	8	4	4	8	8	0	0
Hepatitis	19	3	10	19	18	0	0
Parkinson's	22	5	10	22	2	0	0
Lucas	11	3	3	9	10	11	0
Lucap	143	7	36	121	121	123	0

Table 3. The complexity of each algorithm

Algorithm	Complexity	Remark
FCBF	$O(MNlogN)$	M=number of samples, N= number of features
CFS + SFFS	$< O(N^2)$	N= number of features
PC	$O(|N|^4)$	N= number of features
TPDA	$O(|N|^4)$	N= number of features
SC	$O(2^k \cdot (c+1)! \cdot |J|)$	k = size of candidate set, c = size of the largest separator in cluster tree, J = a family of cluster
HITON	$O(|MB|^3|N|)$	MB = Markov Blanket of the class, N = number of features

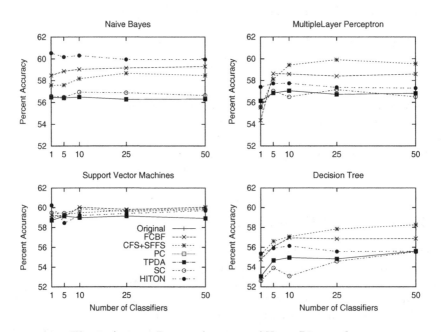

Fig. 1. Average Percent Accuracy of Heart Disease dataset

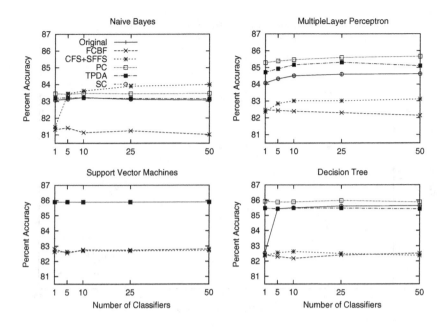

Fig. 2. Average Percent Accuracy of Lucas dataset

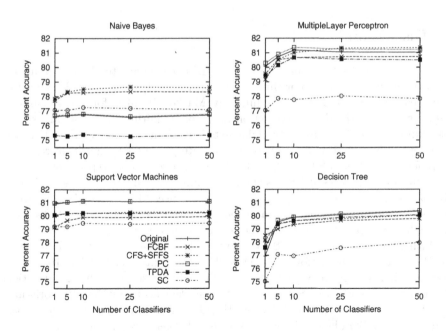

Fig. 3. Average Percent Accuracy of six datasets for each classifier

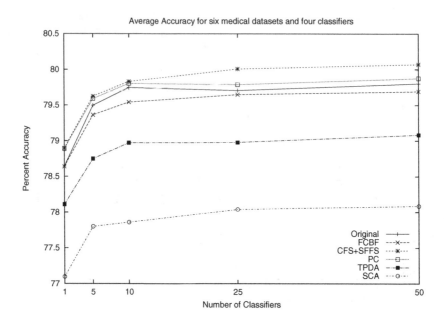

Fig. 4. Average Percent Accuracy of six datasets, four classifiers

5   Discussion

According to table 2, it can be seen that HITON eliminates highest number of features among other algorithms, however, it can define Markov Blanket for only heart disease and does not find any Markov Blanket for the remaining datasets because the data distribution may not be faithful [20] . SC algorithm also eliminates all features in some datasets because it may not identify true set of parents when k parameter is not appropriate [12]. FCBF algorithm removes more features than CFS+SFFS, TPDA and PC algorithms, respectively.

From Table 3, CFS+SFFS has the least complexity among other algorithms. HITON does not have high complexity because it uses Markov Blanket discovery that select only parents, children and spouses of the classes. PC and TPDA have the highest complexity algorithm due to their exhaustive search.

With reference to figure 3, SVM provides less accuracy than MLP because MLP uses back propagation with sigmoid transfer function and has 16 hidden node which is non linear system while SVM uses linear kernel with regularization 0.7. In figure 4, CFS+SFFS provides better average accuracy than PC, original, FCBF, TPDA and SCA, respectively. (HITON algorithm which can select optimal features only in heart disease dataset is not considered in the average graph in order not to bias result.) PC gives the best average accuacy among causal feature selection algorithms, however, it can deal only with few hundred features [16]. FCBF does not provide highest accuracy because its main objective is dealing with high dimensional feature while preserving high accuracy [11].

Although causal feature selection provides slightly less accuracy, more complexity and less number of eliminated features than correlation-based feature selection, it has benefit to learn underlying causal structure of the classes and features as an example shown in figure 5.

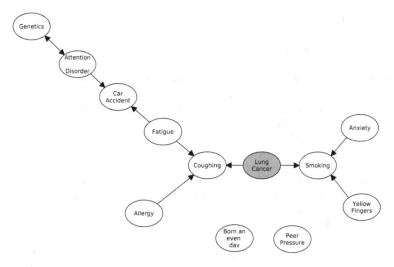

Fig. 5. Causal structure of Lucas dataset from PC algorithm

6   Conclusion

In this paper, we present a comparison analysis between correlation-based and casual feature selection for ensemble classifiers. In conclusion, correlation-based feature selection has slightly higher average accuracy, less complexity and can remove more irrelevant and redundant features than causal feature selection. Nevertheless, causal feature selection can reveal causes and consequence of the classes by defining causal relationship. Ensemble has ability to improve both correlation-based and causal feature selection. The future work will examine the result of causal feature selection from bootstrap dataset and combine result with ensemble classifiers.

References

1. Liu, H., Yu, L.: Toward integrating feature selection algorithms for classification and clustering. IEEE Transactions on Knowledge and Data Engineering 17(4), 491–502 (2005)
2. Saeys, Y., Inza, I., Larranaga, P.: A review of feature selection techniques in bioinformatics. Bioinformatics 23(19), 2507–2517 (2007)
3. Duangsoithong, R., Windeatt, T.: Relevance and Redundancy Analysis for Ensemble Classifiers. In: Perner, P. (ed.) Machine Learning and Data Mining in Pattern Recognition. LNCS, vol. 5632, pp. 206–220. Springer, Heidelberg (2009)

4. Guyon, I., Aliferis, C., Elisseeff, A.: Causal Feature Selection. In: Liu, H., Motoda, H. (eds.) Computational Methods of Feature Selection. Chapman and Hall, Boca Raton (2007)
5. Aliferis, C.F., Tsamardinos, I., Statnikov, A.: HITON, A Novel Markov Blanket Algorithm for Optimal Variable Selection. In: AMIA 2003 Annual Symposium Proceedings, pp. 21–25 (2003)
6. Windeatt, T.: Ensemble MLP Classifier Design, vol. 137, pp. 133–147. Springer, Heidelberg (2008)
7. Windeatt, T.: Accuracy/diversity and ensemble MLP classifier design. IEEE Transactions on Neural Networks 17(5), 1194–1211 (2006)
8. Witten, I.H., Frank, E.: Data Mining: Practical Machine Learning Tools and Techniques, 2nd edn. Morgan Kaufmann, San Francisco (2005)
9. Almuallim, H., Dietterich, T.G.: Learning with many irrelevant features. In: Proceedings of the Ninth National Conference on Artificial Intelligence, pp. 547–552. AAAI Press, Menlo Park (1991)
10. Hall, M.A.: Correlation-based feature selection for discrete and numeric class machine learning. In: Proceeding of the 17th International Conference on Machine Learning, pp. 359–366. Morgan Kaufmann, San Francisco (2000)
11. Yu, L., Liu, H.: Efficient feature selection via analysis of relevance and redundancy. J. Mach. Learn. Res. 5, 1205–1224 (2004)
12. Tsamardinos, I., Brown, L.E., Aliferis, C.F.: The max-min hill-climbing Bayesian network structure learning algorithm. Machine Learning 65, 31–78 (2006)
13. Wang, M., Chen, Z., Cloutier, S.: A hybrid Bayesian network learning method for constructing gene networks. Computational Biology and Chemistry 31, 361–372 (2007)
14. Spirtes, P., Glymour, C., Schinese, R.: Causation, Prediction, and search. Springer, New York (1993)
15. Cheng, J., Bell, D., Liu, W.: Learning Belief Networks from Data: An Information theory Based Approach. In: Proceedings of the Sixth ACM International Conference on Information and Knowledge Management, pp. 325–331 (1997)
16. Tsamardinos, I., Aliferis, C.F., Statnikov, A.: Time and Sample Efficient Discovery of Markov Blankets and Direct Causal Relations. In: KDD 2003, Washington DC, USA (2004)
17. Friedman, N., Nachman, I., Peer, D.: Learning of Bayesian Network Structure from Massive Datasets: The "Sparse Candidate" Algorithm. In: Proceedings of the 15th Conference on Uncertainty in Artificial Intelligence (UAI), pp. 206–215. Morgan Kaufmann, Stockholme (1999)
18. Pudil, P., Novovicova, J., Kitler, J.: Floating Search Methods in Feature Selection. Pattern Recognition Letters 15, 1,119–1,125 (1994)
19. Brown, L.E., Tsamardinos, I., Aliferis, C.F.: A Novel Algorithm for Scalable and Accurate Bayesian Network Learning. Medinfo. 11, 711–715 (2004)
20. Brown, L.E., Tsamardinos, I.: Markov Blanket-Based Variable Selection. Technical Report DSL TR-08-01 (2008)
21. Breiman, L.: Bagging predictors. Machine Learning 24(2), 123–140 (1996)
22. Asuncion, A., Newman, D.: UCI machine learning repository (2007), http://www.ics.uci.edu/mlearn/MLRepository.html
23. Guyon, I.: Causality Workbench (2008), http://www.causality.inf.ethz.ch/home.php

A New Monte Carlo-Based Error Rate Estimator

Ahmed Hefny and Amir Atiya

Cairo University
Faculty of Engineering
Computer Engineering Department

Abstract. Estimating the classification error rate of a classifier is a key
issue in machine learning. Such estimation is needed to compare classifiers or to tune the parameters of a parameterized classifier. Several
methods have been proposed to estimate error rate, most of which rely
on partitioning the data set or drawing bootstrap samples from it. Error
estimators can suffer from bias (deviation from actual error rate) and/or
variance (sensitivity to the data set). In this work, we propose an error
rate estimator that estimates a generative and a posterior probability
models to represent the underlying process that generates the data and
exploits these models in a Monte Carlo style to provide two biased estimators whose best combination is determined by an iterative solution.
We test our estimator against state of the art estimators and show that
it provides a reliable estimate in terms of mean-square-error.

1 Introduction

In a typical supervised learning setting, a classifier is trained on a set of patterns
with the goal being to give accurate classification for future unseen patterns. To
obtain the best possible model, one has to test a number of classifiers and select
the best, as typically different classifiers suit different problems. The misclassification probability is the main performance measure used to select the best
classifier, and it is very important to have an accurate estimator of this measure.
An inaccurate error rate estimator can lead to selecting the wrong classification
model or, for the same classification model, selecting detrimental parameter values. The problem is particularly aggravated by the small sample sizes [1]. Small
sample sizes are encountered in many applications, such as microarray classification [2][3] and domain-specific information extraction [4].

In ideal circumstances, to get an exact estimate of the error rate, one could obtain an exact model of the process that generates the patterns and then use this
model to obtain the error rate estimate either analytically or through Monte Carlo
simulations. However, in realistic situations the pattern generation process is unknown and only a representative data set is available. Thus, the error rate estimation method aims at estimating the error rate solely using this available data set.
Moreover, this same data set has to be used for designing the classifier too. Most
of the existing error rate estimators rely on evaluating the classifier using portions
of the data obtained either by partitioning or bootstrapping. The estimation error

F. Schwenker and N. El Gayar (Eds.): ANNPR 2010, LNAI 5998, pp. 37–47, 2010.
© Springer-Verlag Berlin Heidelberg 2010

for a misclassification rate estimator can be decomposed into the two conflicting components of bias and variance. An estimator that is insensitive to the precise locations of the sampled patterns will typically have a low variance and high bias, and the converse is true too. Obtaining an accurate error rate estimator amounts to mastering the right trade-off between bias and variance.

2   Error Rate Estimators

Consider a classification data set $(X, Y) = (x_i, y_i) \sim G, i = 1, 2, ..., N$, where x_i is a p-dimensional feature vector, y_i is its classification label, and G is the distribution from which (X, Y) is drawn. It is required to estimate the error rate of a classification rule $C : \mathbf{X} \mapsto \mathbf{Y}$ derived from the provided data set, where $\mathbf{X}$ is the domain of features and $\mathbf{Y}$ is the set of possible labels.

The most straightforward estimator is the resubstitution estimator, which is obtained by training the classifier on the whole data set and testing it on the same data. Because the same patterns are used for training and testing, resubstitution estimate suffers from severe negative bias (i.e. it severely underestimates the error rate), especially for complex classifiers, and is of little use by itself.

To combat this bias, researchers have typically used the hold-out method. It is based on dividing the data into a training set and a test set (as a rule of thumb the test set is typically taken as one third of the available dataset). The classifier is trained on the training set and evaluated on the test set. The advantage of such estimator is that it bases its estimate on patterns unseen in the training phase. The disadvantage is that it uses only a fraction of the data for training, leading to a disadvantaged classifier and hence a positive bias. Nevertheless, the hold-out method is widely used, especially for larger datasets.

An approach that attempts to make use of most of the data for training is the K-fold cross-validation (CV). It is based on splitting the data into K parts of equal sizes. Then we perform K different training/testing sessions. In the ith session we train the classifier on all parts except the ith part and test the classifier on the ith part. The estimated misclassification rate is the average of the misclassification rates obtained in the K testing sessions. A special case of CV is the leave-one-out cross-validation (LOOCV) where we have N partitions each consisting of a single pattern. The main disadvantage of CV is its large variance. In general, lower values of K have less variance at the expense of upward bias.

The leave-one-out bootstrap (LOOBS) takes B bootstrap samples of the data. Each bootstrap sample is obtained by sampling N patterns with replacement. The misclassification rate for each pattern x is estimated by classifiers trained on bootstrap samples in which x does not appear. The estimated misclassification rate is averaged over all patterns. LOOBS has less variance than CV at the expense of upward bias.

Efron [5] proposed the .632 method to handle the positive bias of the leave-one-out bootstrap by combining it with the negatively biased resubstitution estimate as follows:

$$\hat{E}_{.632} = 0.632 * \hat{E}_{LOOBS} + 0.368 * \hat{E}_{resub} \tag{1}$$

The weights are derived from the fact that the expected number of unique patterns in a bootstrap sample of size N is approximately $(1 - e^{-1})N = 0.632N$. The .632 suffers from a negative bias if the classifier is (nearly) a perfect memorizer, such as the nearest neighbor classifier (1NN). In this case E_{resub} can be extremely negatively biased (for example $\hat{E}_{resub} = 0$ for 1NN) thus biasing the whole estimate.

The .632+ method [6] attempts to alleviate the bias problem of the .632 method in case of perfect memorizers, by having variable combination weights, w, $1 - w$ where w is a function of the estimated degree of overfitting for the considered classifier instead of being fixed at 0.632. More recently, Sima and Dougherty[7] have shown that the optimal combination between LOOBS and resubstitution depends on the classification rule, sample size and data distribution. They have shown, for example, that the optimal value of w increases with increasing Bayes error rate.

A newly proposed method, the so-called bootstrap cross-validation (BCV), proposed by Fu, Carroll and Wang [2], takes B bootstrap samples of size N. The misclassification rate is estimated for each sample using leave-one-out cross validation and then averaged over all samples. An issue with BCV is that the bootstrap samples contain repeated patterns. Consequently, the LOOCV estimate would be negatively biased due to the overlap between training and testing data. Thus, BCV can suffer from downward bias, especially with perfect memorizers.

Jiang and Simon [3] proposed the adjusted bootstrap (ABS) method which is based on assuming an inverse power law relationship between error rate and number of unique patterns in the training set. The method calculates multiple leave-one-out bootstrap estimates with different bootstrap sample sizes and consequently different expectations of the number of unique patterns. These estimates are used to fit the parameters of the inverse power low curve. The error estimate is then calculated as the error rate at N unique patterns, as predicted by the inverse power law. Experiments conducted by Jiang and Simon show that ABS is moderately conservative (it tends to be upward biased).

In the methods described above each pattern is typically evaluated by an indicator function $I(C(x_i) \neq y_i)$ where $C(x_i)$ is the classifier's output for the ith pattern. This is usually adequate for a large number of patterns. However, as the number of patterns becomes small, the discrete nature of this evaluation function will have a detrimental effect on the estimator's performance due to variance. In such situation, each pattern becomes a valuable source of information that has to be used to its utmost. Replacing the indicator function with a continuous evaluation function improves the overall accuracy and, in particular, reduces the variance.

Most continuous evaluation functions proposed in the literature (see [8] for a review) assume a classification method where the output is obtained by thresholding a discriminant function $f(x) : \mathbf{X} \mapsto \mathbf{R}$. These methods utilize the actual value of the discriminant $f(x)$ instead of thresholded value. Fukunaga and Kessel [9] proposed the posterior probability misclassification rate estimator, where the discriminant function is nothing but posterior probability estimates $\hat{P}(y|x)$. This approach has been also analyzed in detail by [10] and [11]). In this approach the evaluation function becomes $1 - \hat{P}(C(x_i)|x_i)$, where $C(x_i)$ is the classification of pattern x_i.

The posterior probability error estimator does not need the labels (except for designing the classifier) and consequently can make use of unlabeled test patterns. To further reduce variance, Hand [12][13] proposed the utlization of the marginal probability $G(x) = \sum_{y \in \mathbf{Y}} G(x, y)$, which can also be estimated using unlabeled data. The *average conditional error rate* can then be estimated as

$$E_{AC} = \int_{\mathbf{X}} (1 - \hat{P}(C(x)|x))\hat{G}(x)dx \tag{2}$$

Typically, $\hat{P}(y|x)$ is different from the true posterior probability given by

$$P(y|x) = \frac{G(x, y)}{\sum_{y_k \in \mathbf{Y}} G(x, y_k)} \tag{3}$$

This makes the posterior error estimation biased and this bias becomes more severe as the data set size decreases because the error in estimating $\hat{P}$ becomes larger [8].

3   Proposed Estimator

The proposed misclassification rate estimator is based on two different proposed estimators that have favorable complementary features. Subsequently, the two estimators are combined in a certain way so as to emphasize their strong aspects.

First, given a random sample $(\hat{X}, \hat{Y}) \sim \hat{G}$ of size $N_G \gg N$, the generative error rate is defined as follows:

$$\hat{E}_G = \frac{1}{N_G} \sum_{j=1}^{N_G} I(C(\hat{x}_j) \neq \hat{y}_j) \tag{4}$$

Since $\hat{G}$ is typically only an approximation of G, we expect E_G to be biased. If $\hat{G}$ is estimated using a method that has enough degrees of freedom to fit the dataset, such as Parzen windows, the generated set $(\hat{X}, \hat{Y})$ will be similar to the provided set (X, Y) and consequently, $\hat{E}_G$ will be downward biased. $\hat{E}_G$ can thus be thought of as a reduced-bias version of the resubstitution estimate.

The other estimate we develop is the Monte Carlo posterior estimate which is calculated as shown in algorithm 1. The rationale for the Monte Carlo posterior approach is that we would like to replicate the whole process of drawing patterns from the distribution and designing a classifier. Relying on just the original patterns (like in the proposed generative method above or like other estimators in the literature), will get us "stuck" with the specific positions the original patterns happen to fall in.

In this work, we estimate the posterior probability $\hat{P}$ using a Gaussian process classifier [14], we choose Gaussian processes for three main reasons. First, the Gaussian process classifier we use is a discriminative estimator. Discriminative methods are more reliable than generative methods for estimating posterior probabilities especially in small samples, since they do not need to estimate the joint

---

Algorithm 1. Monte Carlo posterior error rate estimation

---
for $i = 1$ to B **do**
    $(X_s, Y_s) \sim \hat{G}$, $N_s = N$
    $(X_t, Y_t) \sim \hat{G}$, $N_t = \frac{N}{2}$
    Train C on (X_s, Y_s)
    $\hat{E}_{MCPi} = \frac{1}{N_t} \sum_{t=1}^{N_t} (1 - \hat{P}(C(x_i)|x_i))$
end for
$\hat{E}_{MCP} = \frac{1}{B} \sum_{i=1}^{B} \hat{E}_{MCPi}$

---

probability $G(x, y)$. Second, Gaussian processes are kernel-based, which means they can easily be applied on non-Euclidean spaces. Third, Gaussian processes are based on a formal probabilistic framework.

To derive a good combination of $\hat{E}_G$ and $\hat{E}_{MCP}$ we need to observe how they are related to the actual error rate which depends on the degree of separation between patterns belonging to different classes based on available features. In a problem where the features clearly separate classes, the true error rate E and $\hat{E}_G$ tend to be minimum and in this case $\hat{E}_G$ tends to be unbiased. On the other hand, in a problem with almost non-informative features, E tends to the error rate of a totally random classifier (0.5) while the E_G estimate will typically be less than 0.5 since the test is similar to the set fitted by the classifier. Consequently, $\hat{E}_G$ will be significantly down-biased in low class separation problems.

The relation between the bias of $\hat{E}_{MCP}$ and the degree of class separation can be explained by the effect of replacing the indicator evaluation function $z_i = I(C(x_i) \neq y_i)$, where $z_i \in \{0, 1\}$, with the posterior estimate $p_i = 1 - \hat{P}(C(x_i)|x_i)$, where $p_i \in [0, 1]$. Since p_i is more smoothed than z_i, we expect that $p_i < 1$ if $z_i = 1$ and $p_i > 0$ if $z_i = 0$. In other words, p_i is more optimistic about misclassified patterns and more pessimistic about correctly classified patterns. Consequently, at low error rates (high separation), $\hat{E}_{MCP}$ becomes upward-biased. This bias decreases as error rate increases and reaches its minimum at error rates near 0.5, where the effects of misclassified and correctly classified patterns balance out.

To test the validity of the above argument, we conducted Monte Carlo simulation experiments on different data sets and classifiers. For each data set, we tried different degrees of class separation by translating and/or scaling each pattern depending on its class. For each degree of class separation we select a set S of N patterns ($N = 20$) from the transformed data set, train the classifier on them and estimate the true error rate E by counting the number of misclassified patterns when testing the classifier on the patterns not selected in S. We use S to estimate E_G and E_{MCP}. This process is repeated K times ($K = 100$).

Figure 1 shows the results of one of these simulations, where a Naive Bayesian classifier is tested on patterns generated from a 10-variate Gaussian distribution with $\sum = I$, $\mu_1 = \underline{0}$ and $\mu_{2i} = d \, \forall \, i : 1 \leq i \leq 10$. Degree of separation is controlled by changing d, which is shown as the x-axis. The figure shows that both E_{MCP} and E_G tend to be more pessimistic (or less optimistic) as the degree of separation increases. It shows that E_{MCP} is a good estimator when there is

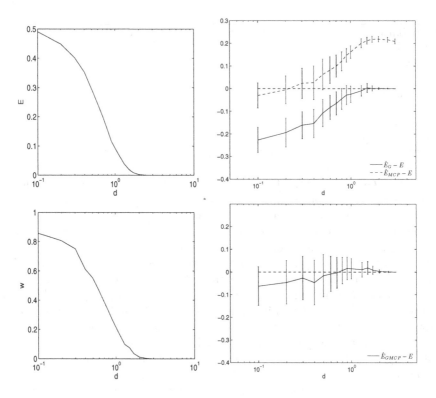

Fig. 1. Effect of degree of class separation on error rate estimation. **Top-left**: Estimation of the true error rate E. **Top-right**: Error estimation bias ($\hat{E}_G - E$ and $\hat{E}_{MCP} - E$). **Bottom-left**: Combination weight as calculated by iteratively applying equations 5 and 6 for 10 iterations. **Bottom-right**: Bias of the final error estimate ($\hat{E}_{GMCP} - E$) All depicted quantities are averaged over 100 trials.

low separation and the error rate is high while E_G is a good estimator when there is high separation and the error rate is low.

This suggests a linear combination of the form

$$\hat{E}_{GMCP} = w\hat{E}_{MCP} + (1-w)\hat{E}_G \quad (0 \le w \le 1) \qquad (5)$$

where

$$w = min(1, 2E) \qquad (6)$$

and E is the error rate which depends on the degree of class separation. Thus w tends to 1 as E tends to 0.5 (low separation) and tends to 0 as E tends to 0 (high separation). Since E is unknown a priori we replace it in 6 with $\hat{E}_{GMCP}$ and iteratively apply 5 and 6 starting with $w = 0.5$. Figure 1 shows that w converges to nearly $2E$. It is worth noting that the combination of E_G and E_{MCP} reduced the resultant bias at the expense of increased variance, a clear example of the bias/variance tradeoff. Similar behavior was consistently observed in the experiments performed on other data sets and other classifiers.

4   Experimental Setup

To test the comparative performance of the proposed estimator, we have performed extensive numerical simulations on four 2-class real-world data sets different from those used to develop our estimator.

For each data set we performed K runs (we set K to 100), where in each run we randomly sample N patterns from the data set. These represent the training set, and they will be used to estimate the classifier error rate for each competing estimator. Table 1 shows the details of the data sets used [1].

The estimators we tested are given in Table 2. For the estimation of $\hat{P}(y|x)$, we used Gaussian processes with the neural network covariance function since our development experiments showed that it has more discriminative power for small samples compared to other functions such as the standard error isotropic covariance function. For the estimation of $G(x, y)$ we used Parzen windows with kernel width estimated using Silverman's rule[16].

For each setup we estimate bias, variance and mean square error (MSE), which are calculated as follows

$$Bias = \frac{1}{K} \sum_{i=1}^{K} (\hat{E}_i - E_i) \tag{7}$$

$$Variance = \frac{1}{K} \sum_{i=1}^{K} (\hat{E}_i - E_i)^2 - (\frac{1}{K} \sum_{i=1}^{K} (\hat{E}_i - E_i))^2 \tag{8}$$

$$MSE = \frac{1}{K} \sum_{i=1}^{K} (\hat{E}_i - E_i)^2 \tag{9}$$

where E_i and $\hat{E}_i$ are the true and estimated misclassification rate in the ith run respectively. The true error rate is estimated by training the classifier on the N sampled data and testing it on the remaining patterns of the dataset.

Table 1. Datasets

Dataset	Number of patterns	Dimensionality	Percentage of class 1 patterns
wdbc	569	10	37%
wine-red	1599	11	53%
wine-white	4898	11	67%
stock-prediction	7516	3	25%

Table 2. Tested error rate estimation methods

Abbreviation	Definition
LOOCV	Leave-one-out cross-validation
LOOCVP	Leave-one-out cross-validation with Gaussian Process posterior evaluation
LOOBS	Leave-one-out bootstrap (50 bootstrap samples)
632	632 bootstrap (50 bootstrap samples)
632+	632+ bootstrap (50 bootstrap samples)
BCV	Bootstrap cross-validation (50 bootstrap samples)
ABS	Adjusted bootstrap (50 bootstrap samples, $l \in \{0.7, 1, 2, 3, 4, 5, 6\}$)
GMCP	Combined Generative and Monte-Carlo posterior estimates

---

[1] All data sets were obtained from UC Ivrine Machine Learning Repository [15] except the stock prediction data set.

For each dataset we tested three different classifiers: the K-nearest neighbor (KNN) with $K = 3$, the diagonal linear discriminant analysis (DLDA), and the classification and regression trees (CART).

5   Experimental Results

Tables 3, 4 and 5 show the results for 3NN, DLDA and CART classifiers respectively. Table 6 shows the ranks of tested estimators (based on MSE) averaged over

Table 3. Results for 3NN Classifier. For each setup, the first row shows the bias, the second shows the variance and the third shows the MSE.

Data	N	True	LOOCV	LOOCVP	LOOBS	632	632+	BCV	ABS	GMCP
wdbc	30	0.096	0.000	0.086	0.024	-0.002	0.003	-0.017	0.023	0.001
			0.004	0.001	0.003	0.002	0.003	0.002	0.004	0.001
			0.004	0.008	0.003	0.002	0.003	0.002	0.004	0.001
wine-red	30	0.356	0.005	0.030	0.030	-0.044	0.002	-0.103	0.041	-0.018
			0.010	0.004	0.006	0.004	0.007	0.003	0.010	0.007
			0.010	0.005	0.007	0.006	0.007	0.013	0.012	0.007
wine-white	30	0.357	0.038	0.052	0.049	-0.030	0.024	-0.090	0.049	0.001
			0.010	0.004	0.006	0.004	0.006	0.003	0.009	0.006
			0.012	0.007	0.008	0.005	0.006	0.011	0.011	0.006
stock	30	0.426	0.083	0.062	0.064	-0.027	0.043	-0.099	0.063	0.063
			0.014	0.003	0.007	0.006	0.006	0.004	0.012	0.003
			0.021	0.007	0.011	0.007	0.008	0.014	0.016	0.007
wdbc	40	0.091	0.005	0.073	0.030	0.005	0.009	-0.014	0.031	0.004
			0.002	0.001	0.002	0.002	0.002	0.001	0.003	0.001
			0.003	0.006	0.003	0.002	0.002	0.001	0.004	0.001
wine-red	40	0.353	0.011	0.022	0.035	-0.045	0.004	-0.107	0.022	-0.028
			0.008	0.003	0.004	0.003	0.005	0.002	0.006	0.004
			0.008	0.003	0.005	0.005	0.005	0.013	0.006	0.004
wine-white	40	0.339	0.018	0.054	0.038	-0.037	0.010	-0.094	0.036	0.002
			0.007	0.002	0.004	0.003	0.004	0.002	0.006	0.004
			0.007	0.005	0.005	0.004	0.005	0.011	0.007	0.004
stock	40	0.367	0.053	0.064	0.060	-0.016	0.038	-0.077	0.080	0.073
			0.010	0.002	0.006	0.005	0.007	0.005	0.012	0.003
			0.013	0.006	0.010	0.006	0.008	0.010	0.018	0.008
wdbc	50	0.085	0.005	0.066	0.024	0.000	0.003	-0.017	0.021	0.002
			0.002	0.001	0.002	0.001	0.001	0.001	0.002	0.001
			0.002	0.006	0.002	0.001	0.001	0.001	0.003	0.001
wine-red	50	0.353	-0.004	0.019	0.024	-0.052	-0.008	-0.115	0.018	-0.027
			0.006	0.003	0.004	0.003	0.005	0.002	0.005	0.004
			0.006	0.003	0.004	0.005	0.005	0.015	0.006	0.004
wine-white	50	0.327	0.005	0.036	0.024	-0.044	-0.002	-0.102	0.019	-0.017
			0.006	0.003	0.004	0.003	0.005	0.002	0.005	0.004
			0.006	0.004	0.005	0.005	0.005	0.012	0.006	0.004
stock	50	0.347	0.031	0.045	0.052	-0.022	0.035	-0.083	0.061	0.058
			0.005	0.001	0.003	0.003	0.003	0.002	0.006	0.002
			0.006	0.003	0.006	0.003	0.004	0.009	0.010	0.005

setups using the same N. The results show that GMCP is a reliable error rate estimator in terms of MSE, with 632 and LOOCVP being very close competitors. The combination of generative and posterior makes GMCP, in most setups, less prone to excessive positive bias occasionally encountered by LOOCVP and excessive negative bias occasionally encountered by 632, especially for overfitting classifiers such as CART.

Table 4. Results for DLDA Classifier. For each setup, the first row shows the bias, the second shows the variance and the third shows the MSE.

Data	N	True	LOOCV	LOOCVP	LOOBS	632	632+	BCV	ABS	GMCP
wdbc	30	0.083	0.007	0.106	0.012	0.003	0.004	-0.004	0.010	0.035
			0.003	0.001	0.002	0.002	0.002	0.002	0.002	0.002
			0.003	0.013	0.003	0.002	0.002	0.002	0.003	0.004
wine-red	30	0.318	-0.006	0.033	0.019	-0.028	-0.007	-0.070	0.007	0.009
			0.010	0.006	0.008	0.006	0.009	0.005	0.008	0.007
			0.010	0.007	0.008	0.007	0.009	0.009	0.008	0.007
wine-white	30	0.340	0.032	0.059	0.045	-0.005	0.020	-0.053	0.035	0.036
			0.011	0.005	0.007	0.005	0.007	0.003	0.008	0.006
			0.012	0.008	0.009	0.005	0.008	0.006	0.009	0.007
stock	30	0.484	0.013	0.005	0.010	-0.020	-0.001	-0.058	0.016	0.006
			0.023	0.015	0.016	0.014	0.016	0.016	0.018	0.017
			0.023	0.016	0.017	0.014	0.016	0.020	0.018	0.017
wdbc	40	0.085	-0.012	0.064	-0.008	-0.014	-0.014	-0.021	-0.010	0.010
			0.002	0.001	0.002	0.002	0.002	0.001	0.002	0.001
			0.002	0.005	0.002	0.002	0.002	0.002	0.002	0.002
wine-red	40	0.313	0.005	0.035	0.029	-0.012	0.002	-0.046	0.014	0.017
			0.006	0.003	0.005	0.004	0.005	0.003	0.005	0.004
			0.006	0.004	0.005	0.004	0.005	0.005	0.005	0.004
wine-white	40	0.332	0.014	0.035	0.030	-0.007	0.009	-0.047	0.019	0.023
			0.007	0.004	0.006	0.005	0.006	0.003	0.006	0.005
			0.008	0.005	0.007	0.005	0.006	0.006	0.007	0.005
stock	40	0.495	0.005	-0.004	-0.002	-0.024	-0.012	-0.060	0.010	-0.010
			0.017	0.006	0.011	0.008	0.009	0.011	0.011	0.015
			0.017	0.006	0.011	0.009	0.009	0.015	0.011	0.015
wdbc	50	0.081	0.003	0.063	0.008	0.001	0.001	-0.004	0.006	0.025
			0.002	0.001	0.002	0.002	0.002	0.002	0.002	0.002
			0.002	0.005	0.002	0.002	0.002	0.002	0.002	0.002
wine-red	50	0.302	0.005	0.028	0.026	-0.005	0.005	-0.036	0.012	0.018
			0.004	0.002	0.003	0.003	0.003	0.002	0.003	0.003
			0.004	0.003	0.004	0.003	0.004	0.004	0.003	0.003
wine-white	50	0.333	-0.010	0.014	0.007	-0.022	-0.012	-0.055	-0.002	0.005
			0.004	0.003	0.004	0.003	0.004	0.003	0.004	0.003
			0.004	0.003	0.004	0.004	0.004	0.006	0.004	0.003
stock	50	0.473	0.005	-0.003	0.001	-0.019	-0.010	-0.046	0.006	0.006
			0.013	0.004	0.011	0.007	0.008	0.011	0.010	0.016
			0.013	0.004	0.011	0.008	0.008	0.013	0.010	0.016

Table 5. Results for CART Classifier. For each setup, the first row shows the bias, the second shows the variance and the third shows the MSE.

Data	N	True	LOOCV	LOOCVP	LOOBS	632	632+	BCV	ABS	GMCP
wdbc	30	0.137	0.005	0.069	0.021	-0.030	-0.017	-0.068	0.016	0.074
			0.006	0.003	0.004	0.003	0.004	0.002	0.005	0.003
			0.006	0.007	0.005	0.004	0.004	0.007	0.005	0.008
wine-red	30	0.378	0.026	0.029	0.022	-0.098	-0.010	-0.194	0.020	0.019
			0.017	0.003	0.006	0.004	0.008	0.003	0.008	0.003
			0.018	0.004	0.007	0.014	0.008	0.040	0.009	0.003
wine-white	30	0.388	0.025	0.039	0.031	-0.099	-0.002	-0.194	0.024	0.029
			0.017	0.005	0.006	0.004	0.008	0.003	0.009	0.004
			0.018	0.006	0.007	0.014	0.008	0.041	0.009	0.005
stock	30	0.458	0.033	0.037	0.029	-0.093	0.004	-0.181	0.039	0.036
			0.017	0.004	0.008	0.006	0.007	0.005	0.009	0.004
			0.018	0.006	0.008	0.014	0.007	0.038	0.010	0.006
wdbc	40	0.125	-0.003	0.053	0.014	-0.028	-0.020	-0.063	0.007	0.076
			0.004	0.001	0.003	0.002	0.002	0.001	0.003	0.002
			0.004	0.004	0.003	0.003	0.003	0.005	0.003	0.007
wine-red	40	0.370	-0.023	0.016	0.011	-0.104	-0.030	-0.196	-0.003	0.016
			0.016	0.004	0.005	0.003	0.007	0.002	0.007	0.002
			0.016	0.004	0.005	0.014	0.008	0.041	0.007	0.003
wine-white	40	0.359	0.032	0.055	0.039	-0.077	0.010	-0.173	0.041	0.048
			0.014	0.004	0.004	0.003	0.005	0.003	0.005	0.004
			0.015	0.007	0.006	0.009	0.006	0.033	0.007	0.006
stock	40	0.400	0.061	0.049	0.048	-0.063	0.030	-0.148	0.055	0.063
			0.011	0.003	0.005	0.004	0.005	0.003	0.006	0.002
			0.015	0.005	0.007	0.008	0.006	0.025	0.009	0.006
wdbc	50	0.117	0.012	0.048	0.018	-0.022	-0.016	-0.055	0.013	0.069
			0.005	0.002	0.003	0.002	0.002	0.001	0.003	0.002
			0.005	0.004	0.003	0.002	0.002	0.004	0.003	0.007
wine-red	50	0.373	0.003	0.019	0.016	-0.096	-0.020	-0.189	0.008	0.017
			0.009	0.003	0.003	0.002	0.004	0.002	0.004	0.002
			0.009	0.003	0.004	0.011	0.005	0.038	0.004	0.002
wine-white	50	0.351	-0.017	0.023	0.005	-0.100	-0.024	-0.183	-0.005	0.021
			0.010	0.003	0.003	0.002	0.004	0.002	0.004	0.002
			0.010	0.003	0.003	0.012	0.005	0.035	0.004	0.002
stock	50	0.390	0.027	0.025	0.015	-0.085	0.001	-0.163	0.024	0.046
			0.008	0.003	0.004	0.003	0.004	0.003	0.005	0.002
			0.008	0.003	0.004	0.010	0.004	0.029	0.006	0.004

Table 6. Average ranks of error rate estimators based on MSE

N	LOOCV	LOOCVP	LOOBS	632	632+	BCV	ABS	GMCP
30	7	3.6667	4.3333	2.9167	3.5000	5.7500	5.7500	3.08337
40	7	3.5833	4.3333	3.0833	3.5000	6	5.5833	2.9167
50	5.9167	2.9167	4.5000	3.9167	3.8333	6.5000	5	3.4167
Overall	6.6389	3.3889	4.3889	3.3056	3.6111	6.0833	5.4444	3.1389

6 Conclusions

In this work we developed a complementary pair of error rate estimators that utilize a generative estimator and a posterior estimator. We proposed an iterative combination of the two estimators. The iterative solution was based on the fact that the best combination depends on a hidden parameter (true error rate). We are planning to study the possibility of integrating more visible and hidden parameters such as number of samples and amount of overfitting to get a more reliable estimator.

References

1. Isaksson, A., Wallman, M., Göransson, H., Gustafsson, M.G.: Cross-validation and bootstrapping are unreliable in small sample classification. Pattern Recogn. Lett. 29(14), 1960–1965 (2008)
2. Fu, W.J., Carroll, R.J., Wang, S.: Estimating misclassification error with small samples via bootstrap cross-validation. Bioinformatics 21, 1979–1986 (2005)
3. Jiang, W., Simon, R.: A comparison of bootstrap methods and an adjusted bootstrap approach for estimating prediction error in microarray classification. Statistics in Medicine (2008)
4. Sordo, M., Zeng, Q.T.: On sample size and classification accuracy: A performance comparison. In: Oliveira, J.L., Maojo, V., Martín-Sánchez, F., Pereira, A.S. (eds.) ISBMDA 2005. LNCS (LNBI), vol. 3745, pp. 193–201. Springer, Heidelberg (2005)
5. Efron, B.: Estimating the error rate of a prediction rule: Improvement on cross-validation. Journal of the American Statistical Association 78(382), 316–331 (1983)
6. Efron, B., Tibshirani, R.: Improvements on cross-validation: The .632+ bootstrap method. Journal of the American Statistical Association 92, 548–560 (1997)
7. Sima, C., Dougherty, E.R.: Optimal convex error estimators for classification. Pattern Recogn. 39(9), 1763–1780 (2006)
8. Raudys, S., Jain, A.: Small sample size effects in statistical pattern recognition: Recommendations for practitioners. IEEE Transactions on Pattern Analysis and Machine Intelligence 13(3), 252–264 (1991)
9. Fukunaga, K., Kessel, D.: Application of optimum error-reject functions. IEEE Transaction on Information Theory 19, 814–817 (1972)
10. Ganesalingam, S., McLachlan, G.J.: Error rate estimation on the basis of posterior probabilities. Pattern Recognition 12(6), 405–413 (1980)
11. Lugosi, G., Pawlak, M.: On the posterior-probability estimate of the error rate of nonparametric classification rules. IEEE Transactions on Information Theory 40(2), 475–481 (1994)
12. Hand, D.J.: An optimal error rate estimator based on average conditional error rate: Asymptotic results. Pattern Recogn. Lett. 4(5), 347–350 (1986)
13. Schiavo, R.A., Hand, D.J.: Ten more years of error rate research. International Statistical Review 68, 295–310 (2000)
14. Rasmussen, C.E., Williams, C.: Gaussian Processes for Machine Learning. MIT Press, Cambridge (2006)
15. Asuncion, A., Newman, D.: UCI machine learning repository (2007)
16. Silverman, B.W.: Density Estimation for Statistics and Data Analysis. Chapman & Hall/CRC (April 1986)

Recognition of Sequences of Graphical Patterns

Edmondo Trentin[1], ShuJia Zhang[2], and Markus Hagenbuchner[2]

[1]DII - Università di Siena, V. Roma, 56 Siena, Italy
`trentin@dii.unisi.it`
[2]University of Wollongong, Wollongong, 2522 NSW, Australia
`{sz603,markus}@uow.edu.au`

Abstract. Several real-world problems (e.g., in bioinformatics/proteomics, or in recognition of video sequences) can be described as classification tasks over sequences of structured data, i.e. sequences of graphs, in a natural way. This paper presents a novel machine that can learn and carry out decision-making over sequences of graphical data. The machine involves a hidden Markov model whose state-emission probabilities are defined over graphs. This is realized by combining recursive encoding networks and constrained radial basis function networks. A global optimization algorithm which regards to the machine as a unity (instead of a bare superposition of separate modules) is introduced, via gradient-ascent over the maximum-likelihood criterion within a Baum-Welch-like forward-backward procedure. To the best of our knowledge, this is the first machine learning approach capable of processing sequences of graphs without the need of a pre-processing step. Preliminary results are reported.

Keywords: Hidden Markov model, relational learning, recursive networks.

1 Introduction

This paper introduces a novel hybrid architecture (along with its training algorithm) for learning over sequences of graphs (i.e., sequential structured data). The scenario is an extension of traditional relational learning, where a learning machine is fed with a graph (an algebraic relation), and it is expected to carry out classification (or, regression) over the input structured data [5]. In the present setup, we are faced with a sequence $g_1, \ldots, g_n$ of individual graphs, and the overall sequence has to be modeled and classified. Examples may help focusing on the nature of this scenario. Several problems in bioinformatics concern the classification of proteins, or of subparts of proteins. For instance, prediction of the secondary structure of certain segments within the protein, or the prediction of the binding state of individual cysteines, starting from the primary structure. These problems are usually faced (e.g., using hidden Markov models) relying on a description of the protein as a sequence (string) of amino-acids, each amino-acid being a symbol drawn from a finite and discrete alphabet, or a real-valued feature vector drawn from a multiple-alignment profile. In so doing, the very nature of the amino-acids is overlooked by the machine, which has no opportunity to take benefit from knowledge of the underlying physical and chemical properties of the amino-acids themselves. As a matter of fact, each amino-acid is a molecule, built up from specific atoms (having specific properties) which are implicitly involved in a binary relation defined by the

F. Schwenker and N. El Gayar (Eds.): ANNPR 2010, LNAI 5998, pp. 48–59, 2010.
© Springer-Verlag Berlin Heidelberg 2010

atom-atom chemical bonds. Such molecules may be represented as labeled graphs in a natural manner. Overall, the whole primary structure of the protein can thus be described in terms of a sequence of graphs, each representing the amino-acid that is found at that specific location along the sequence. A machine which is capable of dealing with such a sequence of graphs, and capable of carrying out an automatic segmentation of the protein into relevant sub-sequences (segments), may exploit the knowledge encapsulated in the sub-molecular properties of the polymer.

Another example is found in the area of video processing. Relational learning machines, such as recursive neural nets (RNN) [8], have been widely applied to image classification tasks. In this case, a graphical representation of images is extracted, e.g. in terms of region adjacency graphs (RAG) [3], or multi-resolution trees (MRT) [2]. The idea is that a graphical representation is richer than a traditional, flat feature vector representation. Given that a video consists of a sequence of still images, and hence, it is quite obvious that a video is suitably represented as a sequence of graphs (either RAGs or MRTs). Again, a machine capable of learning and accomplishing classification over sequences of graphs is sought.

More examples can be found in the areas of molecular chemistry, document processing, the World Wide Web, environmental computing, etc. whenever there is a time dependency between structured objects.

This paper introduces a novel machine that fits the framework of sequential graphical pattern recognition. The architecture is defined as follows. The hidden part of a hidden Markov model (HMM) [7] is taken, along with its intrinsic capability of modeling long-term time dependencies, and of performing automatic segmentation of long sequences into sub-sequences. The emission probabilities associated with the states of the HMM are probability density functions (PDF) defined over labeled graphs[1]. Emission PDFs are modeled using a combined artificial neural network (ANN) architecture. An encoding ANN, as in RNNs for graphs [8], is combined with a radial basis function (RBF)-like network that realizes the estimation of the PDF. This probabilistic interpretation is made possible by (i) a description of individual graphs as the random outcomes of a *generalized random graph*, as was formally defined in [9]; and by (ii) a constrained RBF which actually realizes a PDF model which satisfies probability axioms. We stress the fact that the resulting, hybrid machine is not just the aggregation of separate, cooperating architectures. Rather, it shall be thought of as a whole, since a global, joint optimization algorithm is developed, which trains all the model parameters (RBFs parameters, encoding networks weights, HMM initial and transition probabilities) simultaneously in order to increase a shared, overall criterion function, namely the maximum-likelihood (ML) of the model given a sample of training observation sequences. Training takes place within the popular Forward-backward (or, Baum-Welch) procedure for HMMs [7]. Once training is accomplished, the popular Viterbi algorithm [7] can be applied in order to carry out segmentation and classification of sequences of graphical patterns.

The paper is organized as follows: Section 2 gives the fundamental mathematical details of the algorithm. An application to a learning problem is described in Section 3, and some concluding remarks are made in Section 4.

---

[1] A formal notion of PDF over labeled graphs is given in the next Section.

2   Architecture and Training Algorithm

We will use the notion of *generalized random graph* [9] for a formal definition of the probabilistic quantities over sequences of graphs that underly the given framework. Let $\mathcal{V}$ be a given discrete- or continuous-valued set (*vertex universe*), and let Ω be any given *sample space*. We define a Generalized Random Graph (GRG) over $\mathcal{V}$ and Ω as a function $G : \Omega \rightarrow \{(V, E)|V \subseteq \mathcal{V}, E \subseteq V \times V\}$. Let then $\mathcal{G} = \{(V, E)|V \subseteq \mathcal{V}, E \subseteq V \times V\}$. A *probability density function* (PDF) for GRGs over $\mathcal{V}$ is a function $p : \mathcal{G} \rightarrow \Re$ such that: (1) $p(g) \geq 0, \forall g \in \mathcal{G}$, and (2) $\int_{\mathcal{G}} p(g)dg = 1$ (refer to [9] for a discussion on measurability of GRG spaces, i.e. meaning of this integral). Loosely speaking, any function $G(.) = \xi(t)$ which maps time t (either discrete or continuous) onto a GRG is then defined to be a *stochastic graph process*. In turn, a hidden Markov model over graphs (GHMM) is a pair of stochastic processes: a *hidden* Markov chain[2] and an *observable* stochastic graph process which is a probabilistic function of the states of the former. More precisely, a GHMM is defined as a traditional HMM [7] except for the notion of emission probability, namely as:

1. A set S of Q states, $S = \{S_1, \ldots, S_Q\}$, which are the distinct values that the discrete, hidden stochastic process can take.
2. An *initial state* probability distribution, i.e. $\pi = \{Pr(S_i \mid t = 0), S_i \in S\}$, where t is a discrete time index.
3. A probability distribution that characterizes the allowed transitions between states, that is $\mathbf{a}_{ij} = \{Pr(S_j \text{ at time } t + 1 \mid S_i \text{ at time } t), S_i \in S, S_j \in S\}$ where the *transition probabilities* $\mathbf{a}_{ij}$ are assumed to be independent of time t. Note that $\{Pr(S_j \text{ at time } t + 1 \mid S_i \text{ at time } t, S_k \text{ at time } t - 1, \ldots)\} = \mathbf{a}_{ij}$ due to the Markov assumption.
4. A set of PDFs over GRGs (referred to as *emission* probabilities) that describes the statistical properties of the GRGs for each state of the model: $\mathbf{b_G} = \{b_i(g) = p(g \mid S_i), S_i \in S, g \in \mathcal{G}\}$.

Let us assume that a certain sequence $Y = g_1, \ldots, g_T$ of graphs generated by a (hidden) stochastic graph process has been observed, and that it is the expression (*outcome*) of a certain sequence $\mathcal{W} = \omega_1, \ldots, \omega_L$ of states of nature (i.e., classes). Recognition (classification) of the correct class(es) $\mathcal{W}$ relying on the observations Y can be accomplished according to the class(es) posterior probability given Y, yielded by Bayes' theorem: $Pr(\mathcal{W} \mid Y) = p(Y \mid \mathcal{W})Pr(\mathcal{W})/p(Y)$. The quantity $Pr(\mathcal{W})$ is referred to as the prior probability of $\mathcal{W}$. It can be estimated from relative frequencies of classes as in statistical pattern recognition. We propose GHMMs for modeling the class-conditional density $p(Y \mid \mathcal{W})$. Note that this approach deals with "continuous" recognition tasks, that is a sequence of classes $\omega_1, \ldots, \omega_L$ is hidden behind the observations Y, and no prior segmentation of Y into subsequences $Y_1, \ldots, Y_L$ corresponding with the individual classes in known in advance.

The proposed machine relies on a connectionist non-parametric model of the emission probabilities of a GHMM, with gradient-ascent global training techniques over the ML criterion. An ANN is introduced for each state of the GHMM. The output

---

[2] This is a traditional, discrete time random process.

unit of a generic ANN provides an estimate of the corresponding emission probability (that is, a PDF over GRGs) given the current graphical observation in the input graph space. Training of the other probabilistic quantities in the underlying Markov chain, i.e. *initial* and *transition* probabilities, still relies on likelihood maximization via the forward-backward algorithm [7]. The Viterbi algorithm is then applied to the recognition step [7].

The global criterion function to be maximized during training, namely the *likelihood* L of a graphical observation sequence given the model[3], is defined as $L = \sum_{\iota \in \mathcal{F}} \alpha_{\iota,T}$. The sum is extended to the set $\mathcal{F}$ of all possible *final* states [1] within the GHMM corresponding to the current training sequence. The GHMM is supposed to involve Q states, and T is the length of the current observation sequence $Y = g_1, \ldots, g_T$. The *forward* terms $\alpha_{\iota,t} = Pr(q_{\iota,t}, g_1, \ldots, g_t)$ and the *backward* terms $\beta_{\iota,t} = Pr(g_{t+1}, \ldots, g_T | q_{\iota,t})$ for ι-th state at time t can be computed recursively as follows [7]:

$$\alpha_{\iota,t} = b_{\iota,t} \sum_j a_{j\iota} \alpha_{j,t-1} \tag{1}$$

and

$$\beta_{\iota,t} = \sum_j b_{j,t+1} a_{\iota j} \beta_{j,t+1} \tag{2}$$

where $a_{\iota j}$ denotes the transition probability from the ι-th state to the j-th state, $b_{\iota,t}$ denotes the emission probability associated with the ι-th state over the t-th graph g_t, and the sums are extended to all possible states within the GHMM. The initialization of the forward probabilities is accomplished as in HMMs [7], whereas the backward terms at time T are initialized in a slightly different manner, namely:

$$\beta_{\iota,T} = \begin{cases} 1 \text{ if } \iota \in \mathcal{F} \\ 0 \text{ otherwise.} \end{cases} \tag{3}$$

Given a generic parameter θ of an ANN, hill-climbing gradient-ascent over L prescribes a learning rule of the following well-known kind:

$$\Delta\theta = \eta \frac{\partial L}{\partial \theta} \tag{4}$$

where $\eta \in \Re^+$, and η is commonly known as the *learning rate*. Let us observe (from [1]) that the following property can be easily shown to hold true by taking the partial derivatives of the left- and right-hand sides of Equation (1) with respect to $b_{\iota,t}$:

$$\frac{\partial \alpha_{\iota,t}}{\partial b_{\iota,t}} = \frac{\alpha_{\iota,t}}{b_{\iota,t}}. \tag{5}$$

In addition, by borrowing the scheme proposed by [1], the following theorem can be proved to hold true: $\frac{\partial L}{\partial \alpha_{\iota,t}} = \beta_{\iota,t}$, for each $\iota = 1, \ldots, Q$ and for each $t = 1, \ldots, T$.

---

[3] A standard notation is used in the following to refer to quantities involved in HMM training (e.g. [7]). Note that the Greek letter ι (iota) is used to denote the index for a generic state q_ι of the GHMM; it shall not be confused with the index i that will be introduced later.

Given this theorem and Equation (5), repeatedly applying the chain rule we can expand $\frac{\partial L}{\partial \theta}$ by writing:

$$\frac{\partial L}{\partial \theta} = \sum_q \sum_t \frac{\partial L}{\partial b_{q,t}} \frac{\partial b_{q,t}}{\partial \theta} \qquad (6)$$

$$= \sum_q \sum_t \frac{\partial L}{\partial \alpha_{q,t}} \frac{\partial \alpha_{q,t}}{\partial b_{q,t}} \frac{\partial b_{q,t}}{\partial \theta}$$

$$= \sum_q \sum_t \beta_{q,t} \frac{\alpha_{q,t}}{b_{q,t}} \frac{\partial b_{q,t}}{\partial \theta}$$

where the sums are extended over all states q of the GHMM involved in the current training sequence (i.e., all the rows in the current *trellis* [7]), and to all $t = 1, \ldots, T$, respectively. It is seen that all the quantities in the right-hand side of Equation (6) are available upon recursive processing on the standard HMM trellis, except for $\frac{\partial b_{q,t}}{\partial \theta}$. From now on, attention is thus focused on the calculation of $\frac{\partial b_{q,t}}{\partial \theta}$, where $b_{q,t}$ is the output from the corresponding ANN at time t.

Now, for each $\iota = 1, \ldots, Q$ let us assume the existence of an integer d and of two functions, $\phi_\iota : \mathcal{G} \rightarrow \Re^d$ and $p_\iota : \Re^d \rightarrow \Re$, s.t. $b_{\iota,t}$ can be decomposed as:

$$b_{\iota,t} = p_\iota(\phi_\iota(g_t)). \qquad (7)$$

We call $\phi_\iota(.)$ the *encoding* for ι-th state of the GHMM, while $p_\iota(.)$ is simply referred to as the "emission" associated with that state[4]. Again, we assume parametric forms $\phi_\iota(g_t|\boldsymbol{\theta}_{\phi_\iota})$ and $p_\iota(\mathbf{x}|\boldsymbol{\theta}_{p_\iota})$ for the encoding and for the emission, respectively, and we set $\boldsymbol{\theta}_\iota = (\boldsymbol{\theta}_{\phi_\iota}, \boldsymbol{\theta}_{p_\iota})$. Bearing in mind Equation (7), we propose a state-specific, two-block connectionist/statistical model for $b_{\iota,t}$ as follows. The function $\phi_\iota(g_t|\boldsymbol{\theta}_{\phi_\iota})$ is realized via an *encoding network*, suitable to map graphs g_t into real vectors $\mathbf{x}_t$ for $t = 1, \ldots, T$, as described in [8] for supervised training of RNNs over structured domains. The weights of the encoding network are the parameters $\boldsymbol{\theta}_{\phi_\iota}$. A radial basis functions (RBF)-like neural net is then used to model the emission $p_\iota(\mathbf{x}_t|\boldsymbol{\theta}_{p_\iota})$, where $\boldsymbol{\theta}_{p_\iota}$ are the parameters of the RBF. Basically, for each state ι in the GHMM a state-specific RBF is expected to define a mixture of Normal densities over the state-specific encoding $\phi_\iota(g_t|\boldsymbol{\theta}_{\phi_\iota})$ of t-th input graph. From now on, since (i) HMMs assume that the emission probabilities associated with different states are independent of each other [7], (ii) a separate connectionist model is adopted for each one of the states of the GHMM, and (iii) also individual observations (i.e., graphs) within the input sequence are assumed to be independent of each other given the state [7], we simplify the (cumbersome) notation by dropping the state index ι and the time index t, and we focus on the generic quantity $p(\phi(g|\boldsymbol{\theta}_\phi))|\boldsymbol{\theta}_p)$. Note that, for notational convenience, in the following this quantity may be written in short as $p(g)$.

---

[4] It is seen that there exist (infinite) choices for $\phi(.)$ and $\hat{p}(.)$ that satisfy Equation (7), the most trivial being $\phi(g) = p(g|\boldsymbol{\theta}), \hat{p}(x) = x$.

In order to ensure that a PDF is obtained this way, a standard RBF cannot be used straightforwardly: specific constraints have to be placed on the nature of the Gaussian kernels, as well as on the hidden-to-output connection weights. The training algorithm shall provide us with a likelihood maximization scheme that undergoes such constraints. Three distinct families of adaptive parameters θ of the ANNs have to be considered:

(1) Mixing parameters $c_1, \ldots, c_n$, i.e. the hidden-to-output weight of the RBF network. Constraints have to be placed on these parameters during the ML estimation process, in order to ensure that they are in the range $(0, 1)$ and that they sum to one. A simple way to satisfy the requirements is to introduce n hidden parameters $\gamma_1, \ldots, \gamma_n$, which are unconstrained, and to set

$$c_i = \frac{\varsigma(\gamma_i)}{\sum_{j=1}^{n} \varsigma(\gamma_j)}, i = 1, \ldots, n \tag{8}$$

where $\varsigma(x) = 1/(1 + e^{-x})$. Each γ_i is then treated as an unknown parameter θ to be estimated via ML.

(2) The d-dimensional mean vector μ_i and $d \times d$ covariance matrix Σ_i for each one of the Gaussian kernels $K_i(\mathbf{x}) = N(\mathbf{x}; \mu_i, \Sigma_i)$, $i = 1, \ldots, n$ of the RBF-like network, where $N(\mathbf{x}; \mu_i, \Sigma_i)$ denotes a multivariate Normal PDF having mean vector μ_i, covariance matrix Σ_i, and evaluated over the random vector $\mathbf{x}$. A common (yet effective) simplification is to consider diagonal covariance matrices, i.e. independence among the components of the input vector $\mathbf{x}$. This assumption leads to the following three major consequences: (i) modeling properties are not affected, according to [6]; (ii) generalization capabilities of the overall model may turn out to be improved, since the number of free parameters is reduced to a significant extent; (iii) i-th multivariate kernel K_i may be expressed in the form of a product of d univariate Normal densities as:

$$K_i(\mathbf{x}) = \prod_{j=1}^{d} \frac{1}{\sqrt{2\pi}\sigma_{ij}} exp\left\{ -\frac{1}{2} \left(\frac{x_j - \mu_{ij}}{\sigma_{ij}} \right)^2 \right\} \tag{9}$$

i.e., the free parameters to be estimated are the means μ_{ij} and the standard deviations σ_{ij}, for each kernel $i = 1, \ldots, n$ and for each component $j = 1, \ldots, d$ of the input space.

(3) The weights $\mathcal{U}$ of the encoding network.

In the following, we will derive explicit formulations for $\frac{\partial p(\phi(g|\boldsymbol{\theta}_\phi)|\boldsymbol{\theta}_p)}{\partial \theta}$ for each of the three families of free parameters θ above. These derivatives are then put in place of $\frac{\partial b_{q,t}}{\partial \theta}$ in Equation (6), obtaining the quantity $\frac{\partial L}{\partial \theta}$ and, in turn, the overall learning rule $\Delta\theta = \eta \frac{\partial L}{\partial \theta}$ for the generic parameter θ.

As regards a generic mixing parameter $c_i, i = 1, \ldots, n$, from Equations (7) and (8), and since $p(g) = \sum_{k=1}^{n} c_k K_k(\mathbf{x})$, we can obtain Equation 10.

$$\frac{\partial p(\phi(g|\boldsymbol{\theta}_\phi)|\boldsymbol{\theta}_p)}{\partial \gamma_i} = \sum_{j=1}^{n} \frac{\partial p(g)}{\partial c_j} \frac{\partial c_j}{\partial \gamma_i} \tag{10}$$

$$= \sum_{j=1}^{n} K_j(\mathbf{x}) \frac{\partial}{\partial \gamma_i} \left(\frac{\varsigma(\gamma_j)}{\sum_{k=1}^{n} \varsigma(\gamma_k)} \right)$$

$$= K_i(\mathbf{x}) \left\{ \frac{\varsigma'(\gamma_i) \sum_k \varsigma(\gamma_k) - \varsigma(\gamma_i)\varsigma'(\gamma_i)}{[\sum_k \varsigma(\gamma_k)]^2} \right\} + \sum_{j \neq i} K_j(\mathbf{x}) \left\{ \frac{-\varsigma(\gamma_j)\varsigma'(\gamma_i)}{[\sum_k \varsigma(\gamma_k)]^2} \right\}$$

$$= K_i(\mathbf{x}) \frac{\varsigma'(\gamma_i)}{\sum_k \varsigma(\gamma_k)} - \sum_j K_j(\mathbf{x}) \frac{\varsigma(\gamma_j)\varsigma'(\gamma_i)}{[\sum_k \varsigma(\gamma_k)]^2}$$

$$= K_i(\mathbf{x}) \frac{\varsigma'(\gamma_i)}{\sum_k \varsigma(\gamma_k)} - \left\{ \sum_j c_j K_j(\mathbf{x}) \right\} \frac{\varsigma'(\gamma_i)}{\sum_k \varsigma(\gamma_k)}$$

$$= \frac{\varsigma'(\gamma_i)}{\sum_k \varsigma(\gamma_k)} \{K_i(\mathbf{x}) - p(g)\}$$

Bearing in mind that the calculations were carried out for a connectionist model of the emission probability $b_{\iota,t}$ associated with the generic ι-th state of the GHMM and evaluated over t-th graph in the sequence, and using the symbol $\mathcal{Q}(\iota)$ to denote the subset of the states involved in the current trellis that are instances of the state ι, we can reintroduce the state index and the time index t in the notation, and rewrite Equation (10) as follows:

$$\frac{\partial b_{q,t}}{\partial \gamma_i^{(\iota)}} = \begin{cases} \frac{\varsigma'(\gamma_i^{(\iota)})}{\sum_k \varsigma(\gamma_k^{(\iota)})} \left\{ K_i^{(\iota)}(\mathbf{x}_t^{(\iota)}) - b_{\iota,t} \right\} & \text{if } q \in \mathcal{Q}(\iota) \\ 0 & \text{otherwise} \end{cases} \tag{11}$$

where the writings in the form $\gamma_i^{(\iota)}$, $\mathbf{x}_t^{(\iota)}$ and $K_i^{(\iota)}(.)$ (i.e., those having superscript $^{(\iota)}$ for any value of $\iota = 1, \ldots, Q$) denote the corresponding quantities γ_i, $\mathbf{x}_t$ and $K_i(.)$ within the RBF associated with ι-th state, respectively, according to the previous notation. Substituting Equation (11) into Equation (6) and the latter, in turn, into Equation (4), we obtain the following learning rule for the i-th mixing parameter $\gamma_i^{(\iota)}$ within the ι-th emission model:

$$\Delta \gamma_i^{(\iota)} = \eta \sum_{q \in \mathcal{Q}(\iota)} \sum_t \beta_{q,t} \frac{\alpha_{q,t}}{b_{\iota,t}} \frac{\varsigma'(\gamma_i^{(\iota)})}{\sum_k \varsigma(\gamma_k^{(\iota)})} \left\{ K_i^{(\iota)}(\mathbf{x}_t^{(\iota)}) - b_{\iota,t} \right\} \tag{12}$$

where we implicitly exploited the (obvious) fact that $b_{q,t} = b_{\iota,t}$ for all $q \in \mathcal{Q}(\iota)$.

For the means μ_{ij} and the standard deviations σ_{ij} we proceed as follows. Let θ_{ij} denote the free parameter, i.e. μ_{ij} or σ_{ij}, to be estimated. We can write:

$$\frac{\partial p(\phi(g|\boldsymbol{\theta}_\phi)|\boldsymbol{\theta}_p)}{\partial \theta_{ij}} = \frac{\partial \sum_{k=1}^{n} c_k K_k(\mathbf{x})}{\partial \theta_{ij}} \tag{13}$$

$$= c_i \frac{\partial K_i(\mathbf{x})}{\partial \theta_{ij}}$$

where the calculation of $\frac{\partial K_i(\mathbf{x})}{\partial \theta_{ij}}$ can be accomplished as follows. First of all, let us observe that for any real-valued, differentiable function $f(.)$ this property holds true: $\frac{\partial f(.)}{\partial x} = f(.) \frac{\partial \log[f(.)]}{\partial x}$. As a consequence, from Equation (9) we can write

$$\frac{\partial K_i(\mathbf{x})}{\partial \theta_{ij}} = K_i(\mathbf{x}) \frac{\partial}{\partial \theta_{ij}} \sum_{k=1}^{d} \left\{ -\frac{1}{2} \left[\log(2\pi\sigma_{ik}^2) + \left(\frac{x_k - \mu_{ik}}{\sigma_{ik}} \right)^2 \right] \right\}. \qquad (14)$$

For the means, i.e. $\theta_{ij} = \mu_{ij}$, Equation (14) yields

$$\frac{\partial K_i(\mathbf{x})}{\partial \mu_{ij}} = K_i(\mathbf{x}) \frac{\partial}{\partial \mu_{ij}} \left\{ -\frac{1}{2} \left(\frac{x_j - \mu_{ij}}{\sigma_{ij}} \right)^2 \right\} \qquad (15)$$

$$= K_i(\mathbf{x}) \frac{x_j - \mu_{ij}}{\sigma_{ij}^2}$$

which can be substituted into Equation (13), obtaining

$$\frac{\partial p(\phi(g|\boldsymbol{\theta}_\phi)|\boldsymbol{\theta}_p)}{\partial \mu_{ij}} = c_i K_i(\mathbf{x}) \frac{x_j - \mu_{ij}}{\sigma_{ij}^2}. \qquad (16)$$

Now we can reintroduce the state index ι and the time index t in the notation, and rewrite the Equation as $\frac{\partial b_{\iota,t}}{\partial \mu_{ij}^{(\iota)}} = c_i^{(\iota)} K_i^{(\iota)}(\mathbf{x}_t^{(\iota)}) \frac{x_{tj}^{(\iota)} - \mu_{ij}^{(\iota)}}{\sigma_{ij}^{2(\iota)}}$, where $x_{tj}^{(\iota)}$ denotes the j-th component of the vector $\mathbf{x}_t^{(\iota)}$ which represents the ι-th encoding of t-th input graph g_t within the training sequence, $\sigma_{ij}^{2(\iota)}$ is the j-th component of the diagonal of the covariance matrix associated with the i-th kernel of ι-th emission PDF, while the other symbols have the same meaning as above. Again, since $\frac{\partial b_{q,t}}{\partial \mu_{ij}^{(\iota)}} = 0$ when $q \notin \mathcal{Q}(\iota)$, the expression can be substituted into Equation (6) and the latter, in turn, into Equation (4), obtaining the following learning rule for the j-th component of the mean vector $\mu_{ij}^{(\iota)}$ associated with i-th kernel function within the ι-th emission model:

$$\Delta \mu_{ij}^{(\iota)} = \eta \sum_{q \in \mathcal{Q}(\iota)} \sum_{t} \beta_{q,t} \frac{\alpha_{q,t}}{b_{\iota,t}} c_i^{(\iota)} K_i^{(\iota)}(\mathbf{x}_t^{(\iota)}) \frac{x_{tj}^{(\iota)} - \mu_{ij}^{(\iota)}}{\sigma_{ij}^{2(\iota)}} \qquad (17)$$

where, again, we exploited the fact that $b_{q,t} = b_{\iota,t}$ for all $q \in \mathcal{Q}(\iota)$.

For the covariances, i.e. $\theta_{ij} = \sigma_{ij}$, Equation (14) yields:

$$\frac{\partial K_i(\mathbf{x})}{\partial \sigma_{ij}} = \frac{K_i(\mathbf{x})}{\sigma_{ij}} \left\{ \left(\frac{x_j - \mu_{ij}}{\sigma_{ij}} \right)^2 - 1 \right\} \qquad (18)$$

which can be substituted into Equation (13) obtaining

$$\frac{\partial p(\phi(g|\boldsymbol{\theta}_\phi)|\boldsymbol{\theta}_p)}{\partial \sigma_{ij}} = c_i \frac{K_i(\mathbf{x})}{\sigma_{ij}} \left\{ \left(\frac{x_j - \mu_{ij}}{\sigma_{ij}} \right)^2 - 1 \right\} \qquad (19)$$

that, adopting the notation above for expressing the dependence on the generic ι-th state of the GHMM and on the time index t, can be substituted into Equation (6) and the latter, in turn, into Equation (4), obtaining the following learning rule:

$$\Delta\sigma_{ij}^{(\iota)} = \eta \sum_{q \in \mathcal{Q}(\iota)} \sum_{t} \beta_{q,\iota} \frac{\alpha_{q,t}}{b_{\iota,t}} c_i^{(\iota)} \frac{K_i^{(\iota)}(\mathbf{x}_t^{(\iota)})}{\sigma_{ij}^{(\iota)}} \left\{ \left(\frac{x_{tj}^{(\iota)} - \mu_{ij}^{(\iota)}}{\sigma_{ij}^{(\iota)}} \right)^2 - 1 \right\}. \tag{20}$$

Finally, let us consider the connection weights $\mathcal{U} = \{v_1, \ldots, v_s\}$ within the encoding network. The term $\frac{\partial p(\phi(g|\boldsymbol{\theta}_\phi)|\boldsymbol{\theta}_p)}{\partial v}$ in Equation (6) can be computed as follows. Applying the chain rule yields:

$$\frac{\partial p(\phi(g|\boldsymbol{\theta}_\phi)|\boldsymbol{\theta}_p)}{\partial v} = \frac{\partial p(\phi(g|\boldsymbol{\theta}_\phi)|\boldsymbol{\theta}_p)}{\partial y} \frac{\partial y}{\partial v} \tag{21}$$

where y is the output from the unit (in the encoding net) which is fed from connection v. The quantity $\frac{\partial y}{\partial v}$ can be easily computed by taking the partial derivative of the activation function associated with the unit itself, as usual. In particular, if $v = v_{\ell m}$ is the connection weight between the generic m-th unit in a given layer and ℓ-th unit in the following layer, s.t. the corresponding outputs are y_m and y_ℓ, respectively, we have

$$\frac{\partial y_\ell}{\partial v_{\ell m}} = f_\ell'(a_\ell) y_m \tag{22}$$

where $y_\ell = f_\ell(a_\ell)$ and $y_m = f_m(a_m)$ are the activation functions associated with ℓ-th unit and m-th unit, respectively, and a_ℓ and a_m are the corresponding activations (i.e., inputs), and where $f_\ell'(a_\ell)$ denotes the derivative of the activation function given.

As regards the quantity $\frac{\partial p(\phi(g|\boldsymbol{\theta}_\phi)|\boldsymbol{\theta}_p)}{\partial y}$, we proceed as follows. First of all, let us assume that v feeds the output layer, i.e. it connects a certain hidden unit with j-th output unit of the encoding net. In this case, we have $y = x_j$, and:

$$\frac{\partial p(\phi(g|\boldsymbol{\theta}_\phi)|\boldsymbol{\theta}_p)}{\partial x_j} = \frac{\partial \sum_{i=1}^{n} c_i K_i(\mathbf{x})}{\partial x_j} \tag{23}$$

$$= \sum_{i=1}^{n} c_i K_i(\mathbf{x}) \frac{\partial}{\partial x_j} \sum_{k=1}^{d} \left\{ -\frac{1}{2} \left[\log(2\pi\sigma_{ik}^2) + \left(\frac{x_k w_{ik} - \mu_{ik}}{\sigma_{ik}} \right)^2 \right] \right\}$$

$$= \sum_{i=1}^{n} c_i K_i(\mathbf{x}) \left\{ -\frac{1}{2} \frac{\partial}{\partial x_j} \left(\frac{x_j w_{ij} - \mu_{ij}}{\sigma_{ij}} \right)^2 \right\}$$

$$= -\sum_{i=1}^{n} c_i \frac{K_i(\mathbf{x})}{\sigma_{ij}^2} (x_j w_{ij} - \mu_{ij}) w_{ij}.$$

Equations (22) and (23) can be substituted into Equation (21) obtaining:

$$\frac{\partial p(\phi(g|\boldsymbol{\theta}_\phi)|\boldsymbol{\theta}_p)}{\partial v_{jm}} = -\sum_{i=1}^{n} c_i \frac{K_i(\mathbf{x})}{\sigma_{ij}^2} (x_j w_{ij} - \mu_{ij}) w_{ij} f_j'(a_j) f_m(a_m). \tag{24}$$

By defining the quantity:

$$\delta_j = - \sum_{i=1}^{n} c_i \frac{K_i(\mathbf{x})}{\sigma_{ij}^2} (x_j w_{ij} - \mu_{ij}) w_{ij} f_j'(a_j) \tag{25}$$

for the generic j-th output unit in the encoding network, we can rewrite Equation (24) in the following, compact form:

$$\frac{\partial p(\phi(g|\boldsymbol{\theta}_\phi)|\boldsymbol{\theta}_p)}{\partial v_{jm}} = \delta_j f_m(a_m). \tag{26}$$

When v is a hidden weight (say, $v = v_{m\ell}$ where ℓ and m are the indexes of generic hidden units connected via v), the quantity $\frac{\partial p(\phi(g|\boldsymbol{\theta}_\phi)|\boldsymbol{\theta}_p)}{\partial v_{m\ell}}$ can be obtained applying the usual *backpropagation through structures* (BPTS) algorithm [8], once the *deltas* to be backpropagated have been initialized at the output layer via Equation (25). In so doing, a quantity δ_m can be defined for each hidden unit m such that

$$\frac{\partial p(\phi(g|\boldsymbol{\theta}_\phi)|\boldsymbol{\theta}_p)}{\partial v_{m\ell}} = \delta_m f_\ell(a_\ell). \tag{27}$$

Substituting Equation (26) or Equation (27), respectively, into Equation (6) yields an overall learning rule for any given weight $v_{ij}^{(\iota)}$ within the encoding network associated with ι-th state of the GHMM in the following, common form:

$$\Delta v_{ij}^{(\iota)} = \eta \sum_{q \in \mathcal{Q}(\iota)} \sum_t \beta_{q,t} \frac{\alpha_{q,t}}{b_{\iota,t}} \delta_i^{(\iota)} f_j^{(\iota)}(a_j^{(\iota)}) \tag{28}$$

where the superscript $^{(\iota)}$ has the usual meaning.

3 Demonstration

The proposed machine has been implemented, and preliminary experiments were carried out on a graphical sequence recognition task drawn from the *Policemen* dataset [4]. This dataset features images of synthetically generated policemen, having different color, orientation, position of the arms, etc. Directed ordered acyclic graphs (DOAG) were used for representing the individual images, as explained in [4]. Two-dimensional real-valued labels are associated with the nodes in this DOAG representation. Sequences were generated by taking individual images and creating concatenations of images s.t. a coherent "movement" (as in a cartoon sequence) of the policeman emerged. For instance, the images in a given sequence may represent the policemen gradually rising then lowering his left arm, followed by an analogous movement of the right arm. The task involves 158 sequences overall, belonging to 4 disjoint classes. In turn, each class is further divided into subclasses as follows. Class 1: rotation; subclasses: (1.1) clockwise and (1.2) counter-clockwise. Class 2: shift; subclasses: (2.1) right-left and (2.2) top-down. Class 3: zoom; subclasses: (3.1) zoom-in and (3.2) zoom-out. Class 4: arms movement; subclasses: (4.1) both arms up, (4.2) both arms down, (4.3) right arm up, (4.4) right arm down, (4.5) left arm up, and (4.6) left arm down. Hence, there are 12 classes in total.

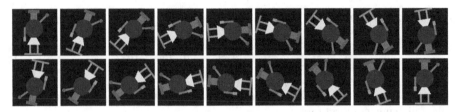

Fig. 1. Sample sequence for subclass 1.1

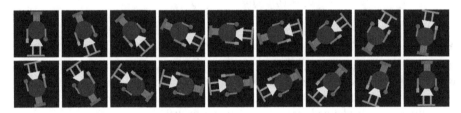

Fig. 2. Sample sequence for subclass 1.2

The length of individual sequences ranged from 10 to 17 graphs. Figures 1 and 2 show two sample sequences from Class 1. Note that the starting and ending frame of some sequences from different classes were identical, and that for sequences from other classes, some frames from within a sequence were identical. Hence, a good classification can only be obtained if the method can encode the given sequences as a whole.

The dataset was split into a training set (72 sequences, chosen by drawing 6 sequences from each subclass at random) and a test set (all the remaining sequences). Separate left-to-right Markov chains were used for each class, each of them having 4 states. Emission probabilities were modeled with 2 multivariate Gaussian kernels RBFs, and recursive encoding networks having 8 sigmoid hidden neurons, 10 state neurons, encoding dimension of 2, and a maximum out-degree of 6 (please refer to [8] for a description of the meaning of such quantities in recursive nets). System parameters were initialized according to a *segmental k-means*-like procedure [7]. In the present, illustrative demonstration the proposed training algorithm was applied for 4 iterations (i.e., epochs) using different learning rates (obtained via cross-validation) for the specific parameters involved in the optimization process. Results are reported in Table 1 in terms of recognition accuracy. Although no direct comparison is possible w.r.t. any other benchmark approaches (since, to the best of our knowledge, the present model is the first attempt to dealing with classification of sequences of graphs), it is seen from Table 1 that: (i) the architecture is indeed suitable to graphical sequence modeling; (ii) the training algorithm, whilst focusing on the maximization of the likelihood (of the model given the training sample) criterion, results also in a significant accuracy in terms of the sequence recognition rate criterion; and (iii) comparison between the accuracy on the training and test sets confirms that learning capability of the machine does not prevent emergence of an appreciable generalization capability.

We found that the residual classification error is attributed to two of the subclasses whose properties are such that a correct classification requires the encoding of the associated data labels. Since the labels provide features which are furthest from the output

Table 1. Recognition accuracy of sequences from the *Policemen* dataset

	Accuracy on training set	Accuracy on test set
Upon initialization	51.39 %	39.53 %
After training	87.50 %	86.05 %

layer, and hence, this implies that the proposed approach priorizes the encoding of structure over the data labels. In general, it can be expected that the classification result will improve further when training is carried out for more iterations, or by adding direct forward links from the labels to the output layer.

4 Conclusion and On-Going Work

The paper introduced a first model for learning and classification over sequences of graphs. The architecture relies on an underlying HMM structure, capable of dealing with long-term dependencies in sequential data of arbitrary length. Emission PDFs over GRGs are estimated by means of a combination of recursive encoding nets and constrained RBF-like nets. A global optimization algorithm, aimed at the maximization of the likelihood of the model given the training observation sequences has been developed. Preliminary results confirm that the architecture and the algorithms are effective, both in terms of learning and generalization capabilities. Current efforts are focused on evaluating and further improving the classification and regression ability of the approach, and on its application to more difficult, real-world tasks.

References

1. Bengio, Y.: Neural Networks for Speech and Sequence Recognition. International Thomson Computer Press, London (1996)
2. Bianchini, M., Maggini, M., Sarti, L.: Object recognition using multiresolution trees. In: Yeung, D.-Y., Kwok, J.T., Fred, A., Roli, F., de Ridder, D. (eds.) SSPR 2006 and SPR 2006. LNCS, vol. 4109, pp. 331–339. Springer, Heidelberg (2006)
3. Di Massa, V., Monfardini, G., Sarti, L., Scarselli, F., Maggini, M., Gori, M.: A comparison between recursive neural networks and graph neural networks. In: World Congress on Computational Intelligence, July 2006, pp. 778–785 (2006)
4. Hagenbuchner, M., Gori, M., Bunke, H., Tsoi, A.C., Irniger, C.: Using attributed plex grammars for the generation of image and graph databases. Pattern Recognition Letters (Special issue on Graph-based Representations) 24(8), 1081–1087 (2002)
5. Haykin, S.: Neural Networks, A Comprehensive Foundation. Macmillan College Publishing Company, Inc., New York (1994)
6. McLachlan, G.J., Basford, K.E. (eds.): Mixture Models: Inference and Applications to Clustering. Marcel Dekker, New York (1988)
7. Rabiner, L.R.: Rabiner. A tutorial on hidden Markov models and selected applications in speech recognition. Proceedings of the IEEE 77(2), 257–286 (1989)
8. Sperduti, A., Starita, A.: Supervised neural networks for the classification of structures. IEEE Transactions on Neural Networks 8(3), 714–735 (1997)
9. Trentin, E., Rigutini, L.: A maximum-likelihood connectionist model for unsupervised learning over graphical domains. In: Alippi, C., Polycarpou, M., Panayiotou, C., Ellinas, G. (eds.) ICANN 2009, Part 1. LNCS, vol. 5768, pp. 40–49. Springer, Heidelberg (2009)

Maximum Echo-State-Likelihood Networks for Emotion Recognition

Edmondo Trentin[1], Stefan Scherer[2], and Friedhelm Schwenker[2]

[1] Dipartimento di Ingegneria dell'Informazione, Università degli studi di Siena,
Siena, Italy
trentin@dii.unisi.it
[2] Institute of Neural Information Processing, Ulm University, Ulm, Germany
{stefan.scherer,friedhelm.schwenker}@uni-ulm.de

Abstract. Emotion recognition is a relevant task in human-computer interaction. Several pattern recognition and machine learning techniques have been applied so far in order to assign input audio and/or video sequences to specific emotional classes. This paper introduces a novel approach to the problem, suitable also to more generic sequence recognition tasks. The approach relies on the combination of the recurrent reservoir of an echo state network with a connectionist density estimation module. The reservoir realizes an encoding of the input sequences into a fixed-dimensionality pattern of neuron activations. The density estimator, consisting of a constrained radial basis functions network, evaluates the likelihood of the echo state given the input. Unsupervised training is accomplished within a maximum-likelihood framework. The architecture can then be used for estimating class-conditional probabilities in order to carry out emotion classification within a Bayesian setup. Preliminary experiments in emotion recognition from speech signals from the WaSeP© dataset show that the proposed approach is effective, and it may outperform state-of-the-art classifiers.

Keywords: Emotion recognition, echo state network, radial basis functions, maximum likelihood, density estimation.

1 Introduction

In the last few years, human-machine interaction (HMI) has been taking a more and more important position in our everyday life. However, up to date expressive and emotional conversation is a matter of human-human communication (HHC) only. Current machines are not capable of neither understanding nor expressing subjective states or emotional expressions. Since this second, implicit channel of communication (containing information about the speaker himself, the situation, and the ongoing interaction) is so important for natural and efficient HHC, it is believed that the only way to render HMI more natural and efficient is to implement the capabilities to recognize, understand, and express these conversational elements in machines. In the present work a novel approach towards recognizing emotional expressions is introduced. The model relies on the combination of an

F. Schwenker and N. El Gayar (Eds.): ANNPR 2010, LNAI 5998, pp. 60–71, 2010.
© Springer-Verlag Berlin Heidelberg 2010

echo state network (ESN) [8] and a constrained radial basis functions (RBF)-like network [1] suitable for density estimation. The basic idea is that the recurrent reservoir of the ESN realizes an encoding of an input sequence (e.g., an acoustic observation sequence obtained from the speech signal of the user whose emotional state has to be recognized) by means of the pattern of activation of its state neurons. The RBF is trained in order to estimate the probability density function (pdf) which characterizes the distribution of the reservoir activation patterns within the encoding space. The model is trained according to an algorithm aimed at the maximization of the likelihood of the encoding (i.e., of the echo-state) given the input sequence. For these reasons, we refer to the overall machine as the maximum echo-state-likelihood network (MESLiN). The training scheme is inherently unsupervised and non-discriminative, along the line of statistical parametric pdf estimation techniques relying on the maximum-likelihood (ML) criterion [2]. Nonetheless, it can be applied in emotion recognition tasks by using a separate MESLiN for estimating the class-conditional pdf [2] for each of the classes involved in the problem. In order to proof the concept of this approach, experiments based on a corpus containing pseudo words spoken in six different emotional prosodies, WaSeP© [17], have been conducted.

The remainder of the paper is organized as follows. In Section 2 the approach is presented, introducing the echo state encoding and the ML estimation algorithm. Section 3 introduces the utilized dataset, the feature extraction process, and reports the performance of humans in a large scale perception test. Sections 4 and 5 report on the achieved classification results for the experiments and conclude the paper, respectively.

2   The Maximum Echo-State-Likelihood Network

The model is introduced in the framework of emotion recognition from speech signals, although it can be applied to several sequence recognition tasks using different feature spaces. As we stated in the previous section, a separate, class-specific, and independent MESLiN is used for each emotion involved in the task. In so doing, in the following we will focus on a generic machine, trained over the corresponding, emotion-specific training sample, with the understanding that the algorithm has to be subsequently applied to as many MESLiNs as the number of classes at hand.

In this perspective, suppose that a sample $\mathcal{T} = \{\mathcal{Y}_1, \ldots, \mathcal{Y}_n\}$ of n acoustic observation sequences has been observed. The pdf estimation problem faced in this paper can be stated as follows: assuming that all the sequences in $\mathcal{T}$ have been independently drawn from a certain pdf $p(\mathcal{Y})$, how can the dataset be used in order to estimate a "reasonable" model of $p(\mathcal{Y})$? We assume that $p(\mathcal{Y})$ is a function having fixed and known parametric form, being determined uniquely by the specific value of a set of *parameters* $\theta = (\theta_1, \ldots, \theta_k)$. To render this dependency on θ in a more explicit manner, we will modify our notation slightly by writing $p(\mathcal{Y})$ as $p(\mathcal{Y}|\theta)$. Given the assumption, the formulation of the question posed above can be restated as: how can we use the sample $\mathcal{T}$ in order to obtain estimates for θ that are meaningful according to a certain "optimality" criterion? A sound answer to the question may be found in the adoption of the ML criterion, along with a suitable method for maximizing the likelihood $p(\mathcal{T}|\theta)$ of the parameters given the sample. Since $\mathcal{Y}_1, \ldots, \mathcal{Y}_n$ are assumed to be i.i.d., the likelihood $p(\mathcal{T}|\theta)$ can be written as $p(\mathcal{T}|\theta) = \prod_{i=1}^{n} p(\mathcal{Y}_i|\theta)$. Before attempting the maximization of the likelihood, it is necessary to specify a well-defined form for the pdf $p(\mathcal{Y}|\theta)$. Let us assume the existence of an integer d and of two functions, $\phi : \{\mathcal{Y}\} \to \Re^d$ (where $\{\mathcal{Y}\}$ is the universe of all possible observation sequences) and $\hat{p} : \Re^d \to \Re$, s.t. $p(\mathcal{Y}|\theta)$ can be decomposed as:

$$p(\mathcal{Y}|\theta) = \hat{p}(\phi(\mathcal{Y})). \tag{1}$$

It is seen that there exist (infinite) choices for $\phi(.)$ and $\hat{p}(.)$ that satisfy Eq. (1), the most trivial being $\phi(\mathcal{Y}) = p(\mathcal{Y}|\theta), \hat{p}(x) = x$. We call $\phi(.)$ the *encoding*, while $\hat{p}(.)$ is simply referred to as the "likelihood". Again, we assume parametric forms $\phi(\mathcal{Y}|\theta_\phi)$ and $\hat{p}(\mathbf{x}|\theta_{\hat{p}})$ for the encoding and for the likelihood, respectively, and we set $\theta = (\theta_\phi, \theta_{\hat{p}})$. For notational convenience, we will sometimes write $p(\mathcal{Y})$ as a shortcut for $p(\mathcal{Y}|\theta)$.

We propose a two-block connectionist/statistical model for $p(\mathcal{Y}|\theta)$ as follows. The function $\phi(\mathcal{Y}|\theta_\phi)$ is realized via an echo state network, suitable to map sequences $\mathcal{Y}$ into real vectors $\mathbf{x}$. The weights of the ESN become the parameters θ_ϕ. A radial basis functions (RBF)-like neural net is then used to model the likelihood function $\hat{p}(\mathbf{x}|\theta_{\hat{p}})$, where $\theta_{\hat{p}}$ are the parameters of the RBF. In order to ensure that a pdf is obtained, constraints have to be placed on the hidden-to-output connection weights of the RBF (assuming that normalized Gaussian kernels are used).

First, let us focus on the ESN-based model for $\phi(\mathcal{Y}|\theta_\phi)$. An ESN [8] is a particular, recent type of recurrent neural network (RNN). Among the advantages of an ESN over common RNNs are the stability towards noisy inputs [14] and the efficient method to adapt the weights of the network [7]. ESNs are applicable in many different tasks such as classification, pattern generation, or controlling tasks [14,15,8,7]. However, in this application the ESN is used to encode the input data sequence within its current state, i.e. the pattern of activation of the neurons in its reservoir. A schematic representation of an ESN is shown in Figure 1. The most important part of the network is the so called reservoir. It is a collection of neurons (typically, from around ten to a few thousand in number), that are loosely connected to each other. Typically, the probability of a connection between neuron a_i and neuron a_j (to be set in the connection matrix W) is around $2\% - 10\%$ and usually decreases with a rising number of neurons within the reservoir, whereas the connections between the input and output layer with the reservoir are all set. This loose connectivity leads, in turn, to several small cliques of neurons that are recursively connected to each other, sensitive to a certain dynamic within the data received through the input and from other connected neurons. If observed as separated from the rest of the network, an individual clique may appear to follow a seemingly random pattern. However, if observed along with all the competing and supporting cliques within the large

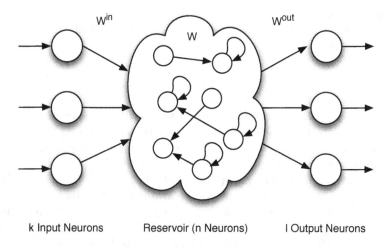

Fig. 1. Schematics of an echo state network with k input neurons, l output neurons, and n neurons in the reservoir. The input is fully connected to the reservoir via W^{in} as well as the reservoir with the output via W^{out}. The topology of the connections within the reservoir, as well as their weights W, are set randomly.

reservoir, the reactions of the clique to the inputs are anything but random. Since there are feedback and recursive connections within the reservoir, not only the input is taken into account for the output but also the current state of each of the neurons, and the history of all previous inputs. Therefore, ESNs are an ideal candidate for encoding dynamic processes, such as emotional expressions or non-verbal utterances [14,15].

In contrast to standard feedforward neural networks, such as multi-layer perceptrons, the ESN incorporates previous features and states into its current state, rendering it an ideal approach for encoding tasks, possibly including the modeling of typical dynamics found in the speech signals (e.g., prosody of emotional expressions). The encoding of an acoustic observation sequence $\mathcal{Y} = \mathbf{y}_1, \ldots, \mathbf{y}_T$ (where T is not fixed, but sequence-specific) for a generic ESN in the system is accomplished as follows:

1. Initialize the states of the ESN randomly.
2. Feed the ESN with the first L acoustic feature vectors $\mathcal{Y}_1, \ldots, \mathcal{Y}_L$ (to minimize the influence of the random initialization)[1].
3. Save the current state $\hat{\mathbf{x}}$ of the ESN as the starting state.
4. Set the ESN in $\hat{\mathbf{x}}$, and sequentially feed the ESN with inputs $\mathcal{Y}$.
5. Let $\mathbf{x}$ denote the *encoding* of $\mathcal{Y}$, i.e. the state of the reservoir at the end of sequence $\mathcal{Y}$.
6. Feed the RBF with $\mathbf{x}$.

---

[1] In this paper we used $L = 5$.

Given the fact that the weights θ_ϕ of the ESN are not optimized, the ML estimation of the RBF parameters $\theta_{\hat{p}}$ given $\mathcal{T}$ requires now to find parameters that maximize the quantity

$$p(\mathcal{T}|\theta_{\hat{p}}) = \prod_{i=1}^{n} \hat{p}(\phi(\mathcal{Y}_i|\theta_\phi)|\theta_{\hat{p}}). \tag{2}$$

A hill-climbing algorithm to carry out ML estimation of the parameters $\theta_{\hat{p}}$ can be obtained as an instance of the gradient-ascent method over $p(\mathcal{T}|\theta_{\hat{p}})$ in two steps: (i) *initialization*, i.e., start with some initial, e.g. "random", assignment of values to the RBF parameters; (ii) *gradient-ascent*, i.e., repeatedly apply a learning rule in the form $\Delta\theta_{\hat{p}} = \eta \nabla_{\theta_{\hat{p}}} \{\prod_{i=1}^{n} p(\phi(\mathcal{Y}_i|\theta_\phi)|\theta_{\hat{p}})\}$ with $\eta \in \Re^+$. This is a batch learning setup. In practice, neural network learning may be simplified, yet even improved, with the adoption of an on-line training scheme that prescribes $\Delta\theta_{\hat{p}} = \eta \nabla_{\theta_{\hat{p}}} \{\hat{p}(\phi(\mathcal{Y}|\theta_\phi)|\theta_{\hat{p}})\}$ upon presentation of each individual training sequence $\mathcal{Y}$. Two distinct families of adaptive parameters θ have to be considered:

(1) Mixing parameters $c_1, \ldots, c_n$, i.e. the hidden-to-output weights of the RBF network. Constraints have to be placed on these parameters during the ML estimation process, in order to ensure that they are in $[0, 1]$ and that they sum to one. A simple way to satisfy the requirements is to introduce n hidden parameters $\gamma_1, \ldots, \gamma_n$, which are unconstrained, and to set

$$c_i = \frac{\varsigma(\gamma_i)}{\sum_{j=1}^{n} \varsigma(\gamma_j)}, i = 1, \ldots, n \tag{3}$$

where $\varsigma(x) = 1/(1 + e^{-x})$. Each γ_i is then treated as an unknown parameter θ to be estimated via ML.

(2) d-dimensional mean vector μ_i and $d \times d$ covariance matrix Σ_i for each of the Gaussian kernels $K_i(\mathbf{x}) = N(\mathbf{x}; \mu_i, \Sigma_i)$, $i = 1, \ldots, n$ of the RBF, where $N(\mathbf{x}; \mu_i, \Sigma_i)$ denotes a multivariate Normal pdf having mean vector μ_i, covariance matrix Σ_i, and evaluated over the random vector $\mathbf{x}$. A common (yet effective) simplification is to consider diagonal covariance matrices, i.e. independence among the components of the input vector $\mathbf{x}$. This assumption leads to the following three major consequences: (i) modeling properties are not affected significantly, according to [10]; (ii) generalization capabilities of the overall model may turn out to be improved, since the number of free parameters is reduced; (iii) i-th multivariate kernel K_i may be expressed in the form of a product of d univariate Normal densities as:

$$K_i(\mathbf{x}) = \prod_{j=1}^{d} \frac{1}{\sqrt{2\pi}\sigma_{ij}} exp\left\{-\frac{1}{2}\left(\frac{x_j - \mu_{ij}}{\sigma_{ij}}\right)^2\right\} \tag{4}$$

i.e., the free parameters to be estimated are the means μ_{ij} and the standard deviations σ_{ij}, for each kernel $i = 1, \ldots, n$ and for each component $j = 1, \ldots, d$ of the input space.

In the following, we will derive explicit formulations for $\frac{\partial \hat{p}(\phi(\mathcal{Y}|\theta_\phi)|\theta_{\hat{p}})}{\partial \theta}$ for the two families of free parameters θ within the proposed model. As regards a generic mixing parameter $c_i, i = 1, \ldots, n$, from Eq. (3), and since $p(\mathcal{Y}) = \sum_{k=1}^{n} c_k K_k(\mathbf{x})$, we have

$$\frac{\partial \hat{p}(\phi(\mathcal{Y}|\theta_\phi)|\theta_{\hat{\mathbf{p}}})}{\partial \gamma_i} = \sum_{j=1}^{n} \frac{\partial p(\mathcal{Y})}{\partial c_j} \frac{\partial c_j}{\partial \gamma_i} \tag{5}$$

$$= \sum_{j=1}^{n} K_j(\mathbf{x}) \frac{\partial}{\partial \gamma_i} \left(\frac{\varsigma(\gamma_j)}{\sum_{k=1}^{n} \varsigma(\gamma_k)} \right)$$

$$= K_i(\mathbf{x}) \left\{ \frac{\varsigma'(\gamma_i) \sum_k \varsigma(\gamma_k) - \varsigma(\gamma_i) \varsigma'(\gamma_i)}{[\sum_k \varsigma(\gamma_k)]^2} \right\} + \sum_{j \neq i} K_j(\mathbf{x}) \left\{ \frac{-\varsigma(\gamma_j) \varsigma'(\gamma_i)}{[\sum_k \varsigma(\gamma_k)]^2} \right\}$$

$$= K_i(\mathbf{x}) \frac{\varsigma'(\gamma_i)}{\sum_k \varsigma(\gamma_k)} - \sum_j K_j(\mathbf{x}) \frac{\varsigma(\gamma_j) \varsigma'(\gamma_i)}{[\sum_k \varsigma(\gamma_k)]^2}$$

$$= K_i(\mathbf{x}) \frac{\varsigma'(\gamma_i)}{\sum_k \varsigma(\gamma_k)} - \left\{ \sum_j c_j K_j(\mathbf{x}) \right\} \frac{\varsigma'(\gamma_i)}{\sum_k \varsigma(\gamma_k)}$$

$$= \frac{\varsigma'(\gamma_i)}{\sum_k \varsigma(\gamma_k)} \{K_i(\mathbf{x}) - p(\mathcal{Y})\}.$$

For the means μ_{ij} and the standard deviations σ_{ij} we proceed as follows. Let θ_{ij} denote the free parameter, i.e. μ_{ij} or σ_{ij}, to be estimated. It is seen that:

$$\frac{\partial \hat{p}(\phi(\mathcal{Y}|\theta_\phi)|\theta_{\hat{\mathbf{p}}})}{\partial \theta_{ij}} = c_i \frac{\partial K_i(\mathbf{x})}{\partial \theta_{ij}} \tag{6}$$

where the calculation of $\frac{\partial K_i(\mathbf{x})}{\partial \theta_{ij}}$ can be accomplished as follows. First of all, let us observe that for any real-valued, differentiable function $f(.)$ this property holds true: $\frac{\partial f(.)}{\partial x} = f(.) \frac{\partial log[f(.)]}{\partial x}$. As a consequence, from Eq. (4) we can write

$$\frac{\partial K_i(\mathbf{x})}{\partial \theta_{ij}} = K_i(\mathbf{x}) \frac{\partial log K_i(\mathbf{x})}{\partial \theta_{ij}} \tag{7}$$

$$= K_i(\mathbf{x}) \frac{\partial}{\partial \theta_{ij}} \sum_{k=1}^{d} \left\{ -\frac{1}{2} \left[log(2\pi\sigma_{ik}^2) + \left(\frac{x_k - \mu_{ik}}{\sigma_{ik}} \right)^2 \right] \right\}.$$

For the means, i.e. $\theta_{ij} = \mu_{ij}$, Eq. (7) yields

$$\frac{\partial K_i(\mathbf{x})}{\partial \mu_{ij}} = K_i(\mathbf{x}) \frac{x_j - \mu_{ij}}{\sigma_{ij}^2}. \tag{8}$$

For the covariances, i.e. $\theta_{ij} = \sigma_{ij}$, Eq. (7) takes the form:

$$\frac{\partial K_i(\mathbf{x})}{\partial \sigma_{ij}} = K_i(\mathbf{x}) \frac{\partial}{\partial \sigma_{ij}} \left\{ -\frac{1}{2} log(2\pi\sigma_{ij}^2) - \frac{1}{2} \left(\frac{x_j - \mu_{ij}}{\sigma_{ij}} \right)^2 \right\} \tag{9}$$

$$= \frac{K_i(\mathbf{x})}{\sigma_{ij}} \left\{ \left(\frac{x_j - \mu_{ij}}{\sigma_{ij}} \right)^2 - 1 \right\}.$$

3   Dataset Description and Feature Extraction

The experiments in this work are based on the "Corpus of spoken words for studies of auditory speech and emotional prosody processing" (WaSeP©) [17], which consists of two main parts: a collection of German nouns and a collection of phonetically balanced pseudo words, which correspond to the phonetical rules of German language. For this study the pseudo words have been chosen as the basis. This pseudo word set consists of 222 words, repeatedly uttered by a male and a female actor in six different emotional prosodies: neutral, joy, sadness, anger, fear, and disgust. Furthermore, a perception test has been conducted with 74 native German listeners, who were asked to rate and name the category or prosody that they were just listening to, resulting in an overall accuracy of 78.53%. It was also observed that the most confused emotion is "disgust", which is conform with the assumptions of Scherer [13].

In the dataset, each of the pseudo words consists of a concatenation of two syllables, including for instance: "hebof", "kebil", or "sepau". The average duration of the speech signals depends on the specific emotion, ranging from 0.75 sec. in the case of the "neutral" prosody, to 1.70 sec. in the case of "disgust". Figure 2 shows a sample waveform (and, spectra) of a signal from the dataset, corresponding to the emotion "joy". The data was recorded using a Sony TCD-D7 DAT-recorder and the Sennheiser MD 425 microphone in an acoustic chamber with a 44.1 kHz sample rate and later down-sampled to 16 kHz with a 16 bit resolution. Relative Spectral Perceptual Linear Predictive Coding (RASTA-PLP) acoustic parameters were used as acoustic features. In [4] perceptual linear predictive speech analysis (PLP) was first introduced as a method to represent speech signals with respect to human perception and with as few parameters as possible. However, PLP is sensitive to steady-state spectral factors caused by transmission channels, e.g. different transducers [6,5]. Therefore, [6] proposes the relative spectral methodology (RASTA) for PLP, rendering it more robust without increasing the computational burden significantly.

The PLP analysis is based on two perceptually and biologically motivated concepts, namely the critical bands, and the equal loudness curves, shown in Figure 3. One line represents the sound pressure (dB) that is required to perceive a sound of any frequency as loud as a reference sound of 1 kHz. The critical band filtering is analogous to the popular Mel frequency scaled cepstral coefficients (MFCC) triangular filtering, but the 21 filters are equally spaced along the Bark scale (instead of applying the Mel scale). The equal loudness curve is approximated by

$$E(w) = 1.151 \cdot \sqrt{\frac{(w^2 + 144 \cdot 10^4) \cdot w^2}{(w^2 + 16 \cdot 10^4) \cdot (w^2 + 961 \cdot 10^4)}},$$

according to [12], and it is applied to the filtered signal. The next steps are specific to the RASTA processing, and they follow the implementation recommendations in [6]. After transforming the spectrum to the logarithmic domain and the application of RASTA filtering, the signal is transformed back using

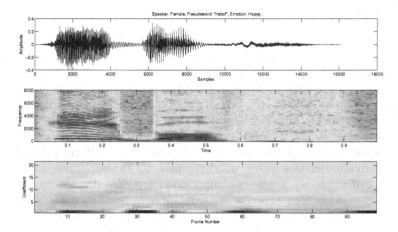

Fig. 2. Original waveform (top), spectrum after RASTA processing (middle), and RASTA-PLP cepstral coefficients (bottom)

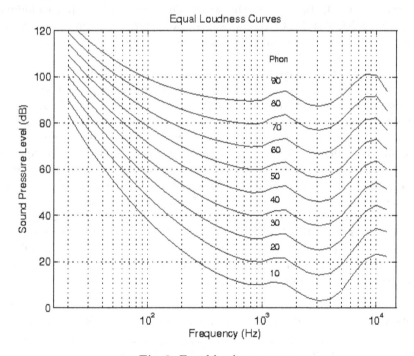

Fig. 3. Equal loudness curves

the exponential function. The last steps are comprised by an estimation of the linear predictive coding (LPC) coefficients as introduced by [11], and by the transformation of the LPC coefficients to cepstral values. In Figure 2 a comparison between the original waveform, the spectrum after RASTA processing and the RASTA-PLP cepstral coefficients are shown.

4   Experiments

Preliminary experiments were carried out using female speech signals from the
WaSeP© dataset, for a total of 1386 variable-length acoustic observation sequences (231 sequences per each class of emotion). A 10-fold crossvalidation
procedure has been adopted for evaluating the classification performance. Each
fold was defined by splitting the overall dataset, at random, into a training sample (1254 sequences) and a test sample (132 sequences). The folds were created
such that (i) the 10 test sets did not overlap with each other (i.e., no sequences
could appear more than once in a test set), and (ii) a uniform prior distribution of individual classes was granted (namely, 22 sequences per class in each
test set). Unfortunately, no direct competition from related work is available for
the dataset used in this experiment. A qualitative baseline is the result from
the human perception test mentioned in Section 3 (average accuracy of 78%).
Furthermore, similar results (around 70%) are available in the literature. For
example, 70% accuracy was achieved in a seven category experiment using three
feature sets in a multi classifier system [16]. In the earlier work by Lee et al.,
similar frame-based features, as in this approach, were used as input to hidden
Markov models (HMM). MFCC-based features were extracted and the classification task was to identify four different emotions. The HMMs reached an overall
accuracy of around 76% [9].

Table 1 reports the recognition accuracies (averaged over the 10 folds) obtained with MESLiN, and other traditional classifiers, in the present setup. All
the traditional classifiers were trained at the acoustic frame level. Classification
on test set was accomplished by averaging over the class-specific scores yielded
by the classifier along the whole observation sequence. The nearest neighbor classifier [2] was used first, in order to fix a baseline. Then, Multi-layer Perceptron
(MLP) networks were trained, having 2-layer topology featuring a hidden layer
with 14 units, sigmoid activation functions in both hidden and output layers.
Training was accomplished by applying 500 training epochs with a learning rate
of 0.3, and a momentum rate equal to 0.2 (these parameters were selected via
preliminary cross-validation). The same MLP architecture, and training parameters, were also applied within a 10-iterations AdaBoost strategy [3], Support
vector machine (SVM) with RBF kernels were used, having $\gamma = 0.1$, $\epsilon = 0.001$
(tolerance of the termination criterion.), and $\nu = 0.5$.

Table 1. Average recognition accuracy of sequences from the WaSeP© dataset

Method	Average accuracy (%)
Nearest Neighbor	33.90
MLP	39.32
AdaBoost	45.87
SVM	48.01
MESLiN	86.39

Finally, MESLiN was evaluated. Due to the intrinsically non-discriminative, ML training setup used in this paper, separate class-specific MESLiNs were trained independently over the training sequences belonging to the corresponding class. Classification was accomplished according to the usual maximum-a-posteriori (MAP) framework, via Bayes decision rule [2]. Since the class prior probabilities are identical in the present setup, the MAP criterion reduces to a direct maximum class-conditional pdf decision rule, i.e. assign a sequence to the emotion whose MESLiN model yields the highest likelihood. The MESLiNs are initialized as follows. Since the output of the feature extraction process described in Section 3 is 21-dimensional, $k = 21$ input neurons are used for the reservoir. Furthermore, they are connected to all the $n = 100$ state neurons (with tanh as transfer function) in the reservoir, featuring randomly initialized topology and values of the connection weights. An overall 10% of the possible unit-to-unit connections in the reservoir are taken, and the weight matrix W is normalized with a spectral width of $\alpha = 1$ [7]. The RBF-like network features $n = 3$ kernels, having mean, covariances, and mixing parameters initialized as follows. The components of the mean vectors were initialized at random (uniformly) over the range $\mathcal{I} = (-0.5, 0.5)$; the components of the diagonal covariance matrices were initialized to a fixed value, namely $\sqrt{|\mathcal{I}|/n}$; the mixing parameters were initialized at random over the interval $(0.0, 1.0)$ such that they sum to 1. Training of the MESLiN was accomplished for 20 epochs, using different, quantity-specific learning rates for the mixing parameters ($\eta_\gamma = 1.0e-06$), the means of the Gaussians ($\eta_\mu = 1.0e-10$), and the corresponding variances ($\eta_\sigma = 1.0e-11$). All the parameters were set during a preliminary cross-validation stage. Results confirm the approach is effective. A concise discussion is given in the next section. Table 2 shows the confusion matrix yielded by MESLiN. It is seen that "disgust" is the most confusable class. This is in line with the same phenomenon reported above for the human listeners. In this case, it is confused with "neutral" most of the times.

Table 2. Confusion matrix (MESLiN)

	neutral	joy	sadness	anger	fear	disgust
neutral	21.90	0.10	0.00	0.00	0.00	0.00
joy	0.50	18.70	0.20	0.00	2.50	0.10
sadness	0.50	0.00	21.40	0.10	0.00	0.00
anger	0.00	1.10	0.30	18.90	0.30	1.40
fear	0.00	0.60	1.60	0.20	19.60	0.00
disgust	14.80	0.90	0.60	0.10	2.00	3.60

5 Conclusions

The paper introduced a novel connectionist approach to the task of sequence modeling and classification. The model combines the reservoir of an ESN with a constrained RBF-like architecture. The former realizes a recurrent encoding of the observation sequence into a pattern of activation of the neurons (i.e., the state of the ESN), whilst the RBF estimates the pdf of such encodings. A ML-based training scheme was developed, which ensures satisfaction of probabilistic axioms (i.e., the estimated model turns out to be a proper pdf). Training is inherently unsupervised (the pdf of the overall training sample is estimated). Nonetheless, the machine can be successfully applied to supervised classification tasks over sequences, by training separate class-conditional models for each one of the classes involved in the problem. In so doing, each class features its own recurrent encoding ESN, and its own pdf model. The algorithm was applied to emotion recognition in speech signals from the WaSeP© dataset. Preliminary results confirm the approach is effective, yielding the highest recognition accuracies over a number of state-of-the-art classifiers. Surprisingly enough, the performance of the proposed classifier is even higher than the average accuracy yielded by humans on a similar task. We argue this is due to the following reasons: (i) the system was trained on data having the same nature as the data used for test, whilst humans do not undergo any data-specific training, but they assign test utterances to an emotional class according to generic, prior knowledge of their (subjective) concept of specific emotions; (ii) in particular, from the humans' point if view, no hard distinctions can be made between certain emotions; and (iii) the audio recordings in the dataset are not real-world utterances, since they were performed by actors. It is likely that this introduces a significant bias, such that humans cannot easily recognize the emotion from the acted expression, while the machine learns from the training sample how actors tend to give a certain interpretation of the same emotion (e.g., affecting specific features) that, later, can be easily recognized in the test data. In the light of these considerations, future work is focusing on a fully unsupervised application of the proposed model: sequences belonging to all classes are merged in a single training set, and the ML training is applied in order to let the machine discover its own "clusters" of emotions (i.e., how the emotions are distributed and concentrated in the encoding space yielded by the ESN).

Acknowledgment

The collaboration of this work would not have been possible without the generous funding of the Vigoni program, supported by DAAD, MIUR and by Ateneo Italo-Tedesco (AIT). Furthermore, the presented work was developed within the Transregional Collaborative Research Centre SFB/TRR 62 "Companion-Technology for Cognitive Technical Systems" funded by the German Research Foundation (DFG).

References

1. Bishop, C.M.: Neural Networks for Pattern Recognition. Oxford University Press, Oxford (1995)
2. Duda, R.O., Hart, P.E.: Pattern Classification and Scene Analysis. Wiley, New York (1973)

3. Freund, Y., Schapire, R.E.: Experiments with a new boosting algorithm. In: Proceedings of the Thirteenth International Conference on Machine Learning, San Francisco, pp. 148–156 (1996)
4. Hermansky, H., Hanson, B., Wakita, H.: Perceptually based linear predictive analysis of speech. In: IEEE International Conference on Acoustics, Speech, and Signal Processing, ICASSP 1985, April 1985, vol. 10, pp. 509–512 (1985)
5. Hermansky, H., Morgan, N., Bayya, A., Kohn, P.: Rasta-plp speech analysis. Technical report, ICSI Technical Report TR-91-069 (1991)
6. Hermansky, H., Morgan, N., Bayya, A., Kohn, P.: Rasta-plp speech analysis technique. In: IEEE International Conference on Acoustics, Speech, and Signal Processing, ICASSP 1992, vol. 1, pp. 121–124 (1992)
7. Jaeger, H.: Tutorial on training recurrent neural networks, covering bppt, rtrl, ekf and the echo state network approach. Technical Report 159, Fraunhofer-Gesellschaft, St. Augustin Germany (2002)
8. Jaeger, H., Haas, H.: Harnessing nonlinearity: Predicting chaotic systems and saving energy in wireless communication. Science 304, 78–80 (2004)
9. Lee, C.M., Yildirim, S., Bulut, M., Kazemzadeh, A., Busso, C., Deng, Z., Lee, S., Narayanan, S.S.: Emotion recognition based on phoneme classes. In: Proceedings of ICSLP 2004 (2004)
10. McLachlan, G.J., Basford, K.E. (eds.): Mixture Models: Inference and Applications to Clustering. Marcel Dekker, New York (1988)
11. Rabiner, L.R.: Fundamentals of Speech Recognition. Prentice-Hall, Englewood Cliffs (1993)
12. Robinson, D.W., Dadson, R.S.: A re-determination of the equal-loudness relations for pure tones. British Journal of Applied Physics 7(5), 166–181 (1956)
13. Scherer, K.R., Johnstone, T., Klasmeyer, G.: Vocal expression of emotion. In: Davidson, R.J., Scherer, K.R., Goldsmith, H.H. (eds.) Handbook of Affective Sciences, Affective Science, pp. 433–456. Oxford University Press, Oxford (2003)
14. Scherer, S., Oubbati, M., Schwenker, F., Palm, G.: Real-time emotion recognition from speech using echo state networks. In: Prevost, L., Marinai, S., Schwenker, F. (eds.) ANNPR 2008. LNCS (LNAI), vol. 5064, pp. 205–216. Springer, Heidelberg (2008)
15. Scherer, S., Schwenker, F., Campbell, W.N., Palm, G.: Multimodal laughter detection in natural discourses. In: Proceedings of 3rd International Workshop on Human-Centered Robotic Systems, HCRS 2009 (2009)
16. Scherer, S., Schwenker, F., Palm, G.: Classifier fusion for emotion recognition from speech. In: 3rd IET International Conference on Intelligent Environments 2007 (IE 2007), pp. 152–155. IEEE, Los Alamitos (2007)
17. Wendt, B., Scheich, H.: The magdeburger prosodie korpus - a spoken language corpus for fmri-studies. In: Speech Prosody 2002, SProSIG (2002)

Robustness Analysis of Eleven Linear Classifiers in Extremely High–Dimensional Feature Spaces

Ludwig Lausser[1] and Hans A. Kestler[1,2,*]

[1] Department of Internal Medicine I, University Hospital Ulm, Germany
ludwig.lausser@uni-ulm.de
[2] Institute of Neural Information Processing, University of Ulm, Germany
hans.kestler@uni-ulm.de

Abstract. In this study we address the linear classification of noisy high-dimensional data in a two class scenario. We assume that the cardinality of the data is much lower than its dimensionality. The problem of classification in this setting is intensified in the presence of noise. Eleven linear classifiers were compared on two-thousand-one-hundred-and-fifty artificial datasets from four different experimental setups, and five real world gene expression profile datasets, in terms of classification accuracy and robustness. We specifically focus on linear classifiers as the use of more complex concept classes would make over-adaptation even more likely. Classification accuracy is measured by mean error rate and mean rank of error rate. These criteria place two large margin classifiers, SVM and ALMA, and an online classification algorithm called PA at the top, with PA being statistically different from SVM on the artificial data. Surprisingly, these algorithms also outperformed statistically significant all classifiers investigated with dimensionality reduction.

1 Introduction

Classification is one of the basic tasks in machine learning. Many different classification methods were proposed (see e.g. [1, 2, 3]). In the standard inductive setting, a classifier will be selected according to a set of training examples and its accuracy is tested on a set of test examples. Problems arise if a collected dataset contains more features than samples. In this case even simple classifiers have the complexity to adapt perfectly to a given training set and loose their ability of generalization (overfitting) [4]. Dimensionality reduction methods, like for example PCA, ICA, can antagonize this problem but complicate the interpretation of a classifier in terms of its original input space [5, 6]. The problem of overfitting is increased in real life applications. The single datapoint can be affected by measurement errors and a classifier will adapt to a noisy dataset.

Aim of this investigation is the influence of different types of noise on the performance of linear classifiers for high-dimensional data of low cardinality.

2 Classification

Classification is the task of predicting a categorial label $y \in \mathbf{Y}$ of a datapoint $x \in \mathbf{X}$. A classifier is a mapping $c : \mathbf{X} \to \mathbf{Y}$. In the following we will concentrate on binary

---

* Corresponding author.

F. Schwenker and N. El Gayar (Eds.): ANNPR 2010, LNAI 5998, pp. 72–83, 2010.
© Springer-Verlag Berlin Heidelberg 2010

classification $\mathbf{Y} = \{+1, -1\}$ and real valued input spaces $\mathbf{X} \subseteq \mathbb{R}^n$. A classifier is chosen from a concept class $\mathbf{C}$, a set describing all classifiers fulfilling some model assumptions. The aim is to find the classifier $c^* \in \mathbf{C}$ which minimizes the number of errors over the distribution of all possible labeled pairs $D(x, y)$

$$c^* = \underset{c}{\operatorname{argmin}} \frac{1}{2} \int |c(x) - y| dD(x, y). \tag{1}$$

The distribution of $D(x, y)$ is usually not known. In this case a classifier c is selected (trained) by a learning algorithm $t(\mathbf{C}, \mathbf{S}) \to c$ according to a finite set $\mathbf{S}$ of m examples

$$\mathbf{S} = \mathbf{S}(\mathbf{P}, \mathbf{N}) = \{(x, +1) \mid x \in \mathbf{P}\} \cup \{(x, -1) \mid x \in \mathbf{N}\}. \tag{2}$$

Here $\mathbf{P}$ denotes the set of k (positive) examples of the first class and $\mathbf{N}$ denotes the set of l (negative) examples of the second class. The error rate of a classifier is estimated on an independent (test-) dataset $\mathbf{S}' = \mathbf{S}(\mathbf{P}', \mathbf{N}')$ with $\mathbf{S} \cap \mathbf{S}' = \emptyset$. This estimator can be formalized as

$$f_{err} = \frac{1}{2|\mathbf{S}'|} \sum_{(\mathbf{x}, \mathbf{y}) \in \mathbf{S}'} |c(\mathbf{x}) - \mathbf{y}|. \tag{3}$$

2.1 Linear Classifiers

The concept class of linear classifiers is given by

$$\mathbf{C}_{lin} = \{c(x) = \operatorname{sign}(\omega^T x - \theta) \mid \omega \in \mathbb{R}^n, \theta \in \mathbb{R}\}. \tag{4}$$

The decision boundaries of these classifiers are linear equations of the form

$$\omega^T x = \theta. \tag{5}$$

ω and θ are normally substituted by $\omega := \omega/||\omega||_2$ and $\theta := \theta/||\omega||_2$ in order to gain a unique representation of each classifier. Here $|| \cdot ||_2$ denotes the Euclidian norm. In a geometric interpretation ω can be seen as the norm vector of the line. The threshold θ can be seen as the line's distance to the origin.

In this analysis we focus on datasets with higher dimensionality than cardinality ($m \ll n$), which is the basic scenario in many tasks, like image analysis [7], speech recognition [8] and gene expression analysis [9]. Although linear classifiers are very simple models, they tend to overfit on such datasets. This was shown, for example, by Cover's theorem [10] stating that a database of m-datapoints (in general position) within a n-dimensional space can be separated in an arbitrary way with probability

$$P(m, n) = \left(\frac{1}{2}\right)^{m-1} \sum_{k=0}^{n-1} \binom{m-1}{k}. \tag{6}$$

If the ratio n/m is greater than 0.5, $P(m, n)$ is rapidly increasing towards 1 and a classifier without any training error can be found for an arbitrary dataset of these dimensions.

3   Training Algorithms

This section contains a brief description of the eleven training algorithms that were used in this study. The algorithms are divided into model-based algorithms (3.1), linear and quadratic programming algorithms (3.2) and iterative algorithms (3.3).

3.1   Model Based Classifiers

The algorithms listed here were created with assumptions on the class densities.

*Fisher Linear Discriminant Analysis (**LDA**).* The LDA classifier is built with the assumption, that both class densities are Gaussians with a common covariance Σ. The hyperplane calculated by this algorithm minimizes the error for datapoints chosen according to these class densities. For this the inverse of Σ is needed. On a real dataset the estimate $\hat{\Sigma}$ of Σ has to be used. $\hat{\Sigma}$ will become singular for datasets with higher dimensionality than cardinality. In this case the inverse of $\hat{\Sigma}$ is usually replaced by the Moore-Penrose Inverse. Besides the standard **LDA (mean)**, we have used a variant **LDA (median)**, for which the estimation of the class centroid was done by applying the median feature-wise.

*Nearest Centroid (**NC**).* The nearest centroid algorithm assumes, that both class densities are Gaussians with a common covariance of form $c \cdot \mathbf{I}$, $c \in \mathbb{R}$ (**I** is the identity matrix). In this way the NC can be seen as a special case of LDA. Under these assumptions only the class centroids have to be calculated for the final classification. For a new example the Euclidian distances to all centroids are calculated. The datapoint will receive the label of its nearest centroid.

*Nearest Shrunken Centroid (**NSC**) [11].* The nearest shrunken centroid is a feature reducing version of the NC. Here, additionally the class independent (overall) centroid is calculated. The main idea of the NSC is, that feature dimensions in which a class centroid is near to the overall centroid are not useful for characterizing the class. The class-wise centroids are shrunken feature–wise towards the overall centroid. If a single entry of a centroid gets negative, it is set to zero. The amount of shrinkage is determined by a set of parameters Δ. In this study experiments for $i \in \{1, \ldots, 30\}$ different sets of shrinking parameters $\Delta_{ij} = i/30 * \max\{|d_{0j}|, |d_{1j}|\}$ were done. Here d_{0j} and d_{1j} denote the distances of the class-wise centroids to the overall centroids in feature dimension j.

3.2   Linear and Quadratic Programming Training Algorithms

This section contains algorithms, which optimize an objective function by a linear or quadratic program. In order to handle non-linear separable datasets a penalty term of slack variables ξ_i is added to the objective function. The tradeoff between the penalty term and the original objective function can now be regulated by a cost parameter C.

*Support Vector Machine (**SVM**) [2].* The support vector machine searches for the hyperplane, which maximizes the Euclidian distance between the hyperplane and the datapoints next to it (maximal L2 margin). This can be formulated as a quadratic problem for minimizing the Euclidian norm $||\omega||_2$ of ω.

$$\min_{\omega,\xi} \quad \|\omega\|_2^2 + C\sum_{i=1}^{N}\xi_i$$

$$\text{s.t.} \quad \forall i : y_i(\omega^T x_i) - \theta \geq 1 - \xi_i$$

$$\forall i : \xi_i \geq 0$$

LIKNON *[12].* The LIKNON algorithm can be seen as the L1 variant of the SVM. Minimizing the L1 norm $\|\omega\|_1$ of ω forces many ω_i to zero. The corresponding features of the datapoints will not be used for the final classification. In this way a feature reduction is achieved. The optimization problem of the LIKNON algorithm can be formalized as a linear program.

$$\min_{\omega,b,\xi} \quad \|\omega\|_1 + C\sum_{i=1}^{N}\xi_i$$

$$\text{s.t.} \quad \forall i : y_i(\omega^T x_i) - \theta \geq 1 - \xi_i$$

$$\forall i : \xi_i \geq 0$$

LESS *[13].* The LESS classifier belongs to the group of weighted centroid classifiers. Linear programming is used here to find a weight vector w, which minimizes the trade-off between its L1 norm and the penalization term. Here again a feature selection is implicitly performed.

$$\min_{w,\xi} \quad \|w\|_1 + C\sum_{i=1}^{N}\xi_i$$

$$\text{s.t.} \quad \forall i : y_i \sum_{j=1}^{M} w_j(2x_{ij}(\mu_{0,j} - \mu_{1j}) + (\mu_{0,j}^2 - \mu_{1j}^2)) \geq 1 - \xi_i$$

$$\forall i : \xi_i \geq 0 \qquad \forall j : w_j \geq 0$$

μ_0 and μ_1 denote the class-wise centroids.

3.3   Iterative Training Algorithms

The algorithms in this section adapt the linear model in an iterative way. During each iteration, i.e. presentation of a data point, the classifier will be modified. Many iterative algorithms are designed for the online learning setting. In this scenario new labeled data points will be available one by one. The classifier will be adapted after receiving a new datapoint. In this study the online learning setting was simulated by iterating 10000 times through permuted versions of the original dataset.

Perceptron *[14].* The perceptron algorithm is one of the classical iterative algorithms. The hyperplane will be updated until it separates the data points correctly. No objective function is considered for the choice of the hyperplane.

ALMA *[15] and* **ROMMA** *[16].* The two algorithms ALMA (approximate maximal margin algorithm) and ROMMA (relaxed online maximum margin algorithm), approximate a maximum margin solution of the L2 margin in an iterative way.

Passive Aggressive Algorithm (**PA**) *[17].* The update rule of PA utilizes the hinge loss $l(\omega^*;(x,y)) = \max(0, 1 - y(\omega x - \theta))$. Here ω^* denotes the vector of all classifier parameters $(\omega_1, \ldots, \omega_n, -\theta)^T$. If a datapoint is classified correctly with a margin greater or equal to one, the hinge loss is equal to zero. Otherwise, the loss is increasing according to the distance between this margin and the datapoint. In the linear separable case, an update step of PA has to fulfill the constraint $l(\omega^*;(x_t,y_t)) = 0$. By this constraint not only a correct classification of x_t but also a minimal distance between the classifier and x_t is enforced. In each iteration t the classifier will be selected, which has the minimal modification of ω_t^*. If the classification of x_t fulfills the constraint, no modifications have to be done and PA is passive. Otherwise, PA forces aggressively the correct classification of x_t. For the linear inseparable case, the optimization problem can be formalized as

$$\omega_{t+1}^* = \underset{\omega^* \in \mathbb{R}^{n+1}}{\text{argmin}} \frac{1}{2} ||\omega^* - \omega_t^*||_2^2 + C\xi^2$$
$$\text{s.t.} \quad l(\omega^*;(x_t,y_t)) \leq \xi$$

This is equal to the PA-II variant proposed in [17].

4   Experimental Setup

The first part of this study is an empirical comparison of the classifiers in several artificial noise settings. For all experiments we use different datasets with a dimensionality of $n = 100$ and 25 datapoints for each of the two classes. A graphical visualization of the experimental setup can be found in Figure 1. For all algorithms various parameters settings were tested prior to the results given here. The best found parameter values were chosen and fixed for the results given in the following. We will first introduce some notation used within this section. The vector **1** is the vector, which is equal to 1 at each position. The vector $\mathbf{1}_x$ is equal to 1 in the first x positions and 0 on the other positions. The vector $\bar{\mathbf{1}}_x$ is defined as $\mathbf{1} - \mathbf{1}_x$. The function $d : \mathbb{R}^n \to \mathbb{R}^{n \times n}$ converts a vector $v \in \mathbb{R}^n$ into a $n \times n$ - dimensional diagonal matrix. The main diagonal of this matrix is filled with the elements of v. A Gaussian distribution with mean μ and covariance Σ will be denoted by $\mathcal{N}(\mu, \Sigma)$. We will write $s(k, \Phi)$ to denote a function which creates a set of k datapoints chosen according to distribution Φ.

4.1   Breakdown Experiments

Here the test error of a classifier trained on samples $\mathbf{P} = s(k, \Psi_1)$ and $\mathbf{N} = s(l, \Psi_0)$ is compared to a classifier trained on contaminated samples $\tilde{\mathbf{P}}$ and $\tilde{\mathbf{N}}$. A contaminated version $\tilde{\mathbf{X}} = \tilde{s}(\mathbf{X}, i, \Phi)$ of a sample $\mathbf{X} \in \{\mathbf{P}, \mathbf{N}\}$ is generated by replacing $i \leq |\mathbf{X}|$ examples by new ones, which were chosen according to distribution Φ. The number of contaminated datapoints is increased from 0 to $|\mathbf{X}|$ (class breakdown). For all experiments the test sets are chosen as $\mathbf{P}' = s(k, \Psi_1)$ and $\mathbf{N}' = s(l, \Psi_0)$. Each test was repeated on ten different samples. A table of the concrete experiments can be found in Table 1. The $mean_x$ experiment was done for $x \in \{5, 10, 25, 50\}$. The $sd_{x'}$ experiment for $x' \in \{10^2, 10^3, 10^4\}$.

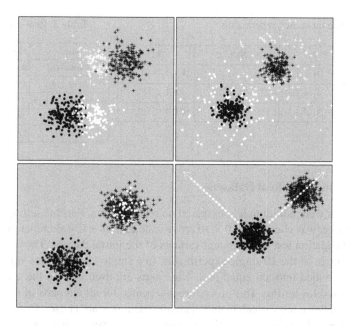

Fig. 1. The four settings of the artificial data experiments: Upper left: *mean* breakdown experiment. The distribution of the noisy datapoints differs from the distribution of the original datapoints in the first x dimensions of their common mean vector. The mean vector of the noisy datapoints is equal to 0 for the first x dimensions. Upper right: *sd* breakdown experiment. The noisy datapoints are chosen according to a Gaussian distribution with higher standard deviation than the standard deviation of the Gaussian of the original one. Lower left: *class* breakdown experiment. In this setting the noisy datapoints are chosen according to the distribution of the other class. Lower right: single outlier experiment. In this experiment a single noisy datapoint is moved in a certain direction. The datapoint is moved either towards the other class (*forward*), or away from the other class (*backwards*), or orthogonal to the other class (*sideways*).

Table 1. Breakdown experiments that were performed

experiment	P/P'	N/N'	$\tilde{P}$	$\tilde{N}$
mean$_x$	$s(25, \mathcal{N}(1, d(1)))$	$s(25, \mathcal{N}(-1, d(1)))$	$\tilde{s}(P, i, \mathcal{N}(\bar{1}_x, d(1)))$	$\tilde{s}(N, i, \mathcal{N}(-\bar{1}_x, d(1)))$
sd$_x$	$s(25, \mathcal{N}(1, d(1)))$	$s(25, \mathcal{N}(-1, d(1)))$	$\tilde{s}(P, i, \mathcal{N}(1, xd(1)))$	$\tilde{s}(N, i, \mathcal{N}(-1, xd(1)))$
class	$s(25, \mathcal{N}(1, d(1)))$	$s(25, \mathcal{N}(-1, d(1)))$	$\tilde{s}(P, i, \mathcal{N}(-1, d(1)))$	$\tilde{s}(N, i, \mathcal{N}(1, d(1)))$

4.2 Single Outlier Experiments

In this test a classifier trained on samples $P = s(k, \Psi_1)$ and $N = s(l, \Psi_0)$ is compared to a classifier trained on samples $P_x^\tau = \tilde{s}(P, x, \tau) = P \cup \{10^\tau x\}$ and N. Here x is a random point from the corresponding unit sphere and $\tau \in \{1, \ldots, 5\}$. For each x, ten different datasets were resampled. Some characteristics of the used datasets are given in Table 2.

Table 2. Single outlier experiments

experiment	P/P'	N/N'	P_x^τ
forward	$s(25, \mathcal{N}(\bar{\mathbf{1}}_1, d(\bar{\mathbf{1}}_1)))$	$s(25, \mathcal{N}(-\bar{\mathbf{1}}_1, d(\bar{\mathbf{1}}_1)))$	$\bar{s}(\mathbf{P}, -\bar{\mathbf{1}}_1/\|-\bar{\mathbf{1}}_1\|, \tau)$
backwards	$s(25, \mathcal{N}(\bar{\mathbf{1}}_1, d(\bar{\mathbf{1}}_1)))$	$s(25, \mathcal{N}(-\bar{\mathbf{1}}_1, d(\bar{\mathbf{1}}_1)))$	$\bar{s}(\mathbf{P}, \bar{\mathbf{1}}_1/\|\bar{\mathbf{1}}_1\|, \tau)$
sideways	$s(25, \mathcal{N}(\bar{\mathbf{1}}_1, d(\bar{\mathbf{1}}_1)))$	$s(25, \mathcal{N}(-\bar{\mathbf{1}}_1, d(\bar{\mathbf{1}}_1)))$	$\bar{s}(\mathbf{P}, \mathbf{1}_1/\|\mathbf{1}_1\|, \tau)$

Table 3. Real data sets

name	#Fea	#Pos	#Neg
Bittner [18]	8067	19	19
Golub [19]	3571	47	25
Buchholz/Kestler [9]	169	37	25
Notterman [20]	7457	18	18
West [21]	7129	25	24

4.3 Experiments on Real Datasets

The classifiers were additionally compared on real data sets. For this setting a 10×10 cross-validation was chosen. A 10×10 cross-validation is a 10-fold repetition of a 10-fold cross-validation test on permuted variants of the initial dataset. The result will be the average error of the 10 single experiments. In a single 10-fold cross-validation test a dataset is divided into ten equal part. Nine parts are used to train the classifier and one part is used for testing. This procedure is repeated for all ten parts of the data. The mean error of these tests is calculated. The used data sets are chosen from the field of gene expression analysis. A list of the used data sets is given in Table 3.

5 Results

mean Breakdown results. The LDA-based algorithms are the only classifiers that are influenced for $mean_5$ and $mean_{10}$ (data not shown). They show error rates between 30% and 50% in these experiments, all other classifiers have zero error. Performance decreases for all other classifiers starting with $mean_{10}$, but is still much better than the LDA classifiers. For 25 and 50 noisy dimensions ($mean_{25}$, $mean_{50}$) performance decreases uniformly to 2% to 50% starting with 15 noisy datapoints. All classifiers are robust to this kind of noise, if the number of noisy datapoints is low. The NC-based classifiers are more robust on an increasing number of noisy datapoints and are only deteriorating for 24 or 25 noisy datapoints. Performance of SVM and LIKNON on lower noise levels is inferior to the iterative algorithms.

sd Breakdown results. The results of the *sd* breakdown experiments can be seen in Figure 2. The single rows contain the results of the sd_{100}, the sd_{1000} and the sd_{10000} breakdown experiment. The classifiers ALMA, PA, LIKNON and SVM are not influenced in any experiment until all datapoints were replaced by noisy datapoints. One noisy datapoint is enough to increase the error rate of the NC-based classifiers. This effect increases with a higher standard deviation. The LDA-based algorithms fluctuate around their initial error rate of about 20%. For all *sd* experiments there is a number of datapoints for which the error rate of LDA (median) is rapidly increasing towards 50%. The LDA (mean) has an error rate of 50% only for the number of 25 noisy datapoints.

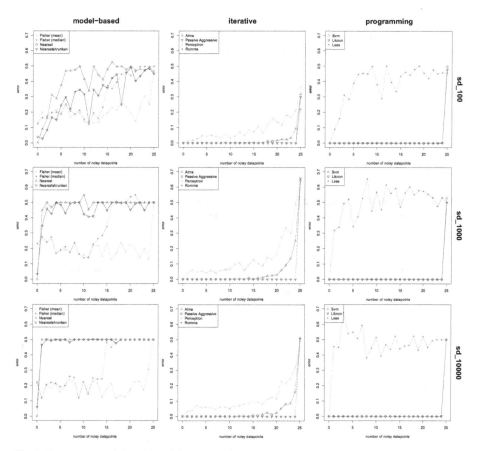

Fig. 2. Error curves of the *sd* breakdown experiment. The mean error of ten repetitions is shown. The number of noisy datapoints per class is given on the horizontal axis. The rows include the results from sd_{100}, sd_{1000}, and sd_{10000} (top to bottom).

class Breakdown results. The results of the *class* breakdown experiment are given in Figure 3. The classifiers show a linear increasing error rate according to the increasing number of noisy datapoints. An exception to this are the model-based classifiers. The classifiers NC, NSC and LESS show a flat error curve until a level of 13 noisy datapoints per class is reached. The LDA-based classifiers fluctuate around the 50% error rate in the range of 5 and 20 noisy datapoints.

Single outlier results. The error curves of the single outlier experiments are given in Figure 4. Only the model based classifiers were influenced in the experiments *backwards* and *sideways*. A exception to this is the NSC in the *sideways* experiment. The other classifiers are only affected in the *forward* experiment.

Average ranking on artificial datasets. The rank over all classifiers was calculated for all single experiments and noise levels. The mean rank is shown in Table 4. The best

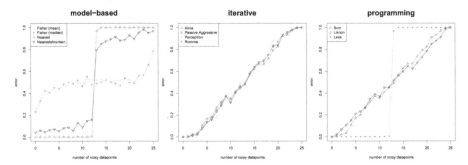

Fig. 3. Error curves of the *class* breakdown experiment over an increasing number of noisy data-points per class. The mean error of ten repetitions is shown.

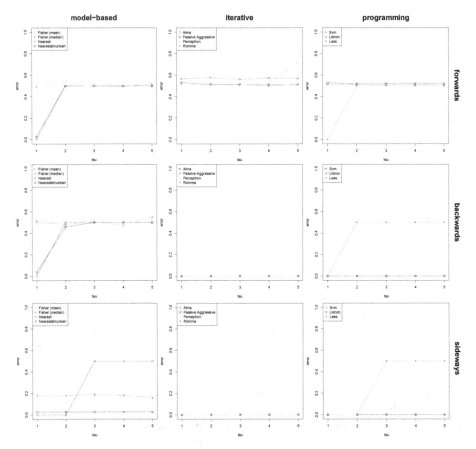

Fig. 4. Error curves of the single outlier experiments. The mean error of ten repetitions is shown. The distance from the outlier to the class centroid is given on the horizontal axis.

Table 4. Average ranks over all experiments on artificial datasets

LDA (mean)	LDA (median)	NC	NSC	PER	ROMMA	ALMA	PA	SVM	LIK	LESS
8.62	9.02	6.48	6.91	5.49	4.41	4.62	4.41	4.60	5.06	6.37

Table 5. Results of the real data experiments. The results of the 10×10 cross-validation are given by the mean errors in percent and standard deviations. The last two columns show the average error and the average rank over all datasets.

	Bittner	Golub	Notterman	Buchholz	West	Average error	Average rank
LDA (mean)	46.84 ± 6.88	42.22 ± 6.84	34.17 ± 7.53	45.32 ± 7.35	44.69 ± 6.83	42.65	9.9
LDA (median)	46.84 ± 7.00	44.31 ± 6.56	35.28 ± 7.30	45.48 ± 6.44	47.35 ± 9.47	43.85	10.7
NC	50.53 ± 1.11	2.50 ± 0.59	2.78 ± 0.00	27.26 ± 0.92	39.39 ± 4.41	24.49	6.4
NSC	8.68 ± 2.17	4.17 ± 1.13	5.28 ± 2.43	29.19 ± 2.21	15.10 ± 1.72	12.48	5.4
PER	28.42 ± 3.88	11.39 ± 2.60	5.83 ± 2.05	26.77 ± 3.58	16.33 ± 2.89	17.75	6.9
ROMMA	20.53 ± 7.21	5.83 ± 2.68	6.39 ± 2.64	26.77 ± 3.15	17.96 ± 3.94	15.50	6.8
ALMA	10.53 ± 0.00	2.22 ± 0.97	2.78 ± 0.00	19.03 ± 5.20	9.59 ± 1.38	8.83	1.9
PA	9.74 ± 2.50	2.64 ± 0.44	2.78 ± 0.00	19.52 ± 3.68	10.41 ± 1.79	9.02	2.8
SVM	14.74 ± 2.83	2.64 ± 0.44	2.78 ± 0.00	16.13 ± 3.88	10.61 ± 2.32	9.38	3.2
LIK	13.95 ± 2.79	7.64 ± 1.35	3.06 ± 0.88	24.52 ± 4.08	17.96 ± 4.49	13.43	5.5
LESS	41.58 ± 4.77	2.50 ± 0.59	4.72 ± 1.87	28.23 ± 2.97	30.00 ± 6.24	21.41	6.5

average ranks were achieved by PA (4.41), ROMMA (4.41), SVM (4.60), and ALMA (4.62). We found significant differences between PA and SVM (Wilcoxon rank sum test: $p = 0.0028$) and ROMMA and SVM (Wilcoxon rank sum test: $p = 0.0006$). We also found a significant difference between SVM, PA, ROMMA and all classifiers with dimensionality reduction NSC, LESS, and LIKNON (9 Wilkoxon rank tests, all $p < 0.000014$ after Holm correction for multiple testing).

Cross-validation results on real datasets. The results of the cross–validation experiments can be seen in Table 5. The LDA variants show high error rates for all datasets. Compared to the other centroid based classifiers the NSC has better error rates on the datasets Bittner and West. On these datasets the improvement is better than 10%. Among the classifiers, which try to maximize the margin, ALMA has best error rates on the datasets Bittner, Golub and West. The SVM achieves equal or better result on the other datasets. The results of PA are comparable to the results of SVM and ALMA.

6   Conclusion

In this study a set of eleven linear classifiers were compared in terms of noise robustness and classification rates. The classifiers were tested on real and artificial high dimensional datasets. The artificial datasets fulfilled the model assumptions of the classifiers LDA and NC. Within these tests, a small number of undirected noisy datapoints lead to rapid increasing error rates. NC-like algorithms are more influenced by this effect than the LDA-like ones. This can be seen in the *sd* breakdown experiments and the single outlier experiments. If the noise is directed, the NC-like algorithms perform different to the LDA-like algorithms. In the *class* breakdown experiment, 50% of all datapoints could be replaced by noisy ones. The LDA-based classifiers show a mean error rate of about 50% in this experiment. The centroid based classifiers are superior to the others in the early stages of the *forward* single outlier test. Effects of the noise variant chosen in the *mean* breakdown experiment could only be seen for high values of *x*. In this case, all NC-based classifiers were more robust for higher noise rates. Only the LDA-based algorithms were highly affected of this kind of noise. This is not too surprising and supports the findings of Raudys & Duin [22], that when the total number of learning samples approaches the dimensionality some of the eigenvalues of the sample covariance matrix become extremely large while the others become extremely small. This negatively affects classifier performance. The other classifiers are more robust in the *sd* breakdown experiments. Especially the large margin classifiers are unaffected by this scenario. These classifiers are more sensitive to direct noise. This can be seen by their linear increasing error rates in the *class* breakdown.

The effects of feature selection were different for single algorithms. The NSC obtained equal or better results than NC in the *sd* breakdown experiments and in the *class* breakdown experiments for higher noise rates. LESS is more comparable to NC, but becomes more instable in the *sd* breakdown experiments. LIKNON has lower error rates than the SVM in the *mean* breakdown experiments. The top mean rank over all artificial experiments was gained by PA, ROMMA, ALMA and SVM. Surprisingly, these algorithms also outperformed statistically significant all classifiers with dimensionality reduction (NSC, LESS, LIKNON). This might be due to the problem of finding a meaningful subset of features in these very high-dimensional spaces of low cardinality [23]. Also as volume of the feature space increases exponentially with dimensionality, noise on each of the coordinates does not affect the location of the datapoint too much and thus margin classifiers seem to be superior in this setting. This is also supported by the good performance of the single outlier experiments in which the pure feature selection algorithm NSC scored worse than margin algorithms.

On the real datasets the large margin classifiers were slightly better than the centroid based classifiers. There were two examples among the used datasets (Bittner and West), which could hardly be classified by NC and LESS. The error curves of the NSC has comparable results to the other classifiers on this datasets. Concerning the error rates over all real datasets, the top three classifiers are ALMA, PA and SVM. The overall best performance for these types of high-dimensional data of low cardinality is given be PA, as it scores top on the artificial data and second on the expression profiles.

Acknowledgments. This work is supported by the Stifterverband für die Deutsche Wissenschaft (HAK) and the German Science Foundation, SFB 518, Project C05.

References

1. Breiman, L., Friedman, J., Stone, C., Olshen, R.: Classification and Regression Trees. Chapman & Hall/CRC (1984)
2. Vapnik, V.N.: Statistical Learning Theory. Wiley, New York (1998)
3. Rojas, R.: Neural Networks: A Systematic Introduction. Springer, Heidelberg (1996)

4. Hastie, T., Tibshirani, R., Friedman, J.: The Elements of Statistical Learning. Springer, New York (2001)
5. Pearson, K.: On lines and planes of closest fit to systems of points in space. Philosophical Magazine 2(6), 559–572 (1901)
6. Ans, B., Hérault, J., Jutten, C.: Adaptive neural architectures: Detection of primitives. In: Proceedings of COGNITIVA 1985, pp. 593–597 (1985)
7. Lu, J., Plataniotis, K., Venetsanopoulos, A.: Face recognition using LDA-based algorithms. IEEE Transactions on Neural Networks 14(1), 195–200 (2003)
8. Zolnay, A., Kocharov, D., Schlüter, R., Ney, H.: Using multiple acoustic feature sets for speech recognition. Speech Commun. 49(6), 514–525 (2007)
9. Buchholz, M., Kestler, H.A., Bauer, A., et al.: Specialized DNA arrays for the differentiation of pancreatic tumors. Clinical Cancer Research 11(22), 8048–8054 (2005)
10. Cover, T.M.: Geometrical and Statistical Properties of Systems of Linear Inequalities with Applications in Pattern Recognition. IEEE Transactions on Electronic Computers 14(3), 326–334 (1965)
11. Tibshirani, R., Hastie, T., Narasimhan, B., Chu, G.: Diagnosis of multiple cancer types by shrunken centroids of gene expression. PNAS 99(10), 6567–6572 (2002)
12. Bhattacharyya, C., Grate, L.R., Rizki, A., et al.: Simultaneous classification and relevant feature identification in high-dimensional spaces: application to molecular profiling data. Signal Process. 83(4), 729–743 (2003)
13. Veenman, C.J., Tax, D.M.: Less: A model-based classifier for sparse subspaces. IEEE Transactions on Pattern Analysis and Machine Intelligence 27(9), 1496–1500 (2005)
14. Rosenblatt, F.: The Perceptron: A Probabilistic Model for Information Storage and Organization in the Brain. Psych. Rev. 65(6), 386–407 (1958)
15. Gentile, C.: A new approximate maximal margin classification algorithm. Journal of Machine Learning Research 2 (2001)
16. Li, Y., Long, P.M.: The Relaxed Online Maximum Margin Algorithm. Machine Learning 46(1-3), 361–387 (2002)
17. Crammer, K., Dekel, O., Keshet, J., Shalev-Shwartz, S., Singer, Y.: Online Passive-Aggressive Algorithms. Journal of Machine Learning Research 7, 551–585 (2006)
18. Bittner, M., Meltzer, P., Chen, Y., et al.: Molecular classification of cutaneous malignant melanoma by gene expression profiling. Nature 406(6795), 536–540 (2000)
19. Golub, T., Slonim, D., Tamayo, P., et al.: Molecular Classification of Cancer: Class Discovery and Class Prediction by Gene Expression Monitoring. Science 286(5439), 531–537 (1999)
20. Notterman, D., Alon, U., Sierk, A., Levine, A.: Transcriptional Gene Expression Profiles of Colorectal Adenoma, Adenocarcinoma, and Normal Tissue Examined by Oligonucleotide Arrays. Cancer Research 61(7), 3124–3130 (2001)
21. West, M., Blanchette, C., Dressman, H., et al.: Predicting the clinical status of human breast cancer by using gene expression profiles. PNAS 98(20), 11462–11467 (2001)
22. Raudys, S., Duin, R.: Expected classification error of the fisher linear classifier with pseudo-inverse covariance matrix. Pattern Recognition Letters 19(5), 385–392 (1998)
23. Dougherty, E.R.: Feature-selection overfitting with small-sample classifier design. IEEE Intelligent Systems 20(6), 64–66 (2005)

Global Coordination Based on Matrix Neural Gas for Dynamic Texture Synthesis*

Banchar Arnonkijpanich[1] and Barbara Hammer[2]

[1] Department of Mathematics, Faculty of Science,
Khon Kaen University, Thailand, 40002
and
Centre of Excellence in Mathematics, The Commission on Higher Education,
Si Ayutthaya Rd., Bangkok 10400, Thailand
[2] University of Bielefeld, CITEC, Bielefeld, Germany

Abstract. Matrix neural gas has been proposed as a mathematically well-founded extension of neural gas networks to represent data in terms of prototypes and local principal components in a smooth way. The additional information provided by local principal directions can directly be combined with charting techniques such that a nonlinear embedding of a data manifold into low dimensions results for which an explicit function as well as an approximate inverse exists. In this paper, we show that these ingredients can be used to embed dynamic textures in low dimensional spaces such that, together with a traversing technique in the low dimensional representation, efficient dynamic texture synthesis can be obtained.

1 Introduction

Neural gas (NG) and topology representing networks as proposed by Martinetz constitute particularly robust methods to represent a given data set and its topology in terms of a lattice of neurons [11,12]. In contrast to the popular self-organizing map [7], no prior lattice structure is specified such that direct visualization of data is not possible. On the contrary, the correct, probably irregular topology of the underlying data manifold can be inferred which accounts for the particular robustness of the approach.

Neural gas is often used for data preprocessing, e.g. data compression or clustering. Recently, extensions of NG have been proposed which also adapt local matrices during training such as to minimize the quantization error [13,1]. This corresponds to local coordinate systems which represent smooth local principal directions of data. It has been demonstrated in [2], that these additional parameters offer sufficient information to extract explicit local coordinate systems from the data which can be further processed to obtain a global nonlinear projection of the underlying manifold e.g. using manifold charting [3]. This way, an explicit mapping together with its approximate inverse is obtained which can map high dimensional data into low dimensional space. In [2], the possibility to use this

* This research is partially supported by Centre of Excellence in Mathematics, the Commission on Higher Education, Thailand.

F. Schwenker and N. El Gayar (Eds.): ANNPR 2010, LNAI 5998, pp. 84–95, 2010.
© Springer-Verlag Berlin Heidelberg 2010

mapping for low dimensional data visualization and representation has been explored in comparison to popular alternative visualization schemes as referenced e.g. in [10]. Unlike popular alternative visualization methods such as locally linear embedding, maximum variance unfolding, or stochastic neighbor embedding, manifold charting in combination with matrix neural gas does not only embed the given data points, but it provides an explicit low dimensional embedding of the data manifold and an approximate inverse of the map. Hence additional information is available which can be explicitly used in further applications.

In this contribution, we will demonstrate the suitability of manifold embedding by matrix neural gas for an interesting problem from computer graphics, the efficient synthesis of dynamic texture based on given examples. Dynamic texture synthesis is the process of producing an animation of dynamic textures which preserve the behavior of the system similar to its original appearance. There exist two fundamentally different approaches for dynamic texture synthesis: physics-based methods generate dynamic texture based on mathematical models of the natural phenomena, see e.g. [18,4]. Physics-based models provide a flexible synthesis, since the dynamic texture can be controlled through a few parameters in a mathematical equation system. The drawback of this approach is that each model is appropriate only for a particular texture and cannot be transferred to other domains. As an alternative, image-based models overcome this limitation. They use a global model for different textures and synthesize dynamic textures from a model based on the appearance of the whole texture in a series of images. Different principled approaches can be distinguished such as simple extensions of static texture synthesis to 3D [20], spatiotemporal models based on the pixel level [16], or dynamical models on the image level, such as proposed in [15,6]. The latter approach is particularly promising since it can capture common non-trivial dynamical development such as rotation. This way, dynamic texture synthesis becomes a problem of system identification based on a sequence of image data such that dynamic texture can be directly generated along the trajectory of the system based on given initial conditions.

Typically, image sequences possess a very high dimensionality such that system identification is not possible in the raw image space. Therefore, the approaches presented in [15,6] first project the images onto low dimensional space with a standard principal component analysis (PCA), performing system identification e.g. using classical linear auto-regressive models in the low dimensional projection space, afterwards. Since PCA gives rise to an approximate inverse by means of the pseudo-inverse of the transformation matrix, dynamic textures can be generated this way. The resulting model is rather flexible, but it has the drawback that a global linear embedding is used such that images are not appropriately sampled and represented in particular at points in time with rapid movements (e.g. flapping flag). Therefore, it has been proposed e.g. in [8,9] to use recent nonlinear manifold learning techniques as proposed in machine learning instead of a global linear embedding.

In this contribution, we demonstrate that recent matrix learning schemes for neural gas together with a global coordination technique, manifold charting, give rise to a nonlinear manifold embedding which can successfully be used in this context. This way, neural low-dimensional nonlinear manifold embedding can serve as an essential step in the highly non-trivial application in computer graphics to automated dynamic texture synthesis. Now, we first introduce matrix neural gas which allows us to extract local linear manifold projections from a given data set. These can be combined to a global nonlinear embedding with approximate inverse using manifold charting. We describe the inclusion of this method into the general pipeline for dynamic texture synthesis, afterwards, and we demonstrate its applicability in a variety of examples.

2   Nonlinear Manifold Embedding Based on Matrix Neural Gas

Matrix neural gas

Assume data $\{x_1, \ldots, x_m\} \subset \mathbb{R}^N$ are sampled from a manifold $X \subset \mathbb{R}^N$. The aim of neural gas is to represent the given data in terms of prototypes $\{w_1, \ldots, w_k\} \subset \mathbb{R}^N$ faithfully such that the prototype w_i adequately resembles its receptive field $R_i := \{x \mid i = \operatorname{argmin}_j\{d(x, w_j)\}\}$ where usually the Euclidean distance is used

$$d(x, w_i) = (x - w_i)^t(x - w_i). \tag{1}$$

NG has been derived in [11] as a stochastic gradient descent of the following cost function

$$E_{\mathrm{NG}}(w) \sim \frac{1}{2} \sum_{i=1}^{k} \int h_\sigma(k_i(x)) \cdot d(x, w_i) \, P(dx) \tag{2}$$

where P refers to the probability distribution of the data points x. $k_i(x) \in \{0, \ldots, k-1\}$ constitutes a permutation of prototypes arranged according to the distance, i.e.

$$k_i(x) := |\{w_j \mid d(x, w_j) < d(x, w_i)\}|$$

If distances coincide, ties are broken deterministically. $h_\sigma(t) = \exp(-t/\sigma)$ is an exponential curve with neighborhood range $\sigma > 0$. For vanishing neighborhood $\sigma \to 0$, the standard quantization error is obtained.

As an alternative to online optimization, a batch approach can be taken if data are given in advance. For a discrete finite data set, the cost function (2) can be optimized in a batch scheme repeatedly updating prototypes and rank assignments until convergence [5]. Usually, during training, the neighborhood range σ is annealed to 0 such that the quantization error is recovered in final steps. In intermediate steps, a neighborhood structure of the prototypes is determined by the ranks according to the given training data. This choice accounts for a high robustness of the algorithm with respect to local minima of the quantization error, further, a smooth update of neighbored points is guaranteed this way [12].

Classical NG relies on the Euclidean metric which induces isotropic cluster shapes. More general ellipsoidal shapes can be achieved by the generalized metric form

$$d_{\Lambda_i}(x, w_i) = (x - w_i)^t \Lambda_i(x - w_i) \tag{3}$$

instead of the squared Euclidean metric (1) where $\Lambda_i \in \mathbb{R}^{N \times N}$ is a symmetric positive definite matrix with $\det \Lambda_i = 1$. These constraints are necessary to guarantee that the resulting formula defines a metric which does not degenerate to a trivial form ($\Lambda_i = 0$ constituting an obvious trivial optimum of the cost functions). A general matrix Λ_i can account for correlations and appropriate nonuniform scaling of the data dimensions. The parameters Λ_i in (3) can be optimized during training together with the prototype parameters and assignments. The corresponding cost function (2) which uses (3) instead of (1) can be optimized in batch mode, yielding matrix NG (MNG):

> init $\boldsymbol{w}_i$ randomly
> init Λ_i as identity I
> repeat until convergence
> > set $k_{ij} := k_i(\boldsymbol{x}_j)$
> > set $\boldsymbol{w}_i := \sum_j h_\sigma(k_{ij}) \boldsymbol{x}^j / \sum_j h_\sigma(k_{ij})$
> > set $\Lambda_i := S_i^{-1} (\det S_i)^{1/n}$ where
> > > $S_i := \sum_j h_\sigma(k_{ij}) (\boldsymbol{x}^j - \boldsymbol{w}^i)(\boldsymbol{x}^j - \boldsymbol{w}^i)^t$

It has been shown in [1] that this update scheme converges to a local optimum of the NG cost function under mild conditions.

Local linear projections

The matrix S_i corresponds to the correlation of the data centered at prototype $\boldsymbol{w}_i$ and weighted according to its distance from the prototype. For vanishing neighborhood $\sigma \to 0$, the standard correlation matrix of the receptive field is obtained. The resulting Mahalanobis distance corresponds to a scaling of the principal axes of the data space by the inverse eigenvalues in the eigendirections. Thus, ellipsoidal cluster shapes arise which are aligned according to local principal components of the data. Since neighborhood cooperation is applied to both, prototype adaptation and matrix learning during batch training, a regularization of matrix learning is given and neighbored matrices have a similar form.

Local matrices as learned by MNG provide local linear transformations of the data in the following way: Assume the eigenvalue decomposition

$$\Lambda_i = \Omega_i^t \cdot D_i \cdot \Omega_i$$

is given with a diagonal matrix D_i of eigenvalues and eigenvectors collected in Ω_i. Assume data should be mapped to dimensionality n where, usually, $n < N$. Then we can reduce D_i to only the d smallest eigenvalues (which are the main eigenvalues of S_i, i.e. they belong to the main principal components of the receptive field) getting the $n \times N$ matrix D_i^{red}. The formula

$$A_i : \mathbb{R}^m \to \mathbb{R}^d, \boldsymbol{x} \mapsto D_i^{\text{red}} \cdot \Omega_i^t \cdot (\boldsymbol{x} - \boldsymbol{w}_i) \tag{4}$$

gives the local linear projection of the data points to the main principal components of the receptive field induced by the ith chart of the data manifold. If n is chosen at most 3, every map A_i provides a linear visualization of the manifold which is faithful within the receptive field of prototype $\boldsymbol{w}^i$ because it corresponds to the main eigenvalues of the local chart, as proposed in the contribution [2].

Note that, depending on the dimensionality N of the original data points, full matrix learning in MNG is rather time consuming, requiring matrix inversion of order $\mathcal{O}(N^3)$ in every step. Since only the minor n eigenvalues of the matrix Ω_i are relevant for the local projection, we can priorly reduce the matrix such that the scaling of only the minor d principal components is individually adapted while the scaling remains identical (and nonzero to avoid degeneration) for all other directions. This can be achieved efficiently with any algorithm which extracts the major n principal components of the generalized data correlation matrix, e.g. a generalized Sanger rule as proposed in [13], reducing the matrix determination to $\mathcal{O}(N)$. It has been demonstrated in [13] that the reduction to the largest n principal components can be explicitly included into the metric computation step, such that an overall reduction of the complexity to $\mathcal{O}(N)$ results, assuming independence of the number of iterations for eigenvector learning and neural gas of N.

Global coordination by manifold charting

The projections (4) provide valid local linear transformations of the data in the neighborhood of the respective prototype. Different methods which allow to combine these local projections to a global map have been proposed e.g. [17,3]. We will rely on manifold charting as introduced in [3] which glues the linear pieces together such that a good agreement can be observed at the overlaps.

Assume that linear projections $A_1, \ldots, A_k$ are given by formula (4) which define k local projections of the data points $\boldsymbol{z}_{1i} = A_1(\boldsymbol{x}_i), \ldots, \boldsymbol{z}_{ki} = A_k(\boldsymbol{x}_i)$ of the data points $\boldsymbol{x}_1, \ldots, \boldsymbol{x}_m$. Assume that, in addition, a responsibility value $p_{ij} = p_i(\boldsymbol{x}_j)$ is specified for every data point $\boldsymbol{x}_j$ and chart A_i which defines the responsibility of this chart for the data point, whereby $\sum_i p_{ij} = 1$ for every j. Here, we can use Gaussian bells centered at the prototypes to arrive at these responsibilities. More precisely, set $N_i = |R_i|$ as the number of points in the ith receptive field. Let S_i be the correlation matrix of the ith receptive field as computed in MNG and $\tilde{S}_i = S_i/N_i$ the associated matrix. Then we set the responsibility of the ith receptive field for point $\boldsymbol{x}_j$ as

$$\tilde{p}_{ij} = \tilde{p}_i(\boldsymbol{x}_j) = \frac{N_i}{p} \cdot \frac{1}{(2\pi)^{m/2}\sqrt{|\tilde{S}_i|}} \cdot \exp(-0.5 \cdot (\boldsymbol{x}_j - \boldsymbol{w}_i)^t \cdot \tilde{S}_i^{-1} \cdot (\boldsymbol{x}_j - \boldsymbol{w}_i)) \quad (5)$$

where the prior N_i/p refers to the relative number of points in chart i. The responsibilities p_{ij} are obtained thereof by normalization $p_{ij} = \tilde{p}_{ij}/\sum_i \tilde{p}_{ij}$.

The goal is to combine the local charts A_i by means of local affine mappings $B_i : \mathbb{R}^n \to \mathbb{R}^n$ to a global mapping such that the compositions lead to matching points if more than one chart is responsible for a data point. The mappings B_i are determined in such a way that the following costs are minimized

$$E_{\text{charting}} = \frac{1}{2} \cdot \sum_{i,j,l} p_{ji} p_{li} \| B_j(\boldsymbol{z}_{ji}) - B_l(\boldsymbol{z}_{li}) \|^2 \quad (6)$$

which (as can be seen by a simple algebraic transformation) also give the difference of the globally mapped points and the local affine transformations of the points. As shown in [3] a unique analytic solution of this problem can be found. The final points are then obtained by the formula $x_j \mapsto \sum_i p_{ij} B_i(z_{ij})$.

Note that, depending on the distribution of the prototypes, the assignments will be sparse since p_{ij} will be almost zero for receptive fields i which lead to a high rank w.r.t x_j. To speed up the computation, it is possible to cut off these small assignments and work with sparse matrices. Based on these choices of p_{ij} provided by (5) and z_{ij} provided by the affine transformations (4), MNG can be combined with manifold charting to give a global nonlinear visualization of data. Obviously, this visualization can be described by an explicit mapping of $\mathbb{R}^N \to \mathbb{R}^n$ by means of the formula

$$x \mapsto \sum_i p_i(x) \cdot B_i(A_i(x)) \tag{7}$$

where $p_i(x)$ is computed according to (5), the local linear mappings A_i are given by (4), and the affine transformations B_i to glue the charts together are determined solving equation (6).

Inverse map

To arrive at an approximate inverse map, we take a simple point of view which allows us to compute the inverse algebraically. Note that every local linear projection A_i possesses an approximate inverse A_i^{-1} induced by the pseudo-inverse of $D_i^{\text{red}} \cdot \Omega_i$. Since A_i maps to lower dimensions, this is, of course, no exact inverse but its best approximation in a least squares sense. Further, obviously, the affine transformations B_i can be inverted exactly. Thus, for every $z \in \mathbb{R}^n$, an approximate inverse of the image of (7) can be determined in the following way: for a given x, we determine the inverse under chart i: $A_i^{-1} \circ B_i^{-1}(z)$. From these possibly preimages, we take the one with maximum responsibility according to (5).

3   Dynamic Texture Synthesis by System Traversal

In [8,9], dynamic texture synthesis is modelled as a system identification problem. First, images are nonlinearly mapped to a low-dimensional space. In low dimensions, a method to track temporal developments based on initial conditions is defined. Since every point in the embedding space can be inversely mapped to a point in the original high dimensional space, a sequence of images corresponding to a dynamic texture results. This way, a compressed representation of dynamic textures can be obtained since it is sufficient to store the parameters of the global nonlinear map and its inverse and only the low dimensional projections of the given image sequence which, for $n \ll N$, requires much less space than the original texture sequence. Further, interpolation of texture images as well as generation of texture based on new starting points becomes possible since a model is available to track the dynamics in the low dimensional projection space.

The contribution [9] relies on a mixture of probabilistic principal component analysis (MPPCA) together with global coordination to obtain a global non-linear embedding of the data manifold [19,17]. Here we propose to substitute MPPCA by matrix NG since, as we will demonstrate in experiments, a greater robustness and smoothness of the method can be achieved. Thus, we combine global coordination based on matrix NG as described in the previous section with the tracking dynamics in the low dimensional projection space as intro-duced in [9]. For convenience, we shortly describe the traversal mechanism as proposed in [9].

Assume a dynamic texture is given which, making the temporal depen-dency explicit, is denoted as $x(t)$, $x(t + 1)$, $\ldots \in \mathbb{R}^N$. The corresponding low-dimensional projections are denoted as $z(t)$, $z(t+1)$, $\ldots \in \mathbb{R}^n$. Motion prediction starts from a sequence of at least two points $g(t - 1)$, $g(t)$ in $\mathbb{R}^n$ which are prob-ably obtained as projections of images. Now the temporal successors of $g(t)$ are obtained in six steps based on the low dimensional vectors $z(i)$ as follows (σ_1, α, σ_2 are positive parameters):

1. Sampling neighbors: K nearest neighbors $\mathcal{N}$ of $g(t)$ are sampled from the data $z(i)$ and exponentially weighted according to the distance from $g(t)$ with weight $W_i^1 := \exp(-\|g(t) - z(i)\|^2/\sigma_1^2)$.
2. Temporal smoothness: The similarity of the difference vectors $dz(i) := z(i) - z(i - 1)$ of the neighbors $z(i)$ in $\mathcal{N}$ and the considered trajectory $dg(t) := g(t) - g(t - 1)$ is computed based on the cosine distance and exponentially weighted, yielding weight $W_i^2 := \exp(\alpha(dz(i)^t d(g(t))/(\|dz(i)\| \cdot \|dx(t)\|) - 1))$.
3. Noise perturbation: for every neighbor $z(i)$ in $\mathcal{N}$, noisy successors of $g(t)$ are sampled using the direction of the trajectory at $z(i)$ and a Gaussian noise vector ν_j with components $\sim N(0, \sigma_2)$ leading to possible positions $g^{ij}(t + 1) = g(t) + (z(i + 1) - z(i)) + \nu_j$.
4. Drift prevention: Each candidate is weighted according to its distance from the trajectory leading to the weight $p(g^{ij}(t + 1)) = W_i^1 W_i^2 \sum_l \varphi((g^{ij}(t + 1) - z_k)/h)$ where φ is a window function with window width h.
5. Normalization: These weights are normalized such that $\sum_{ij} p(g^{ij}(t+1)) = 1$.
6. Prediction: The successor is chosen from these points according to the prob-ability $p(g^{ij}(t + 1))$.

This way, the overall direction of the trajectory gives rise to the respective suc-cessor of a given starting position. Slight noise accounts for typical effects when dealing with natural phenomena. An additional neighborhood integration makes sure that the created trajectory does not diverge from the dynamics as deter-mined by the given data set.

4   Experiments

An important example of dynamic texture is given by image sequences of nat-ural phenomena as available in the DynTex database [14]. Each pixel gradually

Table 1. Number of local models used to map the respective dynamic texture into low dimensional space and number of frames included in the dynamic textures

Dataset	no. local models	no. frames
wave	3	200
escalator	4	251
smoke	5	251
fall	2	200
straw	4	251

changes its intensity or color level depending on the kind of image sequence. Examples of natural phenomena used in this work are referred to as wave, escalator, smoke, straw, and fall. All images have an original size of 288 by 352 pixels with RGB color codes. Because of the high dimensionality, we resized the images to 50% with respect to the original size resulting in 144 times 176 pixels. Then, each image corresponds to a 76,032 dimensional vector formed by the RGB values of the pixels. Before applying matrix NG, data were projected to 100 dimensions using simple principal component analysis.

We compare the result of matrix NG and mixtures of probabilistic component analysis as described in [19] together with global manifold charting and trajectory traversal as described above. The dimensionality n has been chosen as 40. The number k of local components is chosen such that, on average, about 50 frames are represented by one local model. The lengths of the considered dynamic textures and the number of local models is shown in Tab. 1.

We evaluate the method by the mean absolute distance of the generated images and the original images averaged over time, as shown in Tab. 2. Further, we exemplarily show a visual comparison of the images as obtained by MNG and MPPCA in comparison to the original image in Figs. 2 - 5. The synthesis results indicate that manifold embedding based on MNG is able to generate high-quality video, while charting based on MPPCA generates lower visual quality of video over time. The synthesized image sequences by MNG are smooth with respect to temporal evolution due to the included neighborhood cooperation, and sharp features are better preserved in the single images. MPPCA contains blurring in single images and a larger trend towards discontinuities when generating image sequences. This manifests in a larger error of the generated images. The development of the absolute error over time per pixel is depicted in Fig. 1. Obviously, for MPPCA, the error is not uniformly distributed but it accumulates at points in time such that errors are clearly observable for MPPCA. In comparison, the error of MNG is very smooth such that no abrupt changes in the visual appearance can be observed.

Both methods, dynamic tecture generation based on MPPCA or MNG, rely on prototypes which represent parts of the data space. Commonly, these prototypes are computed as averages, such that both methods have the drawback that they somehow smooth details in the images. Since high contrast features are of particular relevance for the human observer, deviations from the original images

Table 2. Average absolute reconstruction errors averaged over the number of frames on the given image sequences and standard deviations

Dataset	MPPCA	MNG
wave	33.7319 ± 2.7425	31.3815 ± 0.5540
escalator	30.8114 ± 2.0043	24.5836 ± 0.1290
smoke	14.2246 ± 0.5515	12.5402 ± 0.2278
fall	48.4483 ± 0.8815	47.3001 ± 0.0537
straw	37.3818 ± 0.5610	36.7924 ± 0.1960

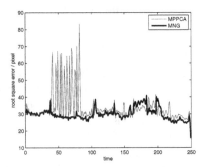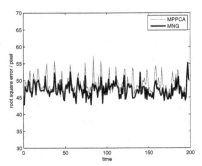

Fig. 1. Absolute error per pixel between the generated images and the true image sequence over time for MPPCA and MNG for the wave texture (left) and fall texture (right). Obviously, the error obtained by MPPCA is large for some time points in which a low quality of the reconstructed texture can visually be observed.

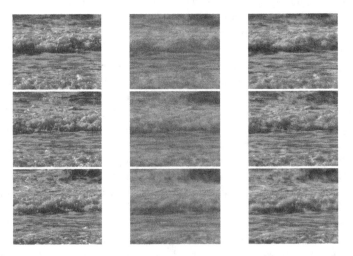

Fig. 2. This figure shows reconstructed image sequence of waves. The first column represents the original reference frames. The second and third columns demonstrate the frames reconstructed by manifold charting based on MPPCA and matrix NG, respectively.

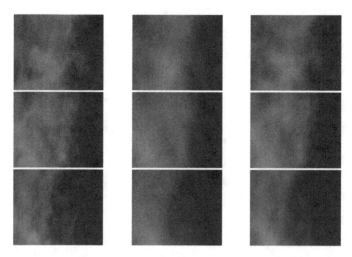

Fig. 3. This figure shows reconstructed image sequence of smoke. The first column represents original reference frames. The second and third columns demonstrate the frames reconstructed by manifold charting based on MPPCA and matrix NG, respectively.

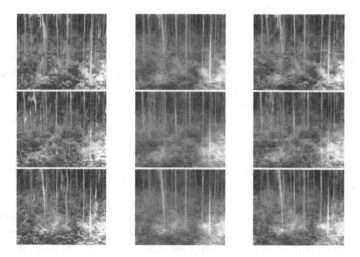

Fig. 4. This figure shows reconstructed image sequence of fall. The first column represents original reference frames. The second and third columns demonstrate the frames reconstructed by manifold charting based on MPPCA and matrix NG, respectively.

can clearly be observed by humans, albeit the error is small, as can be seen for the examples 'fall' and 'straw' which include high contrast or lots of details, respectively. This drawback could be prevented by substituting the averaged prototypes by features with more contrast or details, respectively, obtained e.g. by appropriate postprocessing.

Fig. 5. This figure shows reconstructed image sequence of straw. The first column represents original reference frames. The second and third columns demonstrate the frames reconstructed by manifold charting based on MPPCA and matrix NG, respectively.

5 Discussion

We have introduced a method which allows the global nonlinear embedding of complex manifolds into low dimensions based on matrix neural gas, leading to an explicit embedding function as well as its approximate inverse. Thereby, an essential part is given by matrix neural gas as presented in this paper, which extracts prototypes and local principle directions from the data in a robust and topology preserving way such that a set of smooth local linear maps is obtained. These can be combined using charting techniques such that a global smooth embedding arises. In comparison to alternatives such as mixtures of probabilistic principal components, topology preservation has the beneficial effect that manifold charting deals with smooth maps which can be glued together more easily, and the reconstruction error by means of the approximate inverse shows good agreement to the original manifold also at the borders of the local linear pieces. We have demonstrated that this technique can successfully be used in a complex task from computer graphics, namely the synthesis of dynamic textures based on given image sequences.

References

1. Arnonkijpanich, B., Hammer, B., Hasenfuss, A.: Local matrix adaptation in topographic neural maps, Technical Report IfI-08-07, Department of Computer Science, Clausthal University of Technolgy (September 2008)
2. Arnonkijpanich, B., Hasenfuss, A., Hammer, B.: Local matrix learning in clustering and applications for manifold visualization. In: Neural Networks (to appear, 2010)

3. Brand, M.: Charting a manifold. In: Advances in Neural Information Processing Systems, vol. 15, pp. 961–968. MIT Press, Cambridge (2003)
4. Chuang, Y.Y., Goldman, D.B., Zheng, K.C., Curless, B., Salesin, D.H., Szeliski, R.: Animating pictures with stochastic motion textures. ACM Transactions on Graphics 24(3), 853–860 (2005)
5. Cottrell, M., Hammer, B., Hasenfuss, A., Villmann, T.: Batch and median neural gas. Neural Networks 19, 762–771 (2006)
6. Doretto, G., Chiuso, A., Wu, Y., Soatto, S.: Dynamic Textures. Int. Journal of Computer Vision 51(2), 91–109 (2003)
7. Kohonen, T.: Self-Organizing Maps. Springer, Heidelberg (1995)
8. Liu, C.B., Lin, R.S., Ahuja, N.: Modeling Dynamical Textures Using Subspace Mixtures. In: IEEE International Conference on Multimedia and Expo 2005, pp. 1378–1381 (2005)
9. Liu, C.B., Lin, R.S., Ahuja, N., Yang, M.H.: Dynamic Textures Synthesis as Nonlinear Manifold Learning and Traversing. In: British Machine Vision Conference 2006, vol. 2, pp. 859–868 (2006)
10. van der Maaten, L.J.P., Postma, E.O., van den Herik, H.J.: Matlab Toolbox for Dimensionality Reduction. In: MICC, Maastricht University (2007)
11. Martinetz, T., Berkovich, S.G., Schulten, K.J.: 'Neural-gas' network for vector quantization and its application to time-series prediction. IEEE Transactions on Neural Networks 4, 558–569 (1993)
12. Martinetz, T., Schulten, K.: Topology representing networks. Neural Networks 7, 507–522 (1994)
13. Möller, R., Hoffmann, H.: An Extension of Neural Gas to Local PCA. Neurocomputing 62, 305–326 (2004)
14. R. Peteri, M. Huiskes, and S. Fazekas, The DynTex database Homepage, http://old-www.cwi.nl/projects/dyntex/database.html, Centre for Mathematics and Computer Science, Amsterdam
15. Soatto, S., Doretto, G., Wu, Y.N.: Dynamic Textures. In: IEEE International Conference on Computer Vision, vol. 2, pp. 439–446 (2001)
16. Szummer, M., Picard, R.W.: Temporal texture modeling. In: IEEE International Conference on Image Processing, vol. 3, pp. 823–826 (1969)
17. The, Y.W., Roweis, S.: Automatic alignment of local representations. Advances in Neural Information Processing Systems 15, 841–848 (2003)
18. Treuille, A., McNamara, A., Popovic, Z., Stam, J.: Keyframe control of smoke simulations. ACM Transactions on Graphics 22(3), 716–723 (2003)
19. Tipping, M.E., Bishop, C.M.: Mixtures of probabilistic principal component analyzers. Neural Computation 11, 443–482 (1999)
20. Wei, L.-Y., Levoy, M.: Fast texture synthesis using tree-structured vector quantization. In: Proceedings of ACM SIGGRAPH 2000, pp. 479–488 (2000)

SIC-Means: A Semi-fuzzy Approach for Clustering Data Streams Using C-Means

Amr Magdy and Mahmoud K. Bassiouny

Computer and Systems Engineering, Alexandria University, Egypt
amr.magdy@alex.edu.eg, eng.mkb@gmail.com

Abstract. In recent years, data streaming has gained a significant importance. Advances in both hardware devices and software technologies enable many applications to generate continuous flows of data. This increases the need to develop algorithms that are able to efficiently process data streams. Additionaly, real-time requirements and evolving nature of data streams make stream mining problems, including clustering, challenging research problems. Fuzzy solutions are proposed in the literature for clustering data streams. In this work, we propose a *S*oft *I*ncremental *C-Means* variant to enhance the fuzzy approach performance. The experimental evaluation has shown better performance for our approach in terms of Xie-Beni index compared with the pure fuzzy approach with changing different factors that affect the clustering results. In addition, we have conducted a study to analyze the sensitivity of clustering results to the allowed fuzziness level and the size of data history used. This study has shown that different datasets behave differently with changing these factors. Dataset behavior is correlated with the separation between clusters of the dataset.

Keywords: Fuzz Clustering, Soft Clustering, Data Streams, C-Means, Clustering Data Streams.

1 Introduction

In recent years, the data stream model has gained a significant importance. Advances in both hardware devices and software technologies enable many applications to generate continuous flows of data. This has increased the sources of streaming information and thus the need to develop algorithms that are able to efficiently process data streams. These algorithms have to consider related scalability issues. Huge volumes of data with continuous and evolving nature of streaming data have introduced two main challenges; (i) infeasibility of storing the entire data (ii) and infeasibility of multiple passes processing. These challenges have motivated data research communities to provide appropriate solutions that are able to overcome these problems. Query processing over data streams [3], sketching of data streams [7] and mining problems [12] are the most studied problems in this context.

Data clustering problem is one of the most studied mining problems in the literature [17,16]. Data clustering techniques can be classified on different bases.

F. Schwenker and N. El Gayar (Eds.): ANNPR 2010, LNAI 5998, pp. 96–107, 2010.
© Springer-Verlag Berlin Heidelberg 2010

One possible distinguishing characteristic is how the technique relates data points to different clusters. The literature presented three types: (i) hard clustering (ii) fuzzy clustering (iii) and soft clustering [24]. In hard clustering, each data point belongs to exactly one cluster. In fuzzy clustering each data point belongs to all clusters with different membership degrees. Soft clustering relaxes the absolute fuzziness constraint so it relates a data point to only a subset of clusters which are the most similar under some threshold of similarity. In this work we present a *S*oft *I*ncremental *C-Means* variant for clustering streaming data; the SIC-Means approach.

2 Related Work

The literature of clustering data streams is fairly mature. Research work on dynamic clustering in incremental basis has started in late eighties in the context of information retrieval [5]. Adapting dynamic clustering techniques to work for data streams didn't show a great success due to the evolving nature of data streams. O'Callaghan et al. [22] proposed a k-median based clustering algorithm for streaming data. Their algorithm receives the streaming data in chunks, applies the clustering process and then frees the memory representing the clustering solution with weighted centroids. They obtain the weighted centroids of the entire stream received so far by applying the same algorithm to the centroids obtained from chunks. They showed that their algorithm outperformed BIRCH [31] in terms of sum of squared distance.

Aggarwal et al. [1] proposed an important framework for clustering data streams. They simply suggest to develop a two phase framework; an online phase and an offline phase. The online phase processes the continuous stream of data building sufficient statistics and collecting enough information to the offline phase to use in extracting data clusters. The importance of this framework is not only gained from the CluStream approach they proposed utilizing this framework, but also from being the basic idea of many important following research work in clustering data streams [23, 6, 18, 26, 27]. Park and Lee [23], Jia et al. [18] and Tu and Chen [26] use the online phase to build a data histogram. They then merge neighboring dense cells in clusters during the offline phase under different measures for density. DenStream algorithm [6] considers a different approache for both online and offline phases. They collect neighborhood density information during the online phase. They then use this information to build clusters in a similar way to conventional DBScan algorithm [11].

Many other approaches in the literature proposed solutions for clustering data streams under different assumptions and from different perspectives [13, 30, 21, 2, 8, 29, 4, 25, 19, 20, 9]. However, all these algorithms proposed hard clustering solutions. Some applications generate data items that can belong to more than one cluster of data. Hore et al. [14] proposed a fuzzy incremental clustering solution for data streams based on C-means algorithm. This solution assumes that all data items belong to all clusters with different degrees of strength. Our research attempts to improve the pure fuzzy approach. We argue that assuming all data items to belong to all clusters may be misleading. Some data items are obviously far from some clusters. In general, we can assume that every data item can belong to one or more clusters under some threshold of similarity. We apply this soft clustering concept to the incremental clustering solution.

3 SIC-Means

Streaming data algorithms in the literature assume that data points arrive and get processed either in batches or one at a time. We consider the former paradigm where n_i points arrive at time instant t_i. Our approach is an improvement to Hore's approach presented in [14] with applying the soft clustering concept [15]. We use a threshold α on allowed fuzziness so that after applying the fuzzy clustering process, a data point p_j belongs to a cluster c_k iff the corresponding degree of membership $u_{kj} \geq \frac{\alpha}{number-of-clusters}$.

The SIC-Means approach, which stands for **S**oft **I**ncremental **C-Means**, takes three input parameters; number of clusters c, size of history used h and fuzziness threshold α. SIC-Means assigns a weight to each data point reflecting its importance. The effect of these weights needs to be incorporated in the objective function. Similar to [10], we define the fuzzy objective function to be minimized as follows:

$$J_m(U,V) = \sum_{k=1}^{c} \sum_{j=1}^{n} u_{kj}^m w_j \|p_j - v_k\| \tag{1}$$

Where
U : fuzzy membership matrix
V : cluster centers vector
u_{kj} : the degree of membership of point p_j in cluster c_k
v_k : the center of cluster c_k
w_j : the weight of point p_j

In our work, we consider $J_{1.5}(U,V)$ which has shown better experimental results in our experiments than other values of fuzzifier m. Furthermore, we use Euclidean distance as a dissimilarity measure. At any time instant t_i, SIC-Means clusters N_i weighted points where:

$$N_i = min(n_i + c * (i-1), n_i + c * h), i \geq 1 \tag{2}$$

New data points are assigned unit weights while cluster centers of stage i has weights w_k's where:

$$w_k = \sum_{j=1}^{N_i} u_{kj}, 1 \leq k \leq c \tag{3}$$

Algorithm 1 outlines the main steps of a single SIC-Means stage. At each stage, the clustering algorithm considers old centers of the last h stages in addition to the newly received data points. If the current stage $i < h$, then we use all the available history up to the moment. After each time instant t_i, only $min(c*i, c*h)$ weighted points are kept in memory representing the cluster centers for the last h stages.

Algorithm 1. Stage i of SIC-Means Algorithm

---

1. Collect the instant data and history centers in a matrix P.
2. Prepare weight matrix W, assign a unit weight to every new data point.
3.
if $i = 1$ **then**
    Select initial cluster centers as $(i - 1)th$ stage cluster centers
else
    Select initial cluster centers arbitrarily
end if
4.
repeat
    4.1 Update the membership matrix U utilizing the formula

$$u_{kj} = \left[\sum_{l=1}^{c} \left(\frac{\|p_j - v_k\|}{\|p_j - v_l\|} \right)^{\frac{2}{m-1}} \right]^{-1}, 1 \leq j \leq N_i, 1 \leq k \leq c$$

    4.2
for all u_{kj} in U **do**
      if $u_{kj} < \frac{\alpha}{c}$ **then**
        $u_{kj} \leftarrow 0$
      end if
    end for
    4.3 Normalize U so that

$$\sum_{k=1}^{c} u_{kj} = 1, 1 \leq j \leq N_i$$

    4.4 Update cluster centers using the formula

$$v_k = \frac{\sum_{j=1}^{N_i} w_j u_{kj}^m p_j}{\sum_{j=1}^{N_i} w_j u_{kj}^m}, 1 \leq k \leq c$$

    4.5 Calculate the objective function using the formula
$$J_m(U, V) = \sum_{k=1}^{c} \sum_{j=1}^{N_i} u_{kj}^m w_k \|p_j - v_k\|$$
until there is no change in the objective function
5. Calculate the centers weighs for future instants.

$$w_k = \sum_{j=1}^{N_i} u_{kj}, 1 \leq k \leq c$$

---

4   Evaluation

We have conducted a set of experiments to evaluate the performance of our approach. The experiments have shown an improvement, over using the pure fuzzy approach, under a standard evaluation scheme for fuzzy clustering algorithms. In this section, we first present the evaluation scheme we use, describe the evaluation datasets and then show the experimental results.

4.1   Evaluation Scheme

In our experiments, we use the popular Xie-Beni index [28] which is designed to internally validate fuzzy clustering solutions. XB index can be calculated using the following equation

$$XB = \frac{1}{n} \frac{\sum_{\forall c_k \in C} \sum_{\forall p_j \in P} u_{kj}^m \|p_j - c_k\|}{min_{\forall c_l, c_k}(\|c_l - c_k\|)} \quad (4)$$

Where
C : is set of all clusters
P : is set of all data points
m : is the fuzzifier

As discussed in [28], this index favors more compact and separate clusters. The lower the value of this index, the better the solution.

4.2 Data

In our evaluation, we use three artificial datasets; each of 100,000 2D points, to evaluate SIC-Means performance. The three datasets are drawn randomly from Gaussian distributions to form globular clusters. We have chosen these datasets for two main reasons. First, Hoer et al. [14] used one of these datasets to evaluate their algorithm. As we will introduce later in this section, we compare the performance of our new method relative to that of Hore et al. so it was appropriate to use the same dataset they used in evaluation. Second, a known limitation for C-means algorithm and its variants with Euclidean distance is to favor discovering globular clusters. In our research, we focus on showing the improvement of applying soft clustering concept to incremental fuzzy approaches. So we have chosen the datasets that serve our focus isolating the known C-means limitations.

The main difference between the three datasets is the separation between clusters. As shown in Figure 1, dataset DS1 clusters are strongly overlapped, dataset DS2 clusters are slightly overlapped and dataset DS3 clusters are well-separated. As soft clustering techniques is concerned with cutting off data points from far clusters, this variation in clusters separation should show the effectiveness of the technique.

4.3 Experimental Results

In our experiments, we have studied the effect of two major factors; fuzziness threshold α and size of history h. We have conducted our experiments using the true number of clusters studying the change in Xie-Beni index value with varying these factors. In these experiments, $\alpha = 0$ represents the pure fuzzy approach which is Hore's approach. Figures 2 - 4 show the results of these experiments.

As shown in Figure 2(a), XB index values on DS1 dataset for $0 < \alpha \leq 0.6$ are lower compared to $\alpha = 0$ for all history sizes (values of h). The upper bound of this range is extended to approach 0.8 at h $\geq$ 2. Figure 2(a) shows that history size greater than 2 does not provide a significant improvement for DS1 dataset; curves of history sizes 3 and 4 are approximately identical. These observations are supported by Figure 2(b) that shows obvious lower values of XB index for different α values with changing history size from 0 to 7. In addition, excluding the pure fuzzy approach, at $\alpha = 0$, in Figure 2(c) shows improvement with increasing history size up to $h = 4$. Increasing history size after that degrades the performance. This can be explained by the evolving nature of data stream. According to the input order of the data points, more history weighted points

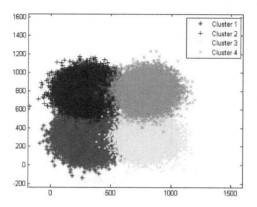

(a) Original distribution of dataset DS1

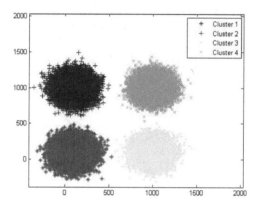

(b) Original distribution of dataset DS2

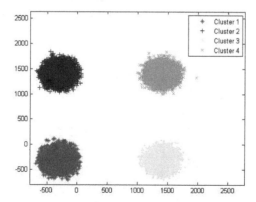

(c) Original distribution of dataset DS3

Fig. 1. Original distribution of evaluation datasets

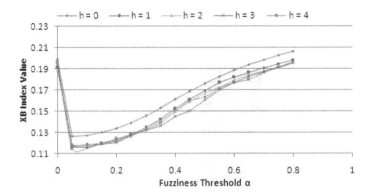

(a) SIC-Means performance for different history sizes

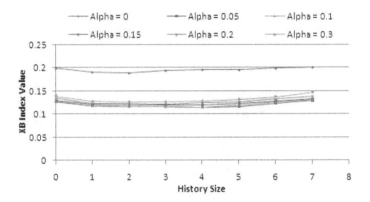

(b) SIC-Means performance for different fuzziness thresholds

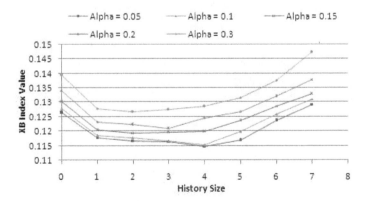

(c) SIC-Means performance for different fuzziness thresholds excluding pure fuzziness

Fig. 2. SIC-Means performance on DS1 dataset

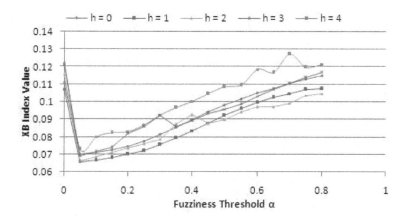

(a) SIC-Means performance for different history sizes

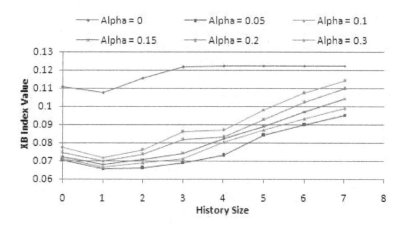

(b) SIC-Means performance for different fuzziness thresholds

Fig. 3. SIC-Means performance on DS2 dataset

can bias new points to the wrong clusters. However, it is obvious that considering more history does not always improve the overall performance.

Figure 3(a) shows similar observations on DS2 dataset. One obvious difference that XB index values are lower on DS2 dataset for wider range of α values, $0 < \alpha \leq 0.8$ than the corresponding value at $\alpha = 0$ for all history sizes < 4. On the other hand, this shows an earlier performance degradation with increasing history size than DS1 dataset. These observations are supported by Figure 3(b).

Figure 4(a) can lead us to conclude that the more the clusters separated, the less the history needed. It shows almost no need to keep history data. Also this can be well observed in Figure 4(b). In almost all cases, the pure fuzzy approach is outperformed for all fuzziness thresholds and history sizes.

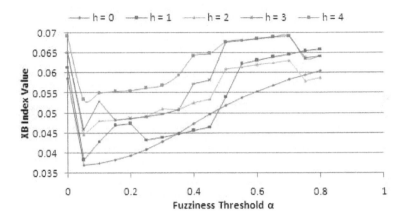

(a) SIC-Means performance for different history sizes

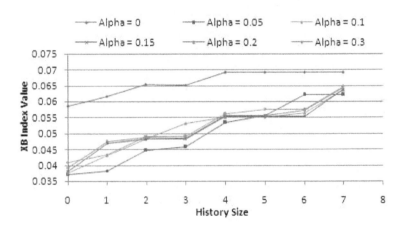

(b) SIC-Means performance for different fuzziness thresholds

Fig. 4. SIC-Means performance on DS3 dataset

5   Conclusion

In this paper, we introduced the combination of incremental and soft clustering concepts. We incrementally cluster batches of data points using weighted C-means variant. At any time instant, we consider a new batch of points with number of old batches of weighted points. These old points represent the cluster centers in earlier stages of the algorithm.

Combining incremental clustering with soft clustering has shown an ability to improve the fuzzy clustering results in terms of Xie-Beni index values. This improvement depends on different factors. From one hand, threshold for allowed fuzziness affect the performance. All thresholds of soft clustering used dominate the pure fuzzy clustering approach. However, considering more history does not always give better results. Evolving nature of data streams may bias new points to join the wrong clusters with increasing the history size above some value. History size value that separates performance improvement from performance degradation differs with changing the clusters separation. The better the clusters are separated, the lower history size needed. This property fits the limited memory constraint imposed by the data stream model.

6   Discussion and Future Work

In this work, we have shown the effectiveness of combining soft clustering technique with C-means based incremental clustering technique. However, more evaluation techniques can be investigated. We may investigate building a fuzzy ground truth for evaluation datasets used. Existence of this ground truth will enable using external validation measures. Also, we may investigate how these clustering solutions tolerate errors in parameter selection; for example how they perform when clustering using wrong number of clusters. Other factors worth investigation are: sensitivity analysis of data input order and effect of using different dissimilarity measures.

Acknowledgment. The authors would like to thank Prof. Dr. M. A. Ismail for his support during Pattern Recognitions graduate course offered during the graduate studies in Computer and Systems Engineering Department, Alexandria University, Egypt. The basics of this work has been conducted as a part of the final project of this course.

References

1. Aggarwal, C.C., Han, J., Wang, J., Yu, P.S.: A framework for clustering evolving data streams. In: VLDB 2003: Proceedings of the 29th international conference on Very large data bases. VLDB Endowment, pp. 81–92 (2003)
2. Aggarwal, C.C., Han, J., Wang, J., Yu, P.S.: A framework for projected clustering of high dimensional data streams. In: VLDB 2004: Proceedings of the Thirtieth international conference on Very large data bases. VLDB Endowment, pp. 852–863 (2004)
3. Babcock, B., Babu, S., Datar, M., Motwani, R., Widom, J.: Models and issues in data stream systems. In: PODS 2002: Proceedings of the twenty-first ACM SIGMOD-SIGACT-SIGART symposium on Principles of database systems, pp. 1–16. ACM, New York (2002)
4. Beringer, J., Hüllermeier, E.: Online clustering of parallel data streams. Data Knowledge Engineering 58(2), 180–204 (2006)
5. Can, F., Ozkarahan, E.: A dynamic cluster maintenance system for information retrieval. In: SIGIR 1987: Proceedings of the 10th annual international ACM SIGIR conference on Research and development in information retrieval, pp. 123–131. ACM, New York (1987)

6. Cao, F., Ester, M., Qian, W., Zhou, A.: Density-based clustering over an evolving data stream with noise. In: Proceedings Sixth SIAM Intl. Conference Data Mining (2006)

7. Cormode, G., Garofalakis, M.: Sketching probabilistic data streams. In: SIGMOD 2007: Proceedings of the 2007 ACM SIGMOD international conference on Management of data, pp. 281–292. ACM, New York (2007)

8. Dai, B.-R., Huang, J.-W., Yeh, M.-Y., Chen, M.-S.: Clustering on demand for multiple data streams. In: ICDM 2004. Fourth IEEE International Conference on Data Mining, November 2004, pp. 367–370 (2004)

9. Dang, X.H., Lee, V.C., Ng, W.K., Ong, K.L.: Incremental and adaptive clustering stream data over sliding window. In: Bhowmick, S.S., Küng, J., Wagner, R. (eds.) DEXA 2009. LNCS, vol. 5690, pp. 660–674. Springer, Heidelberg (2009)

10. Eschrich, S., Ke, J., Hall, L., Goldgof, D.: Fast accurate fuzzy clustering through data reduction. IEEE Transactions on Fuzzy Systems 11(2), 262–270 (2003)

11. Ester, M., Kriegel, H., Sander, J., Xu, X.: A density-based algorithm for discovering clusters in large spatial databases with noise. In: Proc. 2nd Int. Conf. on Knowledge Discovery and Data Mining, Portland, OR, pp. 226–231. AAAI Press, Menlo Park (1996)

12. Gaber, M.M., Zaslavsky, A., Krishnaswamy, S.: Mining data streams: a review. SIGMOD Record 34(2), 18–26 (2005)

13. Guha, S., Meyerson, A., Mishra, N., Motwani, R., O'Callaghan, L.: Clustering data streams: Theory and practice. IEEE Transactions on Knowledge and Data Engineering 15(3), 515–528 (2003)

14. Hore, P., Hall, L., Goldgof, D.: A fuzzy c means variant for clustering evolving data streams. In: ISIC. IEEE International Conference on Systems, Man and Cybernetics, October 2007, pp. 360–365 (2007)

15. Ismail, M.: Soft clustering: algorithms and validity of solutions. Fuzzy Computing, 445–472 (1988)

16. Jain, A.: Data Clustering: 50 Years Beyond K-Means. Pattern Recognition Letters (2009)

17. Jain, A.K., Murty, M.N., Flynn, P.J.: Data clustering: a review. ACM Computing Surveys 31(3), 264–323 (1999)

18. Jia, C., Tan, C., Yong, A.: A grid and density-based clustering algorithm for processing data stream. In: WGEC 2008: Proceedings of the 2008 Second International Conference on Genetic and Evolutionary Computing, pp. 517–521. IEEE Computer Society, Los Alamitos (2008)

19. Liu, Y., Cai, J., Yin, J., Fu, A.: Clustering text data streams. Journal of Computer Science and Technology 23(1), 112–128 (2008)

20. Lühr, S., Lazarescu, M.: Connectivity based stream clustering using localised density exemplars. In: Washio, T., Suzuki, E., Ting, K.M., Inokuchi, A. (eds.) PAKDD 2008. LNCS (LNAI), vol. 5012, pp. 662–672. Springer, Heidelberg (2008)

21. Nasraoui, O., Uribe, C., Coronel, C., Gonzalez, F.: Tecno-streams: tracking evolving clusters in noisy data streams with a scalable immune system learning model. In: ICDM 2003. Third IEEE International Conference on Data Mining, November 2003, pp. 235–242 (2003)

22. Callaghan, L.O., Mishra, N., Meyerson, A., Guha, S., Motwani, R.: Streaming-data algorithms for high-quality clustering. In: Proceedings of the International Conference on Data Engineering, pp. 685–696. IEEE Computer Society Press, Los Alamitos (2002)

23. Park, N.H., Lee, W.S.: Statistical grid-based clustering over data streams. SIGMOD Record 33(1), 32–37 (2004)

24. Selim, S., Ismail, M.: Soft clustering of multidimensional data: a semi-fuzzy approach. Pattern Recognition 17(5), 559–568 (1984)
25. Tasoulis, D.K., Adams, N.M., Hand, D.J.: Unsupervised clustering in streaming data. In: ICDMW 2006: Proceedings of the Sixth IEEE International Conference on Data Mining - Workshops, pp. 638–642. IEEE Computer Society Press, Los Alamitos (2006)
26. Tu, L., Chen, Y.: Stream data clustering based on grid density and attraction. ACM Transactions Knowledge Discovery Data 3(3), 1–27 (2009)
27. Wan, L., Ng, W.K., Dang, X.H., Yu, P.S., Zhang, K.: Density-based clustering of data streams at multiple resolutions. ACM Transactions Knowledge Discovery Data 3(3), 1–28 (2009)
28. Xie, X., Beni, G.: A validity measure for fuzzy clustering. IEEE Transactions on pattern analysis and machine intelligence 13(8), 841–847 (1991)
29. Yang, C., Zhou, J.: Hclustream: A novel approach for clustering evolving heterogeneous data stream. In: ICDMW 2006: Proceedings of the Sixth IEEE International Conference on Data Mining - Workshops, pp. 682–688. IEEE Computer Society, Los Alamitos (2006)
30. Yang, J.: Dynamic clustering of evolving streams with a single pass. In: Proceedings of 19th International Conference on Data Engineering, March 2003, pp. 695–697 (2003)
31. Zhang, T., Ramakrishnan, R., Livny, M.: Birch: an efficient data clustering method for very large databases. SIGMOD Record 25(2), 103–114 (1996)

The Mathematics of Divergence Based Online Learning in Vector Quantization

Thomas Villmann[1,*], Sven Haase[1], Frank-Michael Schleif[2], Barbara Hammer[2], and Michael Biehl[3]

[1] Department of Mathematics/Natural Sciences/Informatics,
University of Applied Sciences Mittweida, 09648 Mittweida, Germany
`thomas.villmann@hs-mittweida.de`
[2] Clausthal University of Technology, Institute of Computer Science,
Clausthal-Zellerfeld, Germany
[3] Rijksuniversity Groningen,
Johann Bernoulli Inst. for Mathematics and Computer Science, The Netherlands

Abstract. We propose the utilization of divergences in gradient descent learning of supervised and unsupervised vector quantization as an alternative for the squared Euclidean distance. The approach is based on the determination of the Fréchet-derivatives for the divergences, wich can be immediately plugged into the online-learning rules. We provide the mathematical foundation of the respective framework. This framework includes usual gradient descent learning of prototypes as well as parameter optimization and relevance learning for improvement of the performance.

Keywords: vector quantization, divergence based learning, information theory, clustering, classification.

1 Introduction

The utilization of non-standard metrics in unsupervised and supervised vector quantization is a challenging topic which has an increasing importance for data processing. Prototype based vector quantization for clustering and classification usually is based on the Euclidean distance like the prominent k-means [18], the self-organizing map (SOM,[15]) or the neural gas (NG,[19]) for unsupervised data modeling and learning vector quantization schemes (LVQ,[15]) or support vector machines (SVM,[29]) in case of supervised learning.

However, the standard Euclidean metric may be not appropriate for faithful data processing [25],. Therefore, recent developments extend the standard approaches by incorporating advanced dissimilarity measures for the data modelling. Examples are in the area of functional data processing and visualization [17],[22],[33] or more generally – kernelized metrics [21],[12], bilinear forms for dissimilarities [28] or general dissimilarities [3],[5]. These dissimilarity measures take into account the structure of the data and, therefore, realize a data adequate processing, which may lead to better results.

In this paper we concentrate on a special data type – positive measures $p(\mathbf{x})$. Positive measures are supposed to be positive functions $p(\mathbf{x})$ for the support

---

[*] Corresponding author.

F. Schwenker and N. El Gayar (Eds.): ANNPR 2010, LNAI 5998, pp. 108–119, 2010.
© Springer-Verlag Berlin Heidelberg 2010

$\mathbf{x} \in \Omega$. If further $\int_\Omega p\,(\mathbf{x})\,d\mathbf{x} = 1$ holds, p is called a density measure, or simply density for short. Density data play an important role in many research areas: For example, spectral data occurring in mass-spectrometry or remote sensing usually are positive measures or densities [34], [32]. The dissimilarity between densities (positive measures) is naturally judged by (generalized) divergences. First vector quantization approaches using divergences apply the batch mode of learning [2],[13] by means of the expectation-maximization methodology. In this scheme, all data have to be available at hand, which is not assumed in the online learning mode of the respective algorithms. Gradient descent leaning usually is realized as stochastic gradient descent optimization. However, this learning mode requires the calculation of the derivatives, which determine the adaptation rule for the prototypes. Thus, we concentrate in this paper, how divergences can be incorporated into gradient based supervised and unsupervised prototype-based learning schemes. For this purpose, we have to investigate the derivatives of divergences, which turn out to be *functional derivatives* mathematically known as *Fréchet-derivatives*.

The paper is organized as follows: We first briefly reconsider SOM/NG and generalized LVQ (GLVQ,[26]) as widely used representatives for the families of gradient based unsupervised and supervised vector quantization algorithms to explain, how the derivatives of the underlying dissimilarity measure come into play. Thereafter we give the *Fréchet-derivatives* for several divergence families, which then can immediately plugged in. Further, we explain some parameter optimization strategies for parametrized divergences, which are related to hyperparameter optimization [27] and relevance learning [10], respectively.

2 Prototype Based Vector Quantization

2.1 Unsupervised Vector Quantization

Prototype based vector quantization (VQ) is a mapping of data $\mathbf{v} \in V \subseteq \mathbb{R}^n$, distributed according to the data density P, onto a set $\mathbf{W} = \{\mathbf{w_r} \in \mathbb{R}^n\}_{r \in A}$ of prototypes. The set A is an appropriate index set, D is the input dimension and $N = \#A$ the number of prototypes.

The aim of *unsupervised vector quantization* during learning is to distribute the prototypes in the data space such that they represent the data as good as possible. This property is judged by quantization error

$$E_{\mathrm{VQ}} = \int \xi\left(\mathbf{v}, \mathbf{w_{s(v)}}\right) P\left(\mathbf{v}\right) d\mathbf{v} \tag{1}$$

based on the dissimilarity measure ξ and

$$\mathbf{s}\left(\mathbf{v}\right) = \operatorname*{argmin}_{r \in A} \left[\xi\left(\mathbf{v}, \mathbf{w_r}\right)\right] \tag{2}$$

being the best matching unit (winner). Hence, the quantization error can be seen as the *expectation value* for the mapping error in the winner determination. Robust approximators for optimum unsupervised vector quantizers are the NG and SOM.

For the NG the above cost function E_{VQ} is modified to

$$E_{\mathrm{NG}} = \frac{1}{2C(\lambda)} \sum_{\mathbf{r}} \int P(\mathbf{v}) \, h_\sigma(\mathbf{r}) \, \xi(\mathbf{v}, \mathbf{w_r}) \, d\mathbf{v}$$

with the so-called neighborhood function $h_\sigma(\mathbf{r}) = \exp\left(\frac{-rank(\mathbf{r})}{2\sigma^2}\right)$ and is the rank function counting the number of prototypes $\mathbf{r}'$ for which $\xi(\mathbf{v}, \mathbf{w_{r'}}) \leq \xi(\mathbf{v}, \mathbf{w_r})$ holds [19]. For SOM a cost function can be defined by

$$E_{\mathrm{SOM}} = \int P(\mathbf{v}) \sum_{\mathbf{r}} \delta_{\mathbf{r}}^{\mathbf{s}(\mathbf{v})} \cdot le(\mathbf{v}, \mathbf{r}) \, d\mathbf{v}$$

with local errors $le(\mathbf{v}, \mathbf{r}) = \sum_{\mathbf{r}'} h_\sigma(\mathbf{r}, \mathbf{r}') \xi(\mathbf{v}, \mathbf{w_{r'}})$ and $\delta_{\mathbf{r}}^{\mathbf{s}(\mathbf{v})}$ is the Kronecker-symbol using HESKES' variant [11]. Here, the neighborhood function $h_\sigma(\mathbf{r}, \mathbf{r}') = \exp\left(\frac{-\xi_A(\mathbf{r}, \mathbf{r}')}{2\sigma^2}\right)$ is the distance measured in the index set A.

For SOMs, the index set A is equipped with a topological order usually taken as regular low-dimensional grid. However, compared with standard SOM the winning rule in Heskes-SOM is slightly modified:

$$\mathbf{s}(\mathbf{v}) = \operatorname*{argmin}_{\mathbf{r} \in A} \left[le(\mathbf{v}, \mathbf{r}) \right]. \tag{3}$$

For both algorithms learning is realized as a stochastic gradient with respect to the prototypes $\mathbf{w_r}$:

$$\triangle \mathbf{w_r} = -\varepsilon \frac{\partial E_{\mathrm{NG/SOM}}}{\partial \mathbf{w_r}} \tag{4}$$

which contains as an essential ingredients the derivative $\frac{\partial \xi(\mathbf{v}, \mathbf{w_{r'}})}{\partial \mathbf{w_r}}$.

2.2 Supervised Vector Quantization

The goal of *supervised learning vector quantization* (LVQ) is the optimization of the classification accuracy for given data $\mathbf{v} \in V \subseteq \mathbb{R}^n$ equipped with class labels $c_\mathbf{v}$. Further, a class label $\mathbf{y_r}$ is attached to each prototype. Again, the data $\mathbf{v}$ are mapped onto the winning prototype according to the mapping rule (2). If $c_\mathbf{v} \neq \mathbf{y_{s(v)}}$ a classification error is detected. The overall classification error cannot be optimized directly by gradient descent learning, because it is not differentiable. Therefore, it has to be replaced by an differentiable cost function reflecting essential properties of the classification accuracy. For this purpose the generalized learning vector quantization (GLVQ) scheme was developed [26]. The cost function of GLVQ is given by

$$E_{\mathrm{GLVQ}} = \sum_{\mathbf{v}} \mu(\mathbf{v}) \tag{5}$$

defining the classifier function $\mu(\mathbf{v})$

$$\mu(\mathbf{v}) = \frac{\xi^+ - \xi^-}{\xi^+ + \xi^-} . \tag{6}$$

with $\xi^+ = \xi\left(\mathbf{v}, \mathbf{w}_{s^+(\mathbf{v})}\right)$. The value $s^+(\mathbf{v})$ is the winning prototype with the additional constraint that $\mathbf{c_v} = \mathbf{y}_{\mathbf{s}^+(\mathbf{v})}$ holds. In analogy, $\mathbf{w}_{s^-(\mathbf{v})}$ has has minimum distance $\xi^- = \xi\left(\mathbf{v}, \mathbf{w}_{s^-(\mathbf{v})}\right)$ for all prototypes $\mathbf{w_r}$ with class labels different to $\mathbf{c_v}$, i.e. $\mathbf{y_r} \neq \mathbf{c_v}$. Then the *generalized* LVQ (GLVQ) is derived as gradient descent on the cost function E_{GLVQ} (5) with respect to the prototypes. In each learning step, for a given data point, both $\mathbf{w}_{s^+(\mathbf{v})}$ and $\mathbf{w}_{s^-(\mathbf{v})}$ are adapted in parallel taking the derivatives $\frac{\partial E_{\mathrm{GLVQ}}}{\partial \mathbf{w}_{+(\mathbf{v})}}$ and $\frac{\partial E_{\mathrm{GLVQ}}}{\partial \mathbf{w}_{-(\mathbf{v})}}$:

$$\triangle \mathbf{w}_{s^+(\mathbf{v})} = \epsilon^+ \cdot \theta^+ \cdot \frac{\partial \xi\left(\mathbf{v}, \mathbf{w}_{s^+(\mathbf{v})}\right)}{\partial \mathbf{w}_{s^+(\mathbf{v})}} \text{ and } \triangle \mathbf{w}_{s^-(\mathbf{v})} = -\epsilon^- \cdot \theta^- \cdot \frac{\partial \xi\left(\mathbf{v}, \mathbf{w}_{s^-(\mathbf{v})}\right)}{\partial \mathbf{w}_{s^-(\mathbf{v})}} \quad (7)$$

with the scaling factors

$$\theta^+ = \frac{2 \cdot \xi^-}{(\xi^+ + \xi^-)^2} \text{ and } \theta^- = \frac{2 \cdot \xi^+}{(\xi^+ + \xi^-)^2} . \quad (8)$$

The values ϵ^+ and $\epsilon^- \in (0,1)$ are the learning rates.

3   Divergences as Dissimilarities and Derivatives Thereof

As mentioned in the introduction, frequently the quadratic Euclidean norm is used for the dissimilarity measure ξ in both supervised and unsupervised vector quantization. In the following we show how it can be replaced by divergence measures. Yet, the strategy is straight forward: If the derivative of a divergence is determined it can be plugged into each gradient based vector quantization scheme including the above examples SOMs, NG or GLVQ.

Divergences estimate the dissimilarity between density functions or positive measures. In information theory they are related mutual information [16]. According to the classification given in CICHOCKI ET AL. [4], one can distinguish at least *three* main classes of divergences, the *Bregman*-divergences, the *Csiszár's* f-divergences and the γ-divergences [4]. If a divergence $D(p\|\rho)$ is given, the mathematical framework for the functional derivative with respect to ρ is the concept of *Fréchet-derivatives* or *functional derivatives* $\frac{\delta D(p\|\rho)}{\delta \rho}$ [8],[14]. In the following we will explain the functional derivatives for these divergence classes. Thereby we assume that p and ρ are positive measures in $\mathbf{x} \in \Omega$ and integrals are taken according to support Ω.

3.1   Basic Divergences

Let Φ be a strictly convex real-valued function with the domain $\mathcal{L}$ (the Lebesgue-integrable functions). Further, Φ is assumed to be twice continuously Fréchet-differentiable [14]. Bregman divergences are defined as $D_\Phi^B : \mathcal{L} \times \mathcal{L} \longrightarrow \mathbb{R}^+$ with

$$D_\Phi^B(p\|\rho) = \Phi(p) - \Phi(\rho) - \frac{\delta \Phi(\rho)}{\delta \rho}(p - \rho) \quad (9)$$

whereby $\frac{\delta\Phi(\rho)}{\delta\rho}$ is the Fréchet-derivative of Φ with respect to ρ. For the choice $\Phi(f) = f^2$, the Euclidean distance is obtained. The Fréchet-derivative is

$$\frac{\delta D_\Phi^B(p||\rho)}{\delta\rho} = \frac{\Phi(p)}{\delta\rho} - \frac{\Phi(\rho)}{\delta\rho} - \frac{\delta\left[\frac{\delta\Phi(\rho)}{\delta\rho}(p-\rho)\right]}{\delta\rho} \tag{10}$$

An important subset of Bregman divergences are the β-divergences

$$D_\beta(p||\rho) = \int p \cdot \frac{p^{\beta-1} - \rho^{\beta-1}}{\beta-1} d\mathbf{x} - \int \frac{p^\beta - \rho^\beta}{\beta} d\mathbf{x} \tag{11}$$

with $\beta \neq 1$ and $\beta \neq 0$. The Fréchet-derivative is

$$\frac{\delta D_\beta(p||\rho)}{\delta\rho} = -p \cdot \rho^{\beta-2} + \rho^{\beta-1} \,. \tag{12}$$

In the limit $\beta \to 1$ the divergence $D_\beta(p,\rho)$ becomes the generalized Kullback-Leibler-divergence

$$D_{GKL}(p||\rho) = \int p \log\left(\frac{p}{\rho}\right) d\mathbf{x} - \int p - \rho d\mathbf{x}. \tag{13}$$

with the Fréchet-derivative

$$\frac{\delta D_{GKL}(p||\rho)}{\delta\rho} = -\frac{p}{\rho} + 1 \tag{14}$$

Csiszár's f-divergences are generated by a *convex* function $f : [0,\infty) \to \mathbb{R}$ with $f(1) = 0$ (without loss of generality) as

$$D_f(p||\rho) = \int \rho \cdot f\left(\frac{p}{\rho}\right) d\mathbf{x} \tag{15}$$

with the definitions $0 \cdot f\left(\frac{0}{0}\right) = 0$, $0 \cdot f\left(\frac{a}{0}\right) = \lim_{x\to 0} x \cdot f\left(\frac{a}{x}\right) = \lim_{u\to\infty} a \cdot \frac{f(u)}{u}$ [6] with the famous Hellinger divergence in case of densities p and ρ [30]:

$$D_H(p||\rho) = \int \left(\sqrt{p} - \sqrt{\rho}\right)^2 d\mathbf{x} \tag{16}$$

with the generating function $f(u) = \left(\sqrt{u} - 1\right)^2$ with $u = \frac{p}{\rho}$. The Fréchet-derivative of $D_f(p||\rho)$ writes as

$$\frac{\delta D_f(p||\rho)}{\delta\rho} = f\left(\frac{p}{\rho}\right) + \rho \frac{\partial f(u)}{\partial u} \cdot \frac{-p}{\rho^2} \tag{17}$$

with $u = \frac{p}{\rho}$ which yields $\frac{\delta D_H(p||\rho)}{\delta\rho} = 1 - \sqrt{\frac{p}{\rho}}$. We can identify also an important subset of f-divergences – the so-called α-divergences [4]:

$$D_\alpha(p||\rho) = \frac{1}{\alpha(\alpha-1)} \int \left[p^\alpha \rho^{1-\alpha} - \alpha \cdot p + (\alpha-1)\rho\right] d\mathbf{x} \tag{18}$$

with the generating f-function

$$f(u) = u \frac{\left(u^{\alpha-1} - 1\right)}{\alpha^2 - \alpha} + \frac{1 - u}{\alpha}$$

and $u = \frac{\rho}{p}$. In the limit $\alpha \to 1$ the generalized Kullback-Leibler-divergence D_{GKL} (13) is obtained. The Fréchet-derivative is calculated as

$$\frac{\delta D_\alpha\left(p||\rho\right)}{\delta\rho} = -\frac{1}{\alpha}\left(p^\alpha \rho^{-\alpha} - 1\right) . \qquad (19)$$

The α-divergences are closely related to the generalized *Rényi-divergences* [1],[23],[24]:

$$D_\alpha^{GR}\left(p||\rho\right) = \frac{1}{\alpha - 1}\log\left(\int\left[p^\alpha \rho^{1-\alpha} - \alpha \cdot p + (\alpha - 1)\rho + 1\right]dx\right) \qquad (20)$$

with the Fréchet-derivative

$$\frac{\delta D_\alpha^{GR}\left(p||\rho\right)}{\delta\rho} = -\frac{\alpha}{\int\left[p^\alpha \rho^{1-\alpha} - \alpha \cdot p + (\alpha - 1)\rho + 1\right]dx}\frac{\delta D_\alpha\left(p||\rho\right)}{\delta\rho} . \qquad (21)$$

The very outlier-robust γ-divergence class is defined according to

$$D_\gamma\left(p||\rho\right) = \frac{1}{\gamma + 1}\log\left[\left(\int p^{\gamma+1}d\mathbf{x}\right)^{\frac{1}{\gamma}} \cdot \left(\int \rho^{\gamma+1}d\mathbf{x}\right)\right] - \log\left[\left(\int p \cdot \rho^\gamma d\mathbf{x}\right)^{\frac{1}{\gamma}}\right] \qquad (22)$$

proposed by FUJISAWA&EGUCHI [9]. In the limit $\gamma \to 0$ $D_\gamma\left(\rho||p\right)$ becomes the usual Kullback-Leibler-divergence for normalized densities. For $\gamma = 1$ the *Cauchy-Schwarz*-divergence

$$D_{CS}\left(p||\rho\right) = \frac{1}{2}\log\left(\int \rho^2\left(\mathbf{x}\right)d\mathbf{x} \cdot \int p^2\left(\mathbf{x}\right)d\mathbf{x}\right) - \log\left(\int p\left(\mathbf{x}\right) \cdot \rho\left(\mathbf{x}\right)d\mathbf{x}\right) \qquad (23)$$

is obtained, which was suggested for information theoretic learning by J. PRINCIPE investigating the Cauchy-Schwarz-inequality for norms [20]. The Fréchet-derivative of $D_\gamma\left(p||\rho\right)$ becomes

$$\frac{\delta D_\gamma\left(p||\rho\right)}{\delta\rho} = \frac{\rho^\gamma}{\left(\int \rho^{\gamma+1}dx\right)} - \frac{p\rho^{\gamma-1}}{\left(\int p \cdot \rho^\gamma dx\right)} \qquad (24)$$

Due to the lack of space, the derivation of these results can be found in [31].

If we now identify $\mathbf{v}$ with a vectorial representation of p and the prototypes $\mathbf{w}$ as the respective ρ representation, the obtained derivative can be immediately plugged into gradient learning schemes as above outlined.

In an example application we consider the data vectors $\mathbf{v} \in \mathbb{R}^2$ with $\|\mathbf{v}\| = 1$ and v_1 distributed in $[0,1]$ according to the density $q\left(v_1\right) = 2v_1$. We learned a one-dimensional SOM for α-, β- and γ-divergences with different parameter setting. The resulted prototype distributions are depicted in Fig. 1. Obviously, the influence of the parameter variations is detectable. In particular, the limits to the Kullback-Leibler-divergence setting are clearly observable.

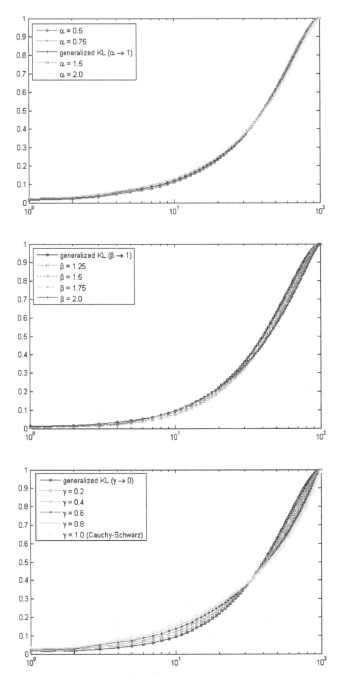

Fig. 1. Illustration of divergence based learning. The w_1-components of the prototypes are depicted for learning α-, β-, γ-divergences (from top to bottom). The horizontal axis is the prottype number. The data distribution was according to $q\left(v_1\right) = 2v_1$ with $v_1 \in [0,1]$, randomly, and $v_2 = 1 - v_1$.

3.2   Parameter Adaptation in Divergence Based Learning

Considering the parametrized divergence families of γ-, α-, and β-divergences, one could further think about the optimal choice of the so-called hyperparameters γ, α, and β as suggested in a similar manner for other parametrized LVQ-algorithms [27]. In case of supervised learning schemes for classification based on differentiable cost functions, the optimization can be handled as an object of a gradient descent based adaptation procedure. Thus, the parameter is optimized in dependence of the classification task at hand.

Suppose, the classification accuracy for a certain approach is given as

$$E = E\left(\xi_\eta, W\right)$$

depending on a *parametrized divergence* ξ_η with parameter η and the set $W = \{\mathbf{w_r}\}$ of prototypes. If E and ξ_η are both differentiable with respect to η according to

$$\frac{\partial E\left(\xi_\eta, W\right)}{\partial \eta} = \frac{\partial E}{\partial \xi_\eta} \cdot \frac{\partial \xi_\eta}{\partial \eta},$$

a gradient based optimization is derived by

$$\triangle \eta = -\varepsilon \frac{\partial E\left(\xi_\eta, W\right)}{\partial \eta} = -\varepsilon \frac{\partial E}{\partial \xi_\eta} \cdot \frac{\partial \xi_\eta}{\partial \eta}$$

depending on the derivative $\frac{\partial \xi_\eta}{\partial \eta}$ for a certain choice of the divergence ξ_η.

We assume in the following that the (positive) measures p and ρ represent the data $\mathbf{v}$ and prototypes $\mathbf{w}$, respectively. If the measures p and ρ are continuously differentiable, then , considering derivatives of parametrized divergences $\frac{\partial \xi_\eta}{\partial \eta}$ with respect to the parameter η, it is allowed to interchange integration and differentiation, under the assumption that the resulting integral exists [7]. Hence, we can differentiate parametrized divergences with respect to their hyperparameter in that case. For the several α-, β-, and γ-divergences characterized in sec. 3.1 we obtain after some elementary calculations [31]:

– β-divergence $D_\beta\left(p||\rho\right)$ from (11)

$$\frac{\partial D_\beta\left(p||\rho\right)}{\partial \beta} = \frac{1}{\beta - 1} \int p \left(p^{\beta-1} \ln p - \rho^{\beta-1} \ln \rho - \frac{\left(p^{\beta-1} - \rho^{\beta-1}\right)}{\left(\beta - 1\right)} \right) d\mathbf{x}$$

$$- \int \left(p^\beta \ln p - \rho^\beta \ln \rho \right) \frac{1}{\beta} - \frac{1}{\beta^2} \left(p^\beta - \rho^\beta \right) d\mathbf{x}$$

– α-divergence $D_\alpha\left(p||\rho\right)$ from (18)

$$\frac{\partial D_\alpha\left(p||\rho\right)}{\partial \alpha} = -\frac{\left(2\alpha - 1\right)}{\alpha^2 \left(\alpha - 1\right)^2} \int \left[p^\alpha \rho^{1-\alpha} - \alpha \cdot p + \left(\alpha - 1\right)\rho \right] d\mathbf{x}$$

$$+ \frac{1}{\alpha \left(\alpha - 1\right)} \int p^\alpha \rho^{1-\alpha} \left(\ln p - \ln \rho \right) - p + \rho dx$$

− generalized Rényi-divergence $D_\alpha^{GR}(p||\rho)$ from (20)

$$\frac{\partial D_\alpha^{GR}(p||\rho)}{\partial \alpha} = -\frac{1}{(\alpha-1)^2} \log\left(\int \left[p^\alpha \rho^{1-\alpha} - \alpha \cdot p + (\alpha-1)\rho + 1\right] dx\right)$$

$$+\frac{1}{\alpha-1} \frac{\int p^\alpha \rho^{1-\alpha}(\ln p - \ln \rho) - p + \rho dx}{\int \left[p^\alpha \rho^{1-\alpha} - \alpha \cdot p + (\alpha-1)\rho + 1\right] dx}$$

− Rényi-divergence $D_\alpha^R(p||\rho)$ from (20) for normalized densities

$$\frac{\partial D_\alpha^{GR}(p||\rho)}{\partial \alpha} = -\frac{1}{(\alpha-1)^2} \log\left(\int p^\alpha \rho^{1-\alpha} dx\right)$$

$$+\frac{1}{\alpha-1} \frac{\int p^\alpha \rho^{1-\alpha}(\ln p - \ln \rho) dx}{\int p^\alpha \rho^{1-\alpha} dx}$$

− γ-divergence $D_\gamma(p||\rho)$ from (22)

$$\frac{\partial D_\gamma(p||\rho)}{\partial \gamma} = -\frac{(2\gamma+1)}{\gamma^2(\gamma+1)^2} \ln\left(\int p^{\gamma+1} dx\right) + \frac{\int p^{\gamma+1} \ln p dx}{(\gamma+1)\gamma \int p^{\gamma+1} dx}$$

$$-\frac{1}{(\gamma+1)^2} \ln\left(\int \rho^{\gamma+1} dx\right) + \frac{\int \rho^{\gamma+1} \ln \rho dx}{(\gamma+1)\int \rho^{\gamma+1} dx}$$

$$+\frac{1}{\gamma^2} \ln\left(\int p \cdot \rho^\gamma dx\right) - \frac{\int p\rho^\gamma \ln \rho dx}{\gamma \int p \cdot \rho^\gamma dx}$$

3.3 Relevance Learning for Positive Measures

Density functions are required to fulfill the normalization condition whereas positive measure are more flexible. This offers the possibility to transfer the idea of relevance learning also to divergence based learning vector quantization. *Relevance learning* in learning vector quantization is weighting the input data dimensions such that classification accuracy is improved [10].

In the framework of divergence based gradient descent learning we multiplicatively weight a positive measure $q(x)$ by $\lambda(x)$ with $0 \leq \lambda(\mathbf{x}) < \infty$ and the regularization condition $\int \lambda(\mathbf{x}) dx = 1$. Incorporating this idea into the above approaches we have to replace in the divergences p by $p \cdot \lambda$ and ρ by $\rho \cdot \lambda$. Doing so we can optimize $\lambda(x)$ during learning for better performance by gradient descent optimization of the GLVQ cost function (5) as it is known from vectorial relevance learning but paying now attention to the utilization of divergences. This leads here, again, to Fréchet-derivatives of the incorporated divergence D but now with respect to the weighting function $\lambda(\mathbf{x}) - \frac{\delta D(\lambda \cdot p||\lambda \cdot \rho)}{\delta \lambda}$.

In particular we obtain for the Bregman divergence

$$\frac{\delta D_\Phi^B(\lambda \cdot p||\lambda \cdot \rho)}{\delta \lambda} = \frac{\Phi(\lambda \cdot p)}{\delta \lambda} - \frac{\Phi(\lambda \cdot \rho)}{\delta \lambda} - \frac{\delta\left[\frac{\delta \Phi(\lambda \cdot \rho)}{\delta \rho}\lambda(p-\rho)\right]}{\delta \lambda} \tag{25}$$

with

$$\frac{\delta\left[\frac{\delta \Phi(\lambda \cdot \rho)}{\delta \rho}\lambda(p-\rho)\right]}{\delta \lambda} = (p-\rho)\left(\frac{\delta^2\left[\Phi(\lambda \cdot \rho)\right]}{\delta \rho\, \delta \lambda}\lambda + \frac{\delta \Phi(\lambda \cdot \rho)}{\delta \rho}\right).$$

This yields for the *generalized* Kullback-Leibler-divergence

$$\frac{\delta D_{GKL}\left(\lambda \cdot p||\lambda \cdot \rho\right)}{\delta \lambda} = p \cdot \log\left(\frac{p}{\rho}\right) - p + \rho \, . \tag{26}$$

Further, for the β-divergences (11) we have

$$\frac{\delta D_\beta\left(\lambda \cdot p||\lambda \cdot \rho\right)}{\delta \lambda} = \frac{\rho \cdot \left(\lambda \cdot p\right)^\beta + \left(\rho \cdot \left(\beta - 1\right) - p \cdot \beta\right) \cdot \left(\lambda \cdot \rho\right)^\beta}{\lambda \rho\left(\beta - 1\right)} \, . \tag{27}$$

For f-divergences (15) we consider with $u = \frac{p}{\rho}$

$$\frac{\delta D_f\left(\lambda \cdot p||\lambda \cdot \rho\right)}{\delta \lambda} = \rho \cdot f\left(\frac{p}{\rho}\right) + \lambda \cdot \rho \frac{\partial f\left(u\right)}{\partial u} \frac{\delta u}{\delta \lambda}$$

$$= \rho \cdot f\left(\frac{p}{\rho}\right) \tag{28}$$

because of $\frac{\delta u}{\delta \lambda} = 0$. The relevance learning of α-divergences (18) follows

$$\frac{\delta D_\alpha\left(\lambda \cdot p||\lambda \cdot \rho\right)}{\delta \lambda} = \frac{1}{\alpha\left(\alpha - 1\right)}\left[\rho \cdot \left(\left(\frac{p}{\rho}\right)^\alpha + \alpha - 1\right) - p \cdot \alpha\right] \, , \tag{29}$$

whereas the respective gradient of generalized Rényi-divergences (20) can be derived from this as

$$\frac{\delta D_\alpha^{GR}\left(\lambda \cdot p||\lambda \cdot \rho\right)}{\delta \lambda} = \frac{\alpha}{\int\left[\lambda \cdot \left(\rho \cdot \left(\frac{p}{\rho}\right)^\alpha - \alpha \cdot p + \left(\alpha - 1\right) \cdot \rho\right) + 1\right]d\mathbf{x}} \frac{\delta D_\alpha\left(\lambda \cdot p||\lambda \cdot \rho\right)}{\delta \lambda} \, . \tag{30}$$

The γ-divergences finally yields

$$\frac{\delta D_\gamma\left(\lambda \cdot p||\lambda \cdot \rho\right)}{\delta \lambda} = \frac{p\left(\lambda \cdot p\right)^\gamma}{\gamma \int\left(\lambda \cdot p\right)^{\gamma+1}d\mathbf{x}} + \frac{\rho\left(\lambda \cdot \rho\right)^\gamma}{\int\left(\lambda \cdot \rho\right)^{\gamma+1}d\mathbf{x}} - \frac{p \cdot \left(\gamma + 1\right) \cdot \left(\lambda \cdot \rho\right)^\gamma}{\gamma \int\left(\lambda \cdot p\right) \cdot \left(\lambda \cdot \rho\right)^\gamma d\mathbf{x}} \, .$$

Again the important special case $\gamma = 1$ is considered: the relevance learning scheme for the Cauchy-Schwarz divergence (23) is derived as

$$\frac{\delta D_{CS}\left(\lambda \cdot p||\lambda \cdot \rho\right)}{\delta \lambda} = \frac{p \cdot \lambda \cdot p}{\int\left(\lambda \cdot p\right)^2 d\mathbf{x}} + \frac{\rho \cdot \lambda \cdot \rho}{\int\left(\lambda \cdot \rho\right)^2 d\mathbf{x}} - \frac{2 \cdot p \cdot \lambda \cdot \rho}{\int \lambda^2 \cdot p \cdot \rho d\mathbf{x}} \, . \tag{31}$$

As before, if we identify p and ρ with the data $\mathbf{v}$ and the prototypes $\mathbf{w}$, the derivatives can be immediately put into a gradiend descent learning scheme.

4   Conclusion

In this article we provide the mathematical foundation for divergence based supervised and unsupervised vector quantization bearing on the derivatives of the applied divergences. For this purpose, we first characterized the main sub-classes of divergences, Bregman-, α-, β-, γ-, and f-divergences following [4]. The mathematical framework of Fréchet-derivatives is then used to calculate the functional divergence derivatives.

We exemplary explain the utilization of this methodology for famous examples of supervised and unsupervised vector quantization including SOM, NG, and GLVQ. Further, we discuss, how a parameter adaptation could be integrated in supervised learning to achieve improved classification results in case of the parametrized α-, β-, and γ-divergences. In the last step we considered a weighting function for generalized divergences based on positive measures. The optimization scheme for this weight function for a given classification task is again obtained by Fréchet derivatives, and one ends up with a relevance learning scheme analogously to relevance learning for usual (Eulidean) supervised learning vector quantization [10].

References

1. Amari, S.-I.: Differential-Geometrical Methods in Statistics. Springer, Heidelberg (1985)
2. Banerjee, A., Merugu, S., Dhillon, I., Ghosh, J.: Clustering with bregman divergences. Journal of Machine Learning Research 6, 1705–1749 (2005)
3. Bezdek, J., Hathaway, R., Windham, M.: Numerical comparison of RFCM and AP algorithms for clustering relational data. Pattern recognition 24, 783–791 (1991)
4. Cichocki, A., Zdunek, R., Phan, A., Amari, S.-I.: Nonnegative Matrix and Tensor Factorizations. Wiley, Chichester (2009)
5. Cottrell, M., Hammer, B., Hasenfuß, A., Villmann, T.: Batch and median neural gas. Neural Networks 19, 762–771 (2006)
6. Csiszár, I.: Information-type measures of differences of probability distributions and indirect observations. Studia Sci. Math. Hungaria 2, 299–318 (1967)
7. Fichtenholz, G.: Differential- und Integralrechnung, 9th edn., vol. II. Deutscher Verlag der Wissenschaften, Berlin (1964)
8. Frigyik, B.A., Srivastava, S., Gupta, M.: An introduction to functional derivatives. Technical Report UWEETR-2008-0001, Dept. of Electrical Engineering, University of Washington (2008)
9. Fujisawa, H., Eguchi, S.: Robust parameter estimation with a small bias against heavy contamination. Journal of Multivariate Analysis 99, 2053–2081 (2008)
10. Hammer, B., Villmann, T.: Generalized relevance learning vector quantization. Neural Networks 15(8-9), 1059–1068 (2002)
11. Heskes, T.: Energy functions for self-organizing maps. In: Oja, E., Kaski, S. (eds.) Kohonen Maps, pp. 303–316. Elsevier, Amsterdam (1999)
12. Hulle, M.M.V.: Kernel-based topographic map formation achieved with an information theoretic approach. Neural Networks 15, 1029–1039 (2002)
13. Jang, E., Fyfe, C., Ko, H.: Bregman divergences and the self organising map. In: Fyfe, C., Kim, D., Lee, S.-Y., Yin, H. (eds.) IDEAL 2008. LNCS, vol. 5326, pp. 452–458. Springer, Heidelberg (2008)
14. Kantorowitsch, I., Akilow, G.: Funktionalanalysis in normierten Räumen, 2nd revised edn. Akademie-Verlag, Berlin (1978)
15. Kohonen, T.: Self-Organizing Maps. Springer Series in Information Sciences, vol. 30. Springer, Heidelberg (1995) (2nd Extended edn. 1997)

16. Kullback, S., Leibler, R.: On information and sufficiency. Annals of Mathematical Statistics 22, 79–86 (1951)
17. Lee, J., Verleysen, M.: Generalization of the l_p norm for time series and its application to self-organizing maps. In: Cottrell, M. (ed.) Proc. of Workshop on Self-Organizing Maps (WSOM) 2005, Paris, Sorbonne, pp. 733–740 (2005)
18. Linde, Y., Buzo, A., Gray, R.: An algorithm for vector quantizer design. IEEE Transactions on Communications 28, 84–95 (1980)
19. Martinetz, T.M., Berkovich, S.G., Schulten, K.J.: Neural-gas network for vector quantization and its application to time-series prediction. IEEE Trans. on Neural Networks 4(4), 558–569 (1993)
20. Principe, J.C., Fisher III, J., Xu, D.: Information theoretic learning. In: Haykin, S. (ed.) Unsupervised Adaptive Filtering. Wiley, New York (2000)
21. Qin, A., Suganthan, P.: A novel kernel prototype-based learning algorithm. In: Proceedings of the 17th International Conference on Pattern Recognition (ICPR 2004), vol. 4, pp. 621–624 (2004)
22. Ramsay, J., Silverman, B.: Functional Data Analysis, 2nd edn. Springer Science+Media, New York (2006)
23. Renyi, A.: On measures of entropy and information. In: Proceedings of the Fourth Berkeley Symposium on Mathematical Statistics and Probability. University of California Press (1961)
24. Renyi, A.: Probability Theory. North-Holland Publishing Company, Amsterdam (1970)
25. Rossi, F., Delannay, N., Conan-Gueza, B., Verleysen, M.: Representation of functional data in neural networks. Neurocomputing 64, 183–210 (2005)
26. Sato, A., Yamada, K.: Generalized learning vector quantization. In: Touretzky, D.S., Mozer, M.C., Hasselmo, M.E. (eds.) Proceedings of the 1995 Conference on Advances in Neural Information Processing Systems, vol. 8, pp. 423–429. MIT Press, Cambridge (1996)
27. Schneider, P., Biehl, M., Hammer, B.: Hyperparameter learning in robust soft LVQ. In: Verleysen, M. (ed.) Proceedings of the European Symposium on Artificial Neural Networks ESANN, pp. 517–522. d-side publications (2009)
28. Schneider, P., Hammer, B., Biehl, M.: Adaptive relevance matrices in learning vector quantization. Neural Computation 21, 3532–3561 (2009)
29. Shawe-Taylor, J., Cristianini, N.: Kernel Methods for Pattern Analysis and Discovery. Cambridge University Press, Cambridge (2004)
30. Taneja, I., Kumar, P.: Relative information of type s, Csiszár's f -divergence, and information inequalities. Information Sciences 166, 105–125 (2004)
31. Villmann, T., Haase, S.: Mathematical aspects of divergence based vector quantization using fréchet-derivatives - extended and revised version. Machine Learning Reports 4(MLR-01-2010), 1–35 (2010), http://www.uni-leipzig.de/~compint/mlr/mlr_01_2010.pdf
32. Villmann, T., Merényi, E., Hammer, B.: Neural maps in remote sensing image analysis. Neural Networks 16(3-4), 389–403 (2003)
33. Villmann, T., Schleif, F.-M.: Functional vector quantization by neural maps. In: Chanussot, J. (ed.) Proceedings of First Workshop on Hyperspectral Image and Signal Processing: Evolution in Remote Sensing (WHISPERS 2009), pp. 1–4. IEEE Press, Los Alamitos (2009)
34. Villmann, T., Schleif, F.-M., Kostrzewa, M., Walch, A., Hammer, B.: Classification of mass-spectrometric data in clinical proteomics using learning vector quantization methods. Briefings in Bioinformatics 9(2), 129–143 (2008)

Cluster Analysis of Cortical Pyramidal Neurons Using SOM

Andreas Schierwagen[1,*], Thomas Villmann[2], Alan Alpár[3], and Ulrich Gärtner[3]

[1] Institute for Computer Science, University of Leipzig, 04109 Leipzig, Germany
schierwa@informatik.uni-leipzig.de
[2] Department of Mathematics/Physics/Computer Sciences,
University of Applied Sciences Mittweida, 09648 Mittweida, Germany
[3] Department of Neuroanatomy, Paul Flechsig Institut for Brain Research,
University of Leipzig, 04109 Leipzig, Germany

Abstract. A cluster analysis using SOM has been performed on morphological data derived from pyramidal neurons of the somatosensory cortex of normal and transgenic mice.

Keywords: Cluster analysis, Kohonen's SOM, transgenic mouse, somatosensory cortex, pyramidal neurons, dendritic morphology.

1 Introduction

In the neurosciences, brain organization is studied at different levels both in ontogenetic and phylogenetic development. At the cellular level, neurons in different brain regions of one species and in the same brain region of different species are compared with respect to their structural and functional properties. We find a great variety in neuronal shapes and cell types, as well as a large variability within neuron classes (Fig. 1, left). Neuronal structure is characterized by elongated processes (neurites); among them, two kinds can be differentiated, the often-branching dendrites and the axon. From a functional point of view, axons and dendrites are conduits for electrical and chemical signals. The shapes of neurites determine both the routes for signal transmission within the nervous system and the way in which electrical signals are processed and transmitted [1].

The formation of a neuron's dendritic and axonal branching patterns is partly determined by genetic factors and partly by interactions with the surrounding tissue. The invention of transgenic mice mutations has provided important means for understanding gene function. In these mutants, gene overexpression may affect several organs and tissues, including the cells and networks of the brain.

In the following, we describe the method of morphological quantification (neuromorphometry) and its use for classification of pyramidal neurons. More specifically, neurons in the same brain region (somatosensory cortex) of two 'species' of mice (wildtype and transgenic type) have been compared with respect to their shape properties. Shape classification was performed with an unsupervised learning method, i.e. Kohonen's self-organizing maps.

---

* Corresponding author.

F. Schwenker and N. El Gayar (Eds.): ANNPR 2010, LNAI 5998, pp. 120–130, 2010.
© Springer-Verlag Berlin Heidelberg 2010

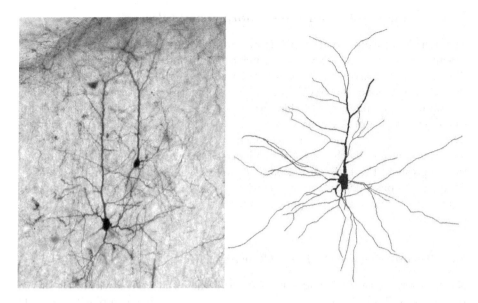

Fig. 1. Pyramidal neurons from layers II/III of *synRas* mouse cortex. Left: light micrograph of retrogradely labelled neurons, right: pyramidal neuron rendered with CVAPP.

2   Materials and Methods

2.1   Experimental Basis and Reconstructions

We used data derived from experiments with three *synRas* mice aged nine months as well as with three wildtype mice of the same age. The details of experimental procedures have been described in [2]. The main deliverables were two sets of retrogradely labelled pyramidal cells (28 cells from wildtype and 28 cells from transgenic mice) which were reconstructed (Fig. 1) using Neurolucida^{TM} (Micro-BrightField, Inc.). The system allowed accurate tracing of the cell processes in all three dimensions and continuous adjustment of the dendritic diameter with a circular cursor. A motorized stage with position encoders enabled the navigation through the section in the xyz axes and the accurate acquisition of the spatial coordinates of the measured structure. All visible dendrites were traced without marking eventual truncation of smaller dendritic sections. This may have led to certain underestimation of the dendritic tree, especially in *synRas* mice with a larger dendritic tree. To gain both optimal transparency for optimal tracing facilites and at the same time a possibly complete neuronal reconstruction, sections of 160 μm thickness were used. Thicker sections allowed only ambiguous tracing of thinner dendritic branches. Shrinkage correction was carried out in the z axis, but not in the xy plane, because shrinkage was negligible in these dimensions [2].

2.2   Editing and Conversion of Morphology Files

The morphology files created in this way were processed with CVAPP [3], a cell viewing, editing and format converting program for morphology files (Fig. 1, right). In particular, CVAPP has been used to edit and convert the Neurolucida ASCII files to the SWC format describing the structure of a neuron in the simplest possible way. In this format, each line encodes the properties of a single neuronal compartment. The format of a line in a SWC file is as follows: $nTxyzRP$. In turn, these numbers mean: (1) an integer label (increasing by one from one line to the next) that identifies the compartment, (2) an integer that represents the type of neuronal segment (0-undefined, 1-soma, 2-axon, 3-dendrite, 4-apical dendrite, etc.); (3)-(5) xyz coordinates of compartment, (6) radius of compartment, (7) parent compartment (defined as -1 for the initial compartment).

2.3   Terminology and Shape Characteristics

Neurons are 3D objects, and the location of their cell bodies within the nerve tissue, as well as the number, spatial extent, branching complexity and 3D embedding of their axonal and dendritic arborizations, are prominent shape characteristics that may differ significantly between cell types. Morphological measurements of these characteristics (neuromorphometry) can be described as follows [4,5].

The dendrites of a neuron are representeded as trees of segments arising from the cell body (soma). A segment is defined as a portion of dendrite extending between two branching points (intermediate segments), or between a node (branching point) and a tip (terminal segments). Dendritic segments are approximated by cylindric sections of length l and diameter d. The distance from the soma to a point on the dendritic tree measured along the course of the segments lying inbetween is the path length which is generally greater than the Euclidean distance between the corresponding points. Measurements of dendritic trees can be differentiated into metrical and topological ones. Measurements include the length and diameter of the segments, path lengths, Euclidean distances of terminal tips from the cell soma and branching angles. A different class of measures is concerned with the spatial embedding in 3D space and focuses on, e.g., the spatial extension, spatial density, spatial orientation, and space filling (fractality) of the structure [6].

Dendritic trees may be categorized by topological type depending on the patterning of segments, independently of metrical and orientation features. The tree is reduced to a skeleton structure of points (branching or terminal points) and segments between these points. Such a skeleton forms a specific tree out of a finite set of possible different topological tree types. The tree-asymmetry index provides a discriminative measure based on asymmetries of pairs of subtrees at bifurcations. Other parameters used are *order* and *degree* (Fig. 2).

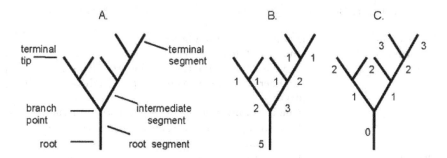

Fig. 2. Representation of dendritic topology. The tree is depicted by a set of connected segments (A) which are labeled by the degree of their subtrees (B) and their centrifugal order (C).

The *order* γ represents the topological distance from the soma. Its value is an integer incremented at every bifurcation (centrifugal order). A value of $\gamma = 0$ is assigned to the primary segments, i.e. those emerging directly from the soma. The *degree* n represents the number of tips of a subtree (or partition) stemming from a segment. In a binary tree, it is related to m, the number of segments of the partition, by $m = 2n - 1$. The dendritic systems of pyramidal cells are divided into basal and apical parts. Each of the subdivisions is considered separately (see Section 3).

All morphologic measurements were extracted from the anatomical files by using the L-Measure software [7]. This software allows the automated extraction of multiple sets of neuroanatomical parameters from reconstructed neurons. Our neuron data included no (or only partial) reconstructions of the axon; thus only the soma and the dendritic trees (basal and apical separately) were considered. L-Measure allows to derive more than 40 neuroanatomical parameters; a number of them, however, are correlated with each other. Thus, we selected 19 independent morphometric parameters (see Table 1) for each of which mean value, standard deviation, minimum and maximum were calculated.

2.4 Feature Space Representation

The shape of a given neuron in three-dimensional space can be characterized by the vector $x = [x_1, x_2, ..., x_N]$ where x_i stands for any measurement. If x allows the original shape to be reconstructed to a specified degree of accuracy, x provides a (more or less) complete representation of the original shape. In the present case, the dimension of the feature space is $19 \times 4 = 76$.

The vector x can be displayed as point in the respective N-dimensional feature space. Neurons belonging to two distinct neural classes can be characterized by clouds of points in the feature space. Provided the measures are sound, vectors defined by similar neurons will be located close to each other in the feature space, while those corresponding to different neurons will tend to be distant from each other. This means, the Euclidean distance in the feature space is related to the

Table 1. Definitions of morphological parameters used in this study

Bifurcations	Number of dendritic bifurcations (nodes)
Segments	Number of dendritic segments
Trees	Number of dendritic trees arising from the soma
Tips	Number of terminal segments per dendritic trees (tree degree)
Segment order	Number of bifurcations between soma and terminal tips
Partition asymmetry	Partition asymmetry of a subtree, defined as $A_p = \frac{|r-l|}{r+l-2}$, with r and l the degrees of the two sub-subtrees
Diameter	Diameter of a segment
Segment length	Length of a dendritic segment
Daughter ratio	Diameter ratio of the two daughter segments at a bifurcation
Parent–child ratio	Diameter ratio between parent and daughter diameter at a bifurcations
Last parent diameter	Diameter of parent segment of the last bifurcation before tips
Rall's power	Best fitting parameter n for Rall's formula: $d_p^n = d_a^n + d_b^n$, where d_p, d_a, d_b, are the parent diameter and the two daughter diameter
Segment taper	Taper rate per segment, calculated as the difference between final and initial diameter divided by the initial diameter at each segment
Unit taper rate	Taper rate per unit length, calculated as the difference between final and initial diameter divided by the segment length
Euclidean distance	Euclidean distance between soma and segment
Path distance	Path distance between soma and segment
Contraction	Ratio between the Euclidean distance and the distance along the path
Local bifurcation angle	Angle between the two daughters at each bifurcation
Remote bifurcation angle	Angle between the tips of the two daughters of each bifurcation

dissimilarity between the shape of the respective cells. It is therefore obvious to approach the problem of classifying nerve cells through a procedure of clustering the respective vectors in the feature space.

2.5   Cluster Analysis

The data classification was realized using the Kohonen Self-Organizing Map (SOM) algorithm [8] to map the complex database on the two-dimensional plane visualizing the *synRas* activation effect, and to designate the relevant variables contributing to the model.

The SOM algorithm is an unsupervised learning procedure which can be summarized as follows. The data are assumed to be n-dimensional real data vectors

$\mathbf{v} \in V$ which are mapped onto a set $\mathbf{W} = \{\mathbf{w_r}\}_{r \in A}$ of prototypes. Thereby, A is an usually two-dimensional discrete lattice and $N = \#A$ is the number of prototypes. SOMs, here taken in the variant of T.HESKES [9], minimize the cost function

$$E = \int P(\mathbf{v}) \sum_r \delta_{\mathbf{r}}^{\mathbf{s}(\mathbf{v})} \sum_{r'} h_\sigma(\mathbf{r}, \mathbf{r}')(\mathbf{v} - \mathbf{w_{r'}})^2 d\mathbf{v} \tag{1}$$

with $\delta_{\mathbf{r}}^{\mathbf{s}(\mathbf{v})}$ being the Kronecker-symbol and $P(V)$ is the data distribution. The neighborhood function

$$h_\sigma(\mathbf{r}, \mathbf{r}') = \exp\left(\frac{-\|\mathbf{r} - \mathbf{r}'\|}{2\sigma^2}\right) \tag{2}$$

describes the learning cooperativeness between the lattice nodes. Learning of the prototypes is realized as a stochastic gradient descent on the cost function E for decreasing neighborhood range σ with respect to the prototypes $\mathbf{w_r}$:

$$\triangle \mathbf{w_r} = -\varepsilon \frac{\partial E}{\partial \mathbf{w_r}} \tag{3}$$

$$= \exp\left(\frac{-\|\mathbf{s}(\mathbf{v}) - \mathbf{r}\|}{2\sigma^2}\right)(\mathbf{v} - \mathbf{w_r}) \tag{4}$$

whereby

$$\mathbf{s}(\mathbf{v}) = \underset{r \in A}{\mathrm{argmin}} \left[\sum_{r'} h_\sigma(\mathbf{r}, \mathbf{r}')(\mathbf{v} - \mathbf{w_{r'}})^2\right]. \tag{5}$$

is the so–called winning node. The mapping rule (5) determines a winner–take–all learning.

SOMs can be taken as a mapping of high-dimensional data onto a low-dimensional lattice [10]. Under certain conditions this mapping is topology preserving, i.e. similar data points are mapped onto neighbored lattice neurons or the same lattice node [11]. Yet, this topographic mapping can not be achieved for any data-lattice-configuration. Growing variants of SOM (GSOM) try to adapt the edge length ratio as well as the dimensionality of the lattice to result a topographic map [12].

Nevertheless, topography has to be judged after SOM-learning by respective quality measures to assure topography. For this purpose, the robust *topographic product* (TP) is frequently applied [13]. It calculates the averaged distortion of distance ratios within the set of prototypes and relates these to the respective node distance ratios within the SOM-lattice. The TP yields approximately zero-values for topology preserving mapping whereas values deviating significantly from zero indicate violations in topographic mapping. For a detailed description we refer to [13]. If the SOM is topology preserving then further investigations like component plane analysis or other can be applied. Component planes picture the value distribution of the prototypes $\mathbf{w_r}$ for a single data dimension. Thereby, the prototypes are arranged according to the position of their assigned lattice nodes $\mathbf{r}$. In this way, the topological ordering of the prototypes can be visualized.

class frequency

SE(5)	WT(2)	SE(4)	WT(7)
SE(4) WT(2)	SE(1) WT(1)	SE(4) WT(2)	WT(3) SE(1)
SE(3) WT(2)	SE(3) WT(2)	SE(3) WT(3)	WT(4)

Fig. 3. SOM frequency map for neuron morphometry data. It shows the distribution of the samples, wildtype (WT) and transgenic neurons (SE), over each SOM node.

One method to detect clusters in the lattice space of SOMs is the U-Matrix method [14]. The U-matrix for a two-dimensional rectangular lattice A of size $n_1 \times n_2$ has the size $(2n_1 - 1) \times (2n_2 - 1)$ and is calculated according the following scheme: For each lattice node $\mathbf{r} = (r_1, r_2)$ the matrix element $u_{2r_1, 2r_2}$ is the mean of the distances of the prototype $\mathbf{w_r}$ to those prototypes $\mathbf{w_{r'}}$ for which $\mathbf{r'}$ is a neighbored node in the rectangular lattice A. The elememt $u_{2r_1-1, 2r_2}$ is the distance to the prototype $\mathbf{w_{r'}}$ with $\mathbf{r'} = (r_1 - 1, r_2)$. The other neighbored matrix elements of the entry $u_{2r_1, 2r_2}$ are calculated accordingly. Thus, matrix entries give an indication of cluster boundaries between the respective nodes in case of a topographic mapping. Therefore, topology preservation is strongly demanded for correct interpretation [10].

The SOM analysis was performed in MATLAB with the publicly available SOM Toolbox [15].

3   Results

For the given data set of 56 data points, each containing the $n = 76$ morphological features, GSOM yielded a 4×3 lattice (Fig. 3). As we can see, each SOM node was hit by pyramidal neuron data. Some nodes were hit either only by transgenic (SE) or by wildtype (WT) neuron data, and others by both. The overall classification accuracy after labeling the map by majority vote is about 77% which refers to non-random partitioning but with class overlap.

After learning the TP was calculated to measure the map quality. It yielded a value of 0.0078 indicating a good topology preservation. This topological ordering can also be seen considering the component planes (CP) of the map (Fig. 4). To make it more accessible, only the first 20 of the 76 parameters were presented as CPs. Shown are the values of only one feature in each map unit (in the original by color-coding). These planes show a clear structure and refer to correlated features if the respective planes look similar in the value distribution.

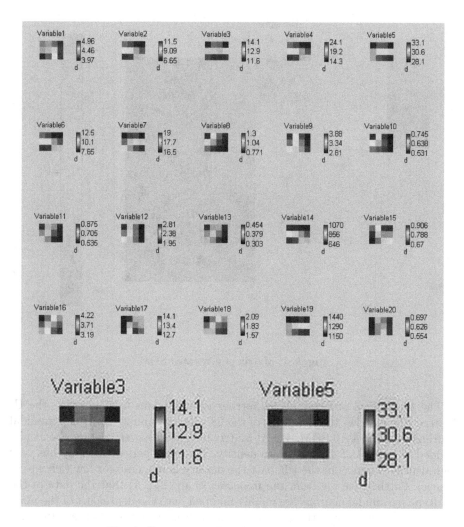

Fig. 4. Component planes of the SOM in Fig. 3

If we take for example the parameters 3 and 5, we see that the CPs of these two features are quite similar, meaning that the distances of these values in the same area of the map correlate. According to the parameter coding list, the two parameters represent the number of bifurcations and the number of branches of the basal dendritic tree. Here a correlation is actually present, because for binary trees a strict relationship holds (see Section 2.3). Comparing the class distribution (Fig. 3) with the component planes one can further detect features which are distinguishing for the classes. For example, low values in the features 8–13 refer with high probability to wildtype pyramidal neurons.

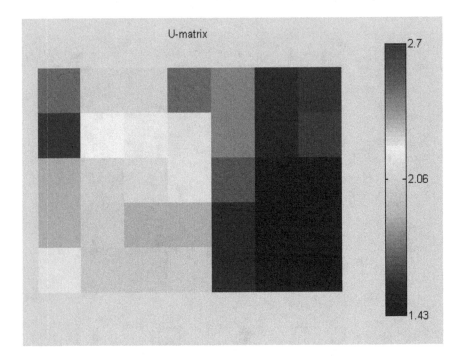

Fig. 5. U-Matrix of generated SOM

The topological ordering allows further investigations by computing the U-Matrix. Fig. 5 shows the U-Matrix of the 28 wildtype and 28 *synRas* pyramidal neurons on a size 4×3 SOM. The U-Matrix indicates that the data space can be divided in regions of different data density. The right part of the map has very high density whereas in the left part the density is moderate or low (left upper corner). Further, we see from the frequency map (Fig. 3) that the data of the wildtype pyramidal neurons are mainly mapped onto the dense part of the map. Hence, the parameter of these data have a low variance whereas the variety for the transgenic pyramidal neurons is higher (upper left corner).

4   Conclusion

In this paper, we described a neuromorphometry protocol and its application for classification of cortical neurons using Kohonen SOM. We employed the visualization capabilities of the SOM method – the U-matrix and the component plane representation – to analyze samples of pyramidal neurons of the somatosensory cortex of wildtype and transgenic mice with respect to their shape properties. Our primary goal was to explore the question of whether pyramidal neurons from layer II/III of the somatosensory cortex of wildtype and of transgenic mice should be considered as members of two different classes or not.

In previous studies, we obtained ambiguous results which needed clarification. In Ref. [2], we found that the volume of the neocortex of synRas mice is expanded up to 25% as compared to wildtype mice. This is due to the dramatically enlarged volume of the cortical pyramidal cells caused mainly by increased dendritic diameter and tree degree, whereas the number of neurons remains unchanged. Changes are generally less prominent in layers II/III than in layer V. For example, topological analyses revealed significant differences between *synRas* and wildtype mice regarding any parameters considered, i.e., number of intersections, branching points (nodes) and tips (leaves), in both basal and apical dendrites of layer V neurons but not of layers II/III neurons. Thus, while pyramidal neurons from layer V of *synRas* and wildtype mice cortex can be clearly considered as members of different classes, the situation has been ambiguous for layer II/III neurons.

Our study has demonstrated that the SOM can be successfully employed for resolving the shape classification problem of layer II/III neurons. As a result of SOM–learning, pooled data of the two samples were clustered with classification accuracy of 77%. This means, the data were non-random partitioned but with class overlap. Both the topological product of 0.0078 and the component planes of the map indicate a good topology preservation.

In particular, the consideration of the component planes suggests that selected features may be sufficient for classification. For this investigation a more detailed analysis is required. In future work, we plan to use the method of supervised learning vector quantization with relevance learning for classification, which automatically provides a classification dependent feature weighting [16].

Acknowledgements

We thank our student C. Roll for technical contributions to the cluster analysis. This work was supported in part by Deutsche Forschungsgemeinschaft (Grant GA 716/1-1).

References

1. Schierwagen, A.: Mathematical and Computational Modeling of Neurons and Neuronal Ensembles. In: Moreno-Díaz, R., Pichler, F., Quesada-Arencibia, A. (eds.) EUROCAST 2009. LNCS, vol. 5717, pp. 159–166. Springer, Heidelberg (2009)
2. Alpár, A., Palm, K., Schierwagen, A., Arendt, T., Gärtner, U.: Expression of constitutively active p21H-rasVal12 in postmitotic pyramidal neurons results in increased dendritic size and complexity. J. Comp. Neurol. 467, 119–133 (2003)
3. Cannon, R.C.: Structure editing and conversion with cvapp (2000), http://www.compneuro.org/CDROM/nmorph/usage.html

4. Van Pelt, J., Schierwagen, A.: Morphological analysis and modeling of neuronal dendrites. Math. Biosciences 188, 147–155 (2004)
5. Schierwagen, A.: Neuronal morphology: Shape characteristics and models. Neurophysiology 40, 366–372 (2008)
6. Schierwagen, A., Costa, L.F., Alpár, A., Gärtner, A.U., Arendt, T.: Neuromorphological Phenotyping in Transgenic Mice: A Multiscale Fractal Analysis. In: Deutsch, A., et al. (eds.) Mathematical Modeling of Biological Systems, vol. II, pp. 191–199. Birkhuser, Boston (2007)
7. Scorcioni, R., Polavaram, S., et al.: L-Measure: a web-accessible tool for the analysis, comparison and search of digital reconstructions of neuronal morphologies. Nat. Protocols 3, 866–876 (2008)
8. Kohonen, T.: Self-Organizing Maps. Springer, Heidelberg (1997)
9. Heskes, T.: Energy functions for Self-Organizing Maps. In: Oja, E., Kaski, S. (eds.) Kohonen Maps, pp. 303–316. Elsevier, Amsterdam (1999)
10. Bauer, H.-U., Herrmann, M., Villmann, T.: Neural Maps and Topographic Vector Quantization. Neural Networks 12, 659–676 (1999)
11. Villmann, T., Der, R., Herrmann, M., Martinetz, T.: Topology Preservation in Self–Organizing Feature Maps: Exact Definition and Measurement. IEEE Transactions on Neural Networks 8, 256–266 (1997)
12. Bauer, H.-U., Villmann, T.: Growing a Hypercubical Output Space in a Self–Organizing Feature Map. IEEE Transactions on Neural Networks 8, 218–226 (1997)
13. Bauer, H.-U., Pawelzik, K.: Quantifying the neighborhood preservation of Self-Organizing Feature Maps. IEEE Transactions on Neural Networks 3, 570–579 (1992)
14. Ultsch, A., Siemon, H.P.: Kohonen's self–organizing feature maps for exploratory data analysis. In: Proceedings of ICNN 1990, International Neural Network Conference, pp. 305–308. Kluwer, Dordrecht (1990)
15. Vesanto, J., Himberg, J., Alhoniemi, E., Parhankangas, J.: SOM Toolbox for Matlab 5. Report A57, April 2000. Helsinki University of Technology, Finland (2000)
16. Hammer, B., Villmann, T.: Generalized Relevance Learning Vector Quantization. Neural Networks 15, 1059–1068 (2002)

Parallelized Kernel Patch Clustering

Stefan Faußer and Friedhelm Schwenker

Institute of Neural Information Processing, University of Ulm, 89069 Ulm, Germany
{stefan.fausser,friedhelm.Schwenker}@uni-ulm.de

Abstract. Kernel based clustering methods allow to unsupervised partition samples in feature space but have a quadratic computation time $O(n^2)$ where n are the number of samples. Therefore these methods are generally ineligible for large datasets. In this paper we propose a meta-algorithm that performs parallelized clusterings of subsets of the samples and merges them repeatedly. The algorithm is able to use many Kernel based clustering methods where we mainly emphasize on Kernel Fuzzy C-Means and Relational Neural Gas. We show that the computation time of this algorithm is basicly linear, i.e. $O(n)$. Further we statistically evaluate the performance of this meta-algorithm on a real-life dataset, namely the Enron Emails.

1   Introduction

Having a large dataset of possibly $100,000$ samples or more imposes several problems on an unsupervised clustering algorithm. While prototype-based clustering algorithms acting in input space, especially k-means, have a linear computation time to cluster the samples they still need to pass multiple times over the samples - one pass per iteration to adapt the prototypes until they converge. This even gets more computationally expensive for Kernel based clustering methods. Kernel based clustering methods allow to unsupervised partition samples in feature space. Often those samples are nonlinear distributed in input space and are easier to partition in a certain feature space which makes the Kernel methods more useful. However Kernel based clustering methods have the main penalties that they have a quadratic computation time and cannot express the prototypes directly but with a convex combination of existing samples. Therefore multiple passes over the dataset gets even more costly for the Kernel based clustering methods.

To reduce the passes over the data, Fahim et al. [1] have observed that certain samples near to their cluster centre stay in their cluster for some iterations and therefore the distance calculation and assignment steps can be ommited for those samples. However these observations only apply to standard k-means, i.e. needs convex cluster shapes and hard sample to cluster assignments. A slightly older but still popular attempt to cluster large datasets is the CLARANS algorithm [2] which approximates a k-medoid method that minimizes the search in the dataset by heuristics and two parameters. Other efforts have been made to parallelize the k-median algorithm (Guha et al. 2000 - 2003) [3] by independently clustering

F. Schwenker and N. El Gayar (Eds.): ANNPR 2010, LNAI 5998, pp. 131–140, 2010.
© Springer-Verlag Berlin Heidelberg 2010

data streams (or Patches) to gain lk weighted cluster (which are represented by samples), where l is the number of parallelized clusterings, and then cluster again the lk samples to finally gain k cluster. First attempts to parallel cluster samples in subsets with k-means and exchange their statistics have been done as well [4] and speeded up the standard k-means algorithm by $O(k/2)$. Later on, Guhas popular method [3] has been extended for other prototype-based clustering algorithms, namely k-means and batch neural gas [5]. Speed ups for the Kernel based methods to cluster large datasets have been done in Kernel K-Means [6] by block-wise calculating and processing the Kernel matrix which represents the similarities between the samples. In a recent scientific paper, Hasenfuss et al [7] have extended Guhas method furthermode with the Relational Neural Gas method (called Patch Relational Neural Gas) but restricted it for the time being to a sequential clustering procedure.

Our contribution in this paper is a generalization of the Patch Relational Neural Gas algorithm [7] to integrate multiple Kernel based clustering methods and call this algorithm simple the Kernel Patch Clustering (KPC) method. We additionally show the integration of Kernel Fuzzy C-Means and Kernel K-Means within the KPC method and the new assignment update formulas. This expands the algorithm generally for soft memberships. Furthermore we describe a meta-algorithm that performs parallelized clusterings of subsets of the samples using the Kernel Patch Clustering method and merges them repeatedly. It can be shown that the computational time is linear regarding the amount of samples. Lastly we statistically evaluate the performance of this meta-algorithm on a real-life dataset, namely the Enron Emails and show that Kernel Fuzzy C-Means with its soft memberships integrated in KPC perform better than Relational Neural Gas with its hard memberships.

2   Kernel Based Methods for Clustering

Typically most divisive clustering methods aim to minimize a common quantization error. Assume that we want to partition $x_1, x_2, \ldots, x_N$ samples in K disjoint sets or cluster and each cluster has a representing prototype c_k then the actual quantization error or intra-cluster variance (ICV) can be written as:

$$E = \sum_{k=1}^{K} \sum_{i=1}^{N} f_k(i) d(c_k, x_i) \tag{1}$$

where $f_k(i)$ is a hard or bounded soft assignment of sample i to cluster k and $d(c_k, x_i)$ is the distance between sample x_i and cluster prototype c_k. If $d(c_k, x_i)$ is measured by the euclidean distance and $f_k(i)$ is a hard assignment then E is the exact quantization error that k-means minimizes. This happens by repeatedly 1. updating $f_k(i)$: assign samples to clusters based on their distance to their nearest prototypes and 2. updating c_k: move prototypes to their cluster centres until the prototypes converges locally. For the euclidean distance the general function to calculate the current prototypes is as follows:

$$c_k = \frac{\sum_{i=1}^{N} f_k(i)x_i}{\sum_{i=1}^{N} f_k(i)}$$

Now suppose that we would transform all samples $x_i \in X$ and prototypes $c_k \in X$ to a (higher dimensional) feature space using the mapping function $\phi : X \to \mathbb{F}$ that maps X from input space to a possible high-dimensional feature space $\mathbb{F}$. This would allow us to calculate the prototypes in feature space. Unfortunately such mapping functions are costly and often unknown. Still we can calculate the distance between such (theoretically) transformed samples using a positive-definite and symmetric kernel $\kappa(x_i, x_j)$ and applying the kernel trick, i.e. define the prototypes as linear combinations of existing transformed samples. Now further assume that we want to weight each sample i with a weight $w(i)$. We can then set up the new weighted distance function $d_{weighted}(\phi(c_k), \phi(x_i))$ in feature space. The distance function $d_{weighted}(\phi(c_k), \phi(x_i))$ can be written as:

$$d_{weighted}(\phi(c_k), \phi(x_i)) = ||\phi(x_i) - \frac{\sum_{j=1}^{N} f_k^{\phi}(j)w(j)\phi(x_j)}{\sum_{j=1}^{N} f_k^{\phi}(j)w(j)}||^2 \qquad (2)$$

$$= \langle \phi(x_i), \phi(x_i) \rangle - 2 \left\langle \phi(x_i), \frac{\sum_{j=1}^{N} f_k^{\phi}(j)w(j)\phi(x_j)}{\sum_{j=1}^{N} f_k^{\phi}(j)w(j)} \right\rangle +$$

$$\left\langle \frac{\sum_{j=1}^{N} f_k^{\phi}(j)w(j)\phi(x_j)}{\sum_{j=1}^{N} f_k^{\phi}(j)w(j)}, \frac{\sum_{j=1}^{N} f_k^{\phi}(j)w(j)\phi(x_j)}{\sum_{j=1}^{N} f_k^{\phi}(j)w(j)} \right\rangle$$

$$= \kappa(x_i, x_i) - \frac{2\sum_{j=1}^{N} f_k^{\phi}(j)w(j)\kappa(x_i, x_j)}{\sum_{j=1}^{N} f_k^{\phi}(j)w(j)} +$$

$$\frac{\sum_{j=1}^{N} \sum_{l=1}^{N} f_k^{\phi}(j)f_k^{\phi}(l)w(j)w(l)\kappa(x_j, x_l)}{[\sum_{j=1}^{N} f_k^{\phi}(j)w(j)]^2}$$

Note that there are many repeatings in the above formula that have to be calculated in the right order to avoid wasting computational time. Contrary to standard k-means where we iteratively update the prototypes we can here only repeatedly update the assignments by calculating and comparing the distances. For Weighted Kernel K-Means the assignment update step is:

$$f_k^{\phi}(i) = \begin{cases} 1, & \text{if} \quad d_{weighted}(\phi(c_k), \phi(x_i)) < d_{weighted}(\phi(c_l), \phi(x_i)), \\ & \quad l = 1, \ldots, K, l \neq k \\ 0, & \text{else} \end{cases} \qquad (3)$$

For Weighted Kernel Batch Neural Gas or Weighted Relational Neural Gas [9] the assignment update step is:

$$f_k^{\phi}(i) = exp\left(\frac{-rank(\phi(c_k), \phi(x_i))}{\lambda}\right) \qquad (4)$$

where $rank(\phi(c_k), \phi(x_i)) = |\{\phi(c_l) \mid d_{weighted}(\phi(c_l), \phi(x_i)) < d_{weighted}(\phi(c_k), \phi(x_i)), l = 1, \ldots, K, l \neq k\}| \in \{0, \ldots, K-1\}$. Lastly the assignment update steps for Weighted Kernel Fuzzy C-Means [8] gets:

$$f_k^{\phi}(i) = \cfrac{1}{\sum_{l=1}^{K} \left[\cfrac{d_{weighted}(\phi(c_k),\phi(x_i))}{d_{weighted}(\phi(c_l),\phi(x_i))} \right]^{\frac{2}{m-1}}} \qquad (5)$$

The output of such a Kernel based clustering method is then the assignments f that determines the hard (Relational Neural Gas, Kernel K-Means) or soft (Kernel Fuzzy C-Means) assignments of the given samples to K cluster. Comparing the three algorithms, Kernel Fuzzy C-Means and Relational Neural Gas are more insensitive to initializations as both algorithms update their indirectly defined prototypes not only by their greedy winner samples but also by other samples determined through neighborhood size λ (Relational Neural Gas) or fuzzifier m (Kernel Fuzzy C-Means). The Kernel Fuzzy C-Means algorithm however has the plus that it uses soft assignments, i.e. gives possibly more information on the data. On the other hand if fuzzifier $m \to 1$ then this algorithms behaves exactly like Kernel K-Means. All three Kernel methods have the same drawback that they have a quadratic computation time $O(N^2)$ which renders them unusable for vast datasets.

3 Kernel Patch Clustering

Assume that we have N samples that we want to separate in K sets but we only can cluster up to M samples with one of the Kernel based clustering methods described above due to the high required computational time. Now suppose that we select instead the first M samples out of N, i.e. set up a Patch, cluster them into K cluster and choose the best k samples per cluster as approximative cluster prototypes. This results in exactly $k \cdot K$ samples that represent K cluster. Now they can be weighted by the number of samples they represent and can be itself clustered with the next M samples. This can be iterated until every N samples have been processed in Patches with a maximum size of M and $k \cdot K$ samples representing the final K prototypes are left. Let us now formulate these thoughts in the following *Kernel Patch Clustering* algorithm:

- **Input:**
  - kernel function κ, samples $x \in X^N$
  - number of patches C, whereas $C = \frac{N}{M}$, $M =$ maximum samples to cluster per patch
  - number of cluster K
  - number of samples per cluster k
  - choose one Kernel clustering method (Kernel K-Means, Kernel Fuzzy C-Means, Relational Neural Gas)
- **Initialize:**
  - $I =$ arbitrary permutation of $\{1, \ldots, N\}$
  - $J_0 = \{\emptyset\}$
  - $w_0 = \{\emptyset\}$

- **for** $t = 1$ to C
    - Construct Patch $P_t \in \mathbb{R}^{|S_t| \times |S_t|}$ with sample indices $S_t = \{J_{(t-1)}, I_{(t-1)\frac{N}{C}}, \ldots, I_{t\frac{N}{C}}\}$ using kernel funktion κ and samples x. This Patch P_t has now the similarities between all current choosen samples $\{I_{(t-1)\frac{N}{C}}, \ldots, I_{t\frac{N}{C}}\}$ plus the last k-approximation of the prototypes $J_{(t-1)}$. Append the vector $\{1\}^{\frac{N}{C}}$ to the weights w_{t-1} so that the whole weight vector w_t has a length of $k \cdot K + \frac{N}{C}$ (or a length of $\frac{N}{C}$ for the first time)
    - Arbitrary initialize cluster assignments $f_{old} \in \mathbb{R}^{K \times |S_t|}$ and perform selected Kernel clustering method using weights w_{t-1} and Patch P_t with K cluster to get new cluster assignments $f_{new} \in \mathbb{R}^{K \times |S_t|}$
    - Select the k best samples for each cluster K out of S_t using the k-approximation (6), the assignments f_{new} and the distance function for the selected clustering method. Fill the sample indices in $J_t \in \mathbb{N}^{K \cdot k}$
    - Calculate the new weights $w_{tl}^{new} = \{\frac{m_l}{k}\} \in \mathbb{R}^k$, for $l = 1, \ldots, K$, where m_l is the number of samples belonging to cluster l. Form the weight vector $w_t = \{w_{t1}^{new}, \ldots, w_{tK}^{new}\}$
- **Output:**
    - k-approximation of final prototypes J_C
    - final cluster assignments f_{new}

Note that the number of Patches C has to be choosen such that it is still computationally possible to cluster $\frac{N}{C} + k \cdot K$ samples per Patch. For the k-approximation we simply determine the k samples that are nearest to their own cluster centre (prototype) and do that for each cluster l (as introduced in [7] for Relational Neural Gas):

$$argmin_{i|g_l^\phi(i)=1}(d(\phi(x_i), \phi(c_l))), l = 1, \ldots, K \tag{6}$$

The above equation has to be repeated k-times to get the k-best approximations of each cluster prototype l, each time removing the winner sample x_i so that it cannot be selected once more. As this requires hard cluster assignments $g_l^\phi(i)$, the soft assignments of Kernel Fuzzy C-Means have to be first converted by defining that a sample i belongs to the cluster l with the highest fuzzy membership $argmax_l(f_l^\phi(i))$. For the most settings the algorithm will be done after having calculated the distances and assignments of the rest of the $(N - k \cdot K)$ samples using the k-approximation of final prototypes J_C and the final cluster assignments f_{new}.

3.1   Parallelized Kernel Patch Clustering

While the Kernel Patch Clustering approach described above clusters the samples in a single pass over the dataset, i.e. if we consider the k-approximations of the prototypes as new samples, it is still a serial process. Easily this can be parallelized by distributing the clustering of the Patches $P_1, \ldots, P_C$ and the calculating of the k-approximations of the resulting prototypes to multiple systems connected by a network or multiple threads on one system possibly handled by multiple processors. This can be formulated in the meta-algorithm to perform parallelized Kernel Patch Clusterings:

- **Input:**
  - maximum samples to cluster per patch M
  - number of maximum parallelizations μ
- **while** not all samples are processed
  1. Construct up to μ patches $P_1, \ldots, P_\mu$, each patch P_i having the next M samples and the last weighted k-approximations of the prototypes J_{last}
  2. Distribute the μ patches to μ systems or threads and cluster the patches parallelized
  3. Calculate the k-approximations of the resulting prototypes J_i and their weights w_i parallelized
  4. Collect all μ k-approximations $J_1, \ldots, J_\mu$, remove any duplicated samples and save it in J_{last}. Divide their corresponding weights by $\frac{1}{\mu}$ as they will be distributed later on in μ parallel clusterings
  5. (optional): Cluster the μ k-approximations in $J_{last} \in \mathbb{N}^{\mu \cdot k \cdot K}$ to get one k-approximation $J_{last} \in \mathbb{N}^{\cdot k \cdot K}$
- **Output:**
  - μ k-approximations of final prototypes J_{last}

Step 2 and 3 can be done completely parallelized, i.e. cluster $M \cdot \mu$ samples parallelized. In step 4, the results of those μ clusterings are collected resulting in $\mu \cdot k \cdot K$ weighted samples, representing k-approximations of K prototypes. As those samples are reinserted in step 1, it is possible that the result of the clusterings in step 4 produces duplicated samples that have to be removed. In step 5 we optionally cluster the final $\mu \cdot k \cdot K$ weighted samples once more to gain exactly $k \cdot K$ k-approximations of K prototypes and to be conform with the output of the serial Kernel Patch Clustering algorithm. This step is recommended for high values of parallelizations μ as without the (re-) clustering of the prototypes this might result in too much samples to cluster in step 1.

Complexity: The complexity of one clustering operation is $O((M + k \cdot K)^2)$ where M is the maximum number of samples per patch, K the number of cluster and k the number of samples per cluster (being the k-approximation of K prototypes). As the size of $M + k \cdot K$ however is bounded and generally independent of N and we have to perform $\frac{N}{M}$ such clustering operations, the summed clustering complexity results in $O((M + k \cdot K)^2 \frac{N}{M}) \sim O(M \cdot N) \sim O(N)$ in terms of N. Moreover this clustering complexity can even be reduced linear through parallelizations by a factor of μ. However the tradeoff is that the complexity of one clustering operation rises to $O((M + \mu \cdot k \cdot K)^2)$ which is bearable for small values of μ, k and K. For higher values of μ, k and K like $\mu \cdot k \cdot K \geq M$, it is advisable to perform an additional clustering step (step 5 in the algorithm above) to get only one k-approximation of the cluster.

4   Experiments and Results

To evaluate the Kernel Patch Clustering method we have conducted experiments with one synthetic (five two-dimensional gaussian distributed cluster), one widely known (Wisconsin Diagnostic Breast Cancer) and one real-life (Enron Emails) dataset. We have compared the performances of Relational Neural Gas (RNG), Kernel Fuzzy C-Means (KFCMEANS) and integrations of those two methods into the Kernel Patch Clustering (KPC). Furthermore we have compared these methods to the basic approach to arbitrarily choose some samples out of all samples and cluster them using KFCMEANS. For all experiments we have set the neighborhood range λ for RNG to be exponentially falling from $\frac{N}{2}$ to 0.01 which are stable standard values (see [9]). The fuzzifier for KFCMEANS had been set to 2.0 for the synthetic dataset and 1.25 for both other datasets to reach more hard than soft memberships. We have done time measurements of these algorithms for the Enron Emails dataset.

4.1   Five Two-Dimensional Gaussian Distributed Cluster

This synthetic dataset had been created by five multivariate (two-dimensional) gaussian distributions with the parameters $\mu_1 = \{0,0\}, \mu_2 = \{1,0\}, \mu_3 = \{0,1\}$, $\mu_4 = \{1,1\}, \mu_5 = \{0.5, 0.5\}$ and a variance of $\sigma^2 = 0.01$. As it can be seen in figure 1, the amount of samples drawn from the distributions are unequal which highly complicates the clustering problem. We have assigned each cluster a class label and have 50 samples in class 1, 100 in class 2, 100 in class 3, 500 in class 4 and 500 in class 5. We have choosen a linear kernel to calculate the similarities between the samples:

$$\kappa(x_i, x_j) = (x_i \cdot x_j)$$

For all clustering methods the aim was to partition the samples in $K = 5$ cluster. We have performed 50 test runs with each algorithm and then have calculated the intra-cluster variance (ICV), i.e. quantization error with hard assignments (same conditions for all methods) and the class prediction score by comparing

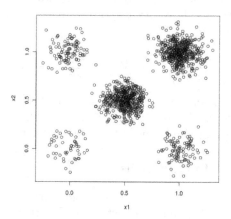

Fig. 1. Five two-dimensional gaussian distributed cluster, $\mu_1 = \{0,0\}, \mu_2 = \{1,0\}, \mu_3 = \{0,1\}, \mu_4 = \{1,1\}, \mu_5 = \{0.5, 0.5\}$, variance $\sigma = 0.01$, 1250 samples (50, 100, 100, 500, 500)

the cluster sets with the class labels. The results can be seen in table 1. For the last method we arbitrarily choose the same size of samples as in one Patch (100). We tried to experiment with the Patch size but have observed that any size between 50 to 500 had not made any statistical difference. Same applies for different number of parallelizations: $1 \leq \mu \leq 6$. However for Patch sizes below 50, the KPC - RNG algorithm had often problems finding $K = 5$ cluster while the KPC - KFCMEANS algorithm still converged. In general the KFCMEANS method performs way better than the RNG method for this dataset, probably because of the soft relaxations of the assignments of samples to cluster. Integrated into the KPC algorithm, both methods still perform nearly as well while being (theoretically) much faster.

Table 1. Cluster validations on five two-dimensional gaussian distributed cluster. All methods had to partition the sample into $K = 5$ cluster. For Kernel Patch Clustering (KPC), 100 samples were processed per Patch, $k = 3$-approximation, $\mu = 1, \ldots, 6$ parallelizations. The values were averaged over 50 runs.

	KFCMEANS	RNG	KPC - KFCMEANS	KPC - RNG	arbitrary choosen KFCMEANS
quantization error	23.43		43.29	23.89	43.77	31.85
class prediction	100%		88.82%	100%	88.8%	97.1%

4.2 Wisconsin Diagnostic Breast Cancer

The Wisconsin Diagnostic Breast Cancer database is a widely known dataset that has been contributed to the UCI machine learning repository [11] in 1995. It consists of 569 samples describing each characteristics of the cell nuclei taken from a Breast mass with 30 real valued features. The task is to classify the cell nuclei either as malignant or benign. We have standardized the samples to zero mean and unit variance and have calculated the similarities between them by a RBF-Kernel with parameter $\sigma = 5$:

$$\kappa(x_i, x_j) = exp(-\frac{||x_i - x_j||^2}{2\sigma^2})$$

Same as for the synthetic dataset we have calculated the quantization error and the class prediction score. The results can be seen in table 2. All algorithms had about the same performance in terms of those two validation criteria.

4.3 Bag of Words – Enron Emails

The Bag of Words dataset at the UCI machine learning repository [11] donated in 2008 consists of five different text collections that each delivers plenty word to document assignments. For our experiments we have choosen the Enron Emails text collection which consists of $39, 861$ documents, $28, 102$ different words and approximately $1, 900, 000$ document to word assignments. To calculate

Table 2. Cluster validations on Wisconsin Diagnostic Breast Cancer database. All methods had to partition the sample into $K = 10$ cluster. For Kernel Patch Clustering (KPC), 50 samples were processed per Patch, $k = 3$-approximation. The values were averaged over 50 runs.

	KFCMEANS	**RNG**	**KPC - KFCMEANS**	**KPC - RNG**
quantization error	221	223	224	222
class prediction	90.5%	90.1%	89.8%	90%

the similarities between the documents we have choosen the Bhattacharyya kernel [10] with a multinomial distribution, setting parameter $X = \frac{1}{8}$:

$$\kappa(p,p') = [\sum_{i=1}^{M}(p_i p'_i)^{\frac{1}{2}}]^X$$

where $p_i = \dfrac{\text{number of words i in document a}}{\sum_j \text{number of words j in document a}}$

and $p'_i = \dfrac{\text{number of words i in document b}}{\sum_j \text{number of words j in document b}}$.

As the complexity of the co-occurence matrix necessary to calculate the kernel rises with the number of words, we could only handle about 200 samples per clustering and had parallelized the clusterings on a four core machine (parameter $\mu = 4$). Furthermore we have performed a serialized clustering on the same machine (parameter $\mu = 1$). As no class labels are known for these documents we can therefore solely evaluate the quantization error. The results with 5 and 10 cluster can be seen in table 3. This time the difference between the simple approach to choose arbitrarily 200 samples and cluster them and the appliance of the KPC algorithm with either KFCMEANS or RNG is clearly visible. Also it can be seen that KPC-KFCMEANS produces slightly better cluster results than KPC-RNG.

Table 3. Cluster validations on the Enron Emails. For Kernel Patch Clustering (KPC), 200 samples were processed per Patch, $k = 3$-approximation. The values were averaged over 20 runs.

	K	μ	quant. error	time[s]
KPC-RNC	5	1	17226	357
KPC-KFCMEANS	5	1	17141	369
KPC-RNC	5	4	17328	126
KPC-KFCMEANS	5	4	17272	131
arbitrary choosen, KFCMEANS	5	-	17596	3
KPC-RNC	10	1	16211	438
KPC-KFCMEANS	10	1	16234	493
KPC-RNC	10	4	16384	155
KPC-KFCMEANS	10	4	16372	203
arbitrary choosen, KFCMEANS	10	-	16780	5

5   Conclusion

We have described a meta-algorithm that performs parallelized clusterings with Kernel based methods and merges the results iteratively. The necessary extensions to the distance calculations and the assignment steps of Kernel Fuzzy C-Means, Kernel K-Means and Relational Neural Gas to include sample weightings have been shown. Experimentally we have observed that the loss of accuracy is rather low and the parallelized KPC algorithm can be used for vast real-life datasets that otherwise could not be clustered. As such a real-life dataset we have choosen the Enron Emails which have approximately $1,900,000$ total words and have shown that the parallelized KPC method performs far better than a simple randomized approach. By the integration of the Kernel Fuzzy C-Means algorithm this parallelized method can determine soft memberships of samples to cluster. The real usefulness of those soft memberships can be better determined by datasets that are naturally fuzzy which might be done in another contribution.

References

1. Fahim, A.M., Salem, A.M., Torkey, F.A., Ramadan, M.A.: An efficient enhanced k-means clustering algorithm. Journal of Zhejiang University SCIENCE A, 1626–1633 (2006) ISSN 1009-3095
2. Ng, R.T., Han, J.: Efficient and Effective Clustering Methods for Spatial Data Mining. In: Proceedings of the 20th VLDB Conference, pp. 286–296. Morgan Kaufmann Publishers, San Francisco (1994)
3. Guha, S., Meyerson, A., Mishra, N., Motwani, R., O'Callaghan, L.: Clustering Data Streams: Theory and Practice. Proceedings of IEEE Transactions on Knowledge and Data Engineering 15(3), 515–528 (2003)
4. Kantabutra, S., Couch, A.L.: Parallel K-means Clustering Algorithm on NOWs. NECTEC Technical Journal 1(6) (2000)
5. Alex, N., Hammer, B.: Parallelizing single patch pass clustering. In: ESANN 2008 (2008) ISBN 2-930307-08-0
6. Zhang, R., Rudnicky, A.I.: A Large Scale Clustering Scheme for Kernel K-Means. In: ICPR 2002, 16th International Conference on Pattern Recognition, vol. 4, p. 40289 (2002)
7. Hasenfuss, A., Hammer, B., Rossi, F.: Patch Relational Neural Gas Clustering of Huge Dissimilarity Datasets. In: Prevost, L., Marinai, S., Schwenker, F. (eds.) ANNPR 2008. LNCS (LNAI), vol. 5064, pp. 1–12. Springer, Heidelberg (2008)
8. Zhang, D.Q., Chen, S.C.: Fuzzy clustering using kernel methods. In: International Conference of Control and Automatation (ICCA 2002), Xiamen, China, pp. 123–128 (2002)
9. Hammer, B., Hasenfuss, A.: Relational Neural Gas. In: Hertzberg, J., Beetz, M., Englert, R. (eds.) KI 2007. LNCS (LNAI), vol. 4667, pp. 190–204. Springer, Heidelberg (2007)
10. Jebara, T., Kondor, R., Howard, A.: Probability Product Kernels. Journal of Machine Learning Research 5, 819–844 (2004)
11. Asuncion, A., Newman, D.J.: UCI Machine Learning Repository (2009), http://www.ics.uci.edu/~mlearn/MLRepository.html

Neural Network Cascade for Facial Feature Localization

Thibaud Senechal, Lionel Prevost, and Shehzad Muhammad Hanif

Universite Pierre and Marie Curie-Paris 6
ISIR, CNRS UMR7222, BC173
4 Place Jussieu, 75252 Paris Cedex 5, France

Abstract. We present here a complete system for the localization of facial features in frontal face images. In the first step, face detection is performed using Viola & Jones state of art algorithm. Then, a cascade of neural networks localizes precisely 28 facial features. The first network performs a coarse detection of three areas in the image corresponding roughly to left and right eyes and mouths. Then, three local networks localize, in these areas, 9 key points per eye and 10 key points on the mouth. Thorough experiments on 3500 images from standard databases (Feret, BioID) show the detector accuracy, its generalization ability and speed.

1 Introduction

Localizing facial features (like mouth and eye corners or eyebrow) is usually the first step in applications like face recognition, expression analysis or action unit identification [1,2]. These key-points are also very useful for model alignment.

Active shape model [3],[4] and active appearance model [5] are commonly used to perform the detection. Unfortunately, they rely on an unstable optimization procedure which depends on hundreds of parameters encoding shape (and texture) variations. Other statistical methods include Neural Networks [6][7], Bayesian Networks [8], Support Vector Machines [9] and Cascade Of Boosted Ensembles (using either Haar filter [10] or Gabor jet [11]). Though most of these algorithms are able to detect precisely a small number of facial features (typically four, including eye centres, mouth and nose), their accuracy on a large number of key-points is rarely stated.

In our previous works (described in [7]), manually cropped images are fed to a neural network trained to output a probability map. Facial feature hypothetic locations (eye centers and mouth corners) corresponded to local maxima in this map.

We present in this paper a fast and an accurate method for precise facial feature localization. A facial detector is used to extract the face in the image. Then, a neural network performs coarse detection, defining 3 regions of interest (left and right eyes and mouth) within the face image. In the second stage, another network is applied on each region to detect 28 points (mouth contour,

F. Schwenker and N. El Gayar (Eds.): ANNPR 2010, LNAI 5998, pp. 141–148, 2010.
© Springer-Verlag Berlin Heidelberg 2010

eye contour and eyebrow). Such a cascade was already explored in [12] but they only detected 10 points.

The paper is organized as follows. Section 2 details the 3 stages: face detection, coarse to fine facial feature localization. Section 3 is devoted to sensitivity analysis and experimental results on several benchmark datasets. Conclusion and prospects are presented in section 4.

2   Overview

2.1   Face Detector

To extract automatically face in images, we use the OpenCV's face detector which provides an implementation of the Viola-Jones algorithm [13]. It uses Haar-like filters as weak classifiers. The AdaBoost algorithm makes a forward selection of the best features and trains the weak learners. To run in real-time, strong classifiers are arranged in a cascade in order of complexity, where each classifier is trained only on examples which pass through the previous classifiers.

2.2   Coarse Localization

This step detects three Regions Of Interest (ROI) corresponding roughly to eyes and mouth in the detection window. To achieve, we train a fully connected multilayer perceptron using back-propagation algorithm to detect five points: eye centers, nose tip and mouth corners.

Neural network inputs can be either the gray levels of sub-sampled extracted faces. To improve the robustness of our detector, we synthesized new face images by translating the detection and modifying the scale factor of the face detector by 10% of the inter-ocular distance (this corresponds to the standard deviation of the eye position in the detection image). To obtain face images invariant to illumination effects, the images are normalized. Statistical mean and standard

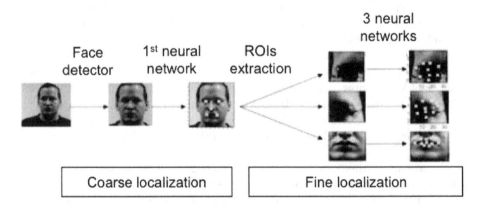

Fig. 1. System overview

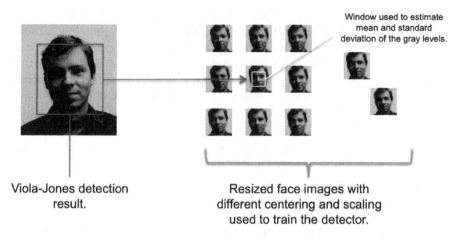

Fig. 2. Train database

deviation are estimated on a window centered on the center of the detection window with a size equal to the half of the detection window (fig. 2. The 10 outputs are the (X,Y) coordinates normalized between -1 and 1 of the five facial features we try lo localize.

2.3   Precise Localization

In this last step, three fully connected multilayer perceptrons are used to deal with eyes and mouth area. They detect 28 points: 9 for each eye and eyebrow contours and 10 for the two lip contours (fig. 1). The eye localizers have 9x2=18 outputs and the mouth localizer has 20 outputs.

Coarse localization determines three ROIs bounding the eye centres and the two mouth corners. We chose the size of these ROIs as a fraction of the detection window size to obtain images including all the points that we try to localize. We made a statistical study on face pattern coordinates in detection images to find the smallest window that contains these points in most of the cases. Finally, to compensate the imprecision of the ROI centre coordinates, we increased ROI dimensions by the mean localization error of the coarse localizer. As before (Section 2.2), to train these three neural networks, we synthesized new images.

3   Setup and Sensitivity Analysis

In order to estimate the parameters of the detector, we manually labeled the ground truth of 320 images from a homemade database (so called ISIR database) containing frontal faces with small expression and natural rotation changes of men and women with different specificities (facial hair, glasses). Training dataset includes 256 peoples and the other 64 images are used to stop training. The size of the training set is artificially expanded to 3700 examples using basic image transformations like small translation, rotation and scaling. For each set of parameters, we performed 3-fold cross-validation. The localization error E for an example is the mean Euclidian distance between the detected feature positions (x_i, y_i) and the true (labelled) feature positions $(\tilde{x}_i, \tilde{x}_i)$, normalized with respect to the inter-ocular distance D. The mean localization error E_m is computed over the whole dataset.

$$E = \frac{1}{N} \sum_{i=1}^{N} \frac{\sqrt{(x_i - \tilde{x}_i)^2 + (y_i - \tilde{y}_i)^2}}{D} \tag{1}$$

3.1 Coarse Localization

We evaluated several image coding schemes: gray-level sub-sampled images of size 20x20 and 30x30 pixels and principal component analysis on images of size 20x20, 30x30 and 60x60 pixels resulting into 70, 90 and 100 eigenvalues respectively (corresponding to 90% of the explained variance). Mean localization results for the 5 points are reported table 1. We get quite similar results using gray level 30x30 images (900 input cells) and eigenvalues (100 input cells) and 100 hidden cells. In both cases, the mean localization error E_{mv} on the validation set is lower than 6% for the five points we try to localize.

Table 1. Mean localization error on training set (E_{mt}) and validation set (E_{mv}) for five points with the coarse detector

Inputs	Number of input cells	E_{mt}	E_{mv}
20x20 images	400	6.0%	7.1%
ACP on 20x20 images	70	8.9%	9.2%
30x30 images	900	4.3%	5.9%
ACP on 30x30 images	90	6.0%	6.9%
ACP on 60x60 images	100	4.3%	5.8%

3.2 Precise Eye Localization

We want to localize 4 points on each eyebrow and 5 points on each eye as shown in fig 1. We have already defined the sizes and position of the ROI in section 2.3. We evaluated several resolutions (6x10, 10x15, 13x20, 16x25, 19x29, 22x34 pixels) for precise eye detection.

Lowest resolution (6x10 pixels) results in a high localization error (higher than 5%) on the training set while the highest resolution (22x34 pixels) generalizes poorly. Best results correspond to the following parameters: 13x20=260 input cells and 10 hidden neurons. The mean localization error $E_m v$ on these 18 feature points is 4.8%.

3.3 Precise Mouth Localization

We have evaluated two sizes for the mouth region of interest: small size (including mouth only) and large size (including the mouth and some parts of nose and chin). For each size, three image resolutions were tested. Small ROIs lead to 4.5% as mean localization error while large regions give better results: the mean localization error is 4%. The sensitivity to resolution is quite low. Best results correspond to the following parameters: 21x23 input cells and 20 hidden neurons. The mean localization error $E_m v$ on these 10 feature points is 4%.

4   Experimental Results

We trained and tune the parameters of the cascade localizer on the ISIR dataset divided into independent training and cross-validation sets. Then, to evaluate the localizer generalization ability, we tested it extensively, without any retraining on several benchmark databases.

4.1   Results on the ISIR Database

Table 2 details performances (mean localization error) for all the face patterns that we want to localize. Center and contour of the eye are localized with a high precision (the mean error is less than 3%). This is partially due to the face detector that produces eye-centered detections. Mouth features are precisely localized too (the mean error is equal to 4%) though mouth position is more variable. Precision on eyebrow is a little poorer (6.4%) but performance evaluation is biased due to ground truth imprecision. The percentage of correctly detected images with a localization error lower than 10% is 98% for the eye and eyebrow regions and 93% for the mouth contour (fig. 3).

Table 2. Mean localization error on training set (E_{mt}) and validation set (E_{mv}) for different facial features compared to the standard deviation of their position in detection image (SD)

	SD	E_{mt}	E_{mv}
Mouth (10 points)	10.5%	3.3%	4.0%
Eye centers (2 points)	6.4%	2.7%	3.0%
Eyes contour (8 points)	6.9%	3.1%	3.3%
Eyebrows (8 points)	9.4%	5.7%	6.4%
28 facial points	8.9%	3.8%	4.4%

Table 3 compares test error of each step of our architecture for 4 points (eyes and mouth corners). First column reports the localization error on eye centers and mouth corners using only the Viola-Jones detector: we use as localization hypothesis the mean position of each feature (estimated on the training dataset) inside the detection windows. Second column reports the mean error after the

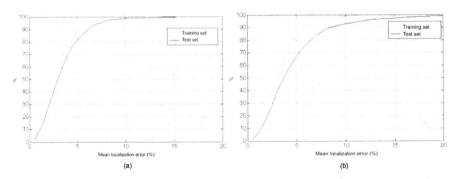

Fig. 3. Percentage of images with a mean localization error lower to the X axis value for eyes points (a) and mouth points (b)

coarse localization. The third column reports the mean error for the same points after the fine localization step. This proves that reducing the research area with a coarse step allows us to increase locally the image resolution without having too many input cell. Moreover it shows that cascading localizers increases drastically the system accuracy.

Table 3. Mean localization error on validation set after each step of the system

	VJ detector	Coarse localization	Fine localization
Eye centers (2 points)	6.4%	4.8%	3.0%
Mouth corners (2 points)	10.5%	7.2%	3.7%

4.2   Results on the Inrialpes, BioID and Feret Test Databases

The system has been successively evaluated on the 60 frontal faces from Inrialpes database, 1500 images from BioID database [14] and the 1918 images FERET Duplicate I dataset [15] without retraining the neural networks. (fig 4) shows some results we obtain on these bases.

Inrialpes database includes frontal faces with poor luminosity conditions and the system performs a mean localization error of 7% (fig 4b).

On BioID database that contains large changes in expressions and has very poor luminosity conditions (emphasis has been laid on "real world" conditions), the system gives a mean localization error of 11% on 28 facial features (fig 4c).

Finally, Feret database contains multi-ethnic subjects with small facial expression changes and some subjects wear glasses. We only have the eyes, mouth centres and the nose tip manually labelled. Although the learning database that we used only includes European subjects without glasses, the mean localization error on eyes and mouth centre is lower than 6% (fig 4a).

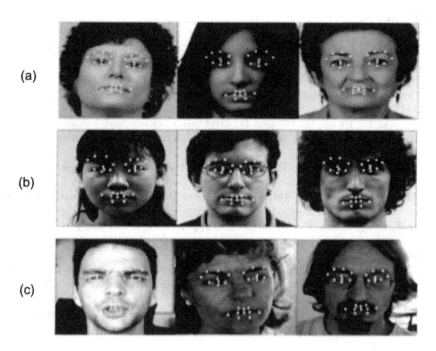

Fig. 4. Localisations on Feret database (a), Inrialpes database (b) and BioID database (c)

5   Conclusion and Prospects

We have presented here a new localizer able to detect precisely the eyes, eyebrows and mouth contours. It uses a cascade of neural networks. The face is first detected using a standard algorithm. The first stage performs a coarse detection of three regions of interest corresponding roughly to eyes and mouth. The second stage localizes precisely 28 facial features on the eye contour, the eyebrow and the mouth contour. The mean localization error is lower than 5% on the validation set. To show the generalization ability, we evaluate the system on three standard databases, namely Inrialpes, bioId and Feret, where faces are sometime slightly expressive, multi-ethnic or poorly illuminated. Results are really encouraging as the overall localization error is lower than 8%. Moreover, the computation speed (including face detection, coarse and fine localization) is nearly 20 images per second. In our previous works [7], we already showed the system ability to localize coarsely facial features in orientation-free images by combining several experts dedicated to each facial pose. So, we can easily combine parallel and cascade approaches to build an orientation-free fine localizer. Other future works include active appearance model initialization and action unit detection for facial expression labelling.

References

1. Zhao, W., Chellappa, R., Phillips, P., Rosenfeld, A.: Face recognition: A literature survey. ACM Computing Surveys 35(4), 399–458 (2003)
2. Pantic, M., Rothkrantz, L.: Automatic analysis of facial expressions: The state of the art. IEEE Transactions on Pattern Analysis and Machine Intelligence, 1424–1445 (2000)
3. Cootes, T., Taylor, C., Cooper, D., Graham, J., et al.: Active shape models-their training and application. Computer Vision and Image Understanding 61(1), 38–59 (1995)
4. Milborrow, S.: Locating Facial Features with Active Shape Models. PhD thesis, Faculty of Engineering, University of Cape Town (2007)
5. Matthews, I., Baker, S.: Active appearance models revisited. International Journal of Computer Vision 60(2), 135–164 (2004)
6. Duffner, S., Garcia, C.: A connexionist approach for robust and precise facial feature detection in complex scenes. In: Image and Signal Processing and Analysis, pp. 316–321 (2005)
7. Hanif, S.M., Prevost, L., Belaroussi, R., Milgram, M.: Real-time facial feature localization by combining space displacement neural networks. Pattern Recognition Letters 29(8), 1094–1104 (2008)
8. Yan, S., Li, M., Zhang, H., Cheng, Q.: Ranking prior likelihood distributions for bayesian shape localization framework. In: International Conference on Computer Vision, pp. 51–58 (2003)
9. Nguyen, M., Perez, J., De la Torre Frade, F.: Facial feature detection with optimal pixel reduction svms. In: International Conference on Automatic Face and Gesture Recognition (2008)
10. Cristinacce, D., Cootes, T.: Facial feature detection using adaboost with shape constraints. In: British Machine Vision Conference, vol. 1, pp. 231–240 (2003)
11. Vukadinovic, D., Pantic, M.: Fully automatic facial feature point detection using gabor feature based boosted classifiers. In: International Conference on Systems, Man and Cybernetics, vol. 2 (2005)
12. Duffner, S., Garcia, C.: A hierarchical approach for precise facial feature detection. In: Compression et Representation des Signaux Audiovisuels (2005)
13. Viola, P., Jones, M.: Robust real-time face detection. International Journal of Computer Vision 57(2), 137–154 (2004)
14. Jesorsky, O., Kirchberg, K., Frischholz, R., et al.: Robust face detection using the hausdorff distance. In: Bigun, J., Smeraldi, F. (eds.) AVBPA 2001. LNCS, vol. 2091, pp. 90–95. Springer, Heidelberg (2001)
15. Phillips, P., Wechsler, H., Huang, J., Rauss, P.: The feret database and evaluation procedure for face-recognition algorithms. Image and Vision Computing 16(5), 295–306 (1998)

A Hidden Markov Model Based Approach for Facial Expression Recognition in Image Sequences

Miriam Schmidt, Martin Schels, and Friedhelm Schwenker

Institute of Neural Information Processing,
University of Ulm, 89069 Ulm, Germany
{miriam.k.schmidt,martin.schels,friedhelm.schwenker}@uni-ulm.de

Abstract. One of the important properties of hidden Markov models is the ability to model sequential dependencies. In this study the applicability of hidden Markov models for emotion recognition in image sequences is investigated, i.e. the temporal aspects of facial expressions. The underlying image sequences were taken from the Cohn-Kanade database. Three different features (principal component analysis, orientation histograms and optical flow estimation) from four facial regions of interest (face, mouth, right and left eye) were extracted. The resulting twelve paired combinations of feature and region were used to evaluate hidden Markov models. The best single model with features of principal component analysis in the region face achieved a detection rate of 76.4 %. To improve these results further, two different fusion approaches were evaluated. Thus, the best fusion detection rate in this study was 86.1 %.

1 Introduction

For over a hundred years people are interested in understanding emotions. One of the most important and pioneering researcher, who has dealt with this issue, was Charles Darwin [6]. By examining the interaction between facial muscles and the associated emotions, he introduced the first rules of emotion recognition. Paul Ekman developed a so-called Facial Action Coding System (FACS) [9] to encode the emotions using the facial muscles. He distinguished six basic emotions: "joy", "anger", "surprise", "disgust", "sadness" and "fear".

Hidden Markov Models (HMMs) are often used in speech recognition [15] and increasingly also for emotion recognition [20,4] because they are able to model temporal dependencies. HMMs are probabilistic models which consist of a countable number of states, transitions and corresponding emissions. HMMs are easy to model, but variable by the parameters that describe them.

Lisetti and Rumelhart [14] proposed a NN-based approach to recognize the facial expressions in which they selected different face regions manually. Lien et al. [13] compared different methods that use optical flow. Lin et al. [7] utilized principle component analysis (PCA) and hierarchical radial basis functions for facial emotion recognition.

F. Schwenker and N. El Gayar (Eds.): ANNPR 2010, LNAI 5998, pp. 149–160, 2010.
© Springer-Verlag Berlin Heidelberg 2010

In this study, HMMs were utilized for recognition of facial expressions in image sequences. The aim was to investigate the behavior of HMMs in facial expression recognition and to demonstrate whether HMMs are capable of recognizing emotions in image sequences sufficiently well. Hence, four regions in the image sequences were selected manually: face, mouth, right and left eye. Afterward, in each region of the image, three different features, namely principal component analysis, orientation histograms and optical flow estimation, were extracted. The resulting twelve paired combinations of feature and region were used to evaluate the HMMs. Numerous experiments to optimally adjust the HMMs were conducted. The optimal number of states and the optimal number of the normal distributions of the Gaussian Mixture Models (GMMs), which were attached to the states, were determined empirically. Additionally, two different model architectures were evaluated. To improve the over-all performance further, two approaches to fuse the results of the twelve individual models were developed.

The rest of the paper is organized as follows: Section 2 provides a brief overview of the HMMs, GMMs and the two classifier fusion approaches. In Sect. 3 the images and the extraction of the features are described. The experiments and the results will be introduced in Sect. 4. Section 5 gives a brief summary of the paper.

2   Stochastic and Functional Principles

2.1   Hidden Markov Models

A Hidden Markov Model (HMM) $\lambda = (Z, V, \pi, A, E)$ is a statistical model which is composed of two random processes [8][16]. The first process is a Markov chain consisting of a defined number of states $Z = (z_1, ..z_n)$ and corresponding state transition probabilities which form the transition matrix $A = (a_{ij})$. The probability a_{ij} designates the probability to change from state z_i to state z_j. The third component of this process is the initial probability vector π which defines the probabilities of the states to be the initial state. The second random process defines the output: it consists of possible observations $V = (v_1 .. v_m)$ and the observation matrix $E = \{e_j(k)\}$ storing the probabilities $e_j(k)$ of observation v_k being produced from the state z_j. The sequence of observations provides information about the sequence of the hidden states.

The topology of the transition matrix defines the structure of the model. A connection between two states z_i and z_j is given, if the corresponding entry a_{ij} is greater than 0. The following two models were utilized in this study: *fully connected model* with $a_{ij} > 0 \; \forall \; i, j$ and a *forward model* with $a_{ij} = 0$ for $i > j$ and $j > i + 1$, i.e. only connections to the state itself and to the following state are allowed.

One property of the HMM is that the next state only depends on the current state. The values at time $t - i$, $i > 1$ have no influence. This is called the Markov property, e.g. for a sequence of states $Q = q_1 ... q_L, q_i \in Z$:

$$P(q_{t+1} = z_j \mid q_t = z_i, q_{t-1} = z_k, ...) = P(q_{t+1} = z_j \mid q_t = z_i) \; . \qquad (1)$$

There are three basic problems associated with HMMs [16]. Each of it can be solved with a specific dynamic programming algorithm:

Decoding Problem. Given the parameters of the model $\lambda = (Z, V, \pi, A, E)$ and a observed output sequence $O = O_1, ..., O_L$. Evaluate the most likely state sequence which could have generated the output sequence.
Solution: *Viterbi algorithm.*

Evaluation Problem. Given the parameters of the model $\lambda = (Z, V, \pi, A, E)$. Compute the probability of an observed output sequence $O = O_1, ..., O_L$.
Solution: *forward algorithm.*

Learning Problem. Given a set of output sequences $O^1, ..., O^L$ and the structure of the model $\lambda = (Z, V)$. Determine the transition matrix A, the observation matrix E and the initial probability vector π so that the probability for this HMM, producing $O^1, ..., O^L$, is the maximum value.
Solution: *Baum-Welch algorithm*[1].

The HMMs were utilized for classification. Therefore, one HMM λ_i, $i = 1, ..., n$ for each emotion class was trained using data of one class. The probabilities $P(O \mid \lambda_i)$ for an unclassified observation O and the i-th HMM were estimated with the forward algorithm. For numerical reasons, the logarithm of the probabilities was used in the computation because they could become very small. The maximum of the n achieved values $P(O \mid \lambda_i)$, one for each HMM, lead to the most likely class. This architecture is called a log-likelihood estimator.

2.2   Gaussian Mixture Models

A Gaussian mixture model (GMM) $g(f_1, ..., f_m)$ is a probabilistic model for estimation of probability density functions [3]. It combines m Gaussian distributions $f_1, ..., f_m$. Each distribution f_i is defined as $f_i = (\mu_i, \Sigma_i, \alpha_i)$. The value μ_i is the expected value and thus the center of the normal distribution. The value Σ_i defines the covariance matrix of the i-th distribution. The third value α_i stands for the weight of the i-th distribution in the probability density function with $\sum_{i=1}^{m} \alpha_i = 1$.

The individual distributions f_i are assumed to be stochastically independent and defined as followed:

$$f_i(x \mid \mu_i, \Sigma_i) = \frac{1}{\sqrt{(2\pi)^{N/2} \mid \Sigma_i \mid}} exp\left(-\frac{1}{2}(x - \mu_i)^T \Sigma_i^{-1}(x - \mu_i) \right) . \qquad (2)$$

This defines the total probability $P(X \mid f_1, ..., f_m)$ with:

$$P(X \mid f_1, ..., f_m) = \sum_{i=1}^{m} \alpha_i f_i(X, \mu_i, \Sigma_i) . \qquad (3)$$

In this study, one GMM in each state of the HMM was utilized to define the observation probabilities E, which were mentioned in Sect. 2.1. It should be

---

[1] The Baum-Welch algorithm is an instance of the expectation-maximization algorithm.

noted that GMMs cannot take temporal dependencies into account as HMMs do. The parameters of the GMM are trained with an expectation maximization algorithm.

2.3   Classifier Fusion

Combining classifiers is a promising approach to improve classifier results. A team of classifiers, which is intended to be fused, needs to be accurate and diverse [11]. While accuracy of a classifier is clear, diversity means that if a sample is classified falsely, not all classifiers should agree on a wrong class-label. In this paper different feature views on the data are constructed to produce diversity and two very simple but intuitive fusion techniques were evaluated.

Vote-Fusion. The results of the log-likelihood estimators lead at best all to the same class, but normally they are different. To nullify this problem, the first fusion method vote the results of selected log-likelihood estimators.

Probability-Fusion. The other fusion approach did not combine the results but the probabilities $P(O \mid \lambda_i)$ of selected log-likelihood estimators. To combine several estimators, the class-wise sums are computed. A summation of the logarithmic values is equivalent to a multiplication of the probabilities. Only then is the maximum determined to obtain the most likely class. This implies statistic independence of the models, which is unfortunately not fully given since the features were generated from the same sequence.

3   Data Collection

The Cohn-Kanade dataset is a collection of image sequences with emotional content [5], which is available for research purposes. It contains image sequences which were recorded in a resolution of 640×480 (sometimes 490) pixels with a temporal resolution of 33 frames per second. Every sequence is played by an amateur actor who is filmed from a frontal view. The sequences always start with a neutral facial expression and end with the full blown emotion which is one of the six categories "fear", "joy", "sadness", "disgust", "surprise" or "anger".

To acquire a suitable label, the sequences were presented to 15 human labelers (13 male and two female). The sequences were presented as a video. After the play-back of a video the last image remained on the screen and the test person was asked to select a label. Thus, a label for every sequence was created as the majority vote of the 15 different opinions. The result of the labeling procedure is given in Table 1 showing the confusion matrix of the test persons according to the majority of all persons. The resulting data collection showed to be highly imbalanced: the class "joy" (105 samples) occurred four times more often than the class "fear" (25 samples) and in addition, this expression could not be identified by the test persons.

In all automatic facial expression recognition systems some relevant features are extracted from the facial image first and these feature vectors are utilized to train some type of classifier to recognize the facial expression. One problem is here how to categorize the emotions: one way is to model emotions through

Table 1. Confusion matrix of the human test persons against the majority of all 15 votes (left). The right column shows the share of the facial expressions in the data set (hardened class labels).

maj.\test pers.	joy	ang.	sur.	disg.	sad.	fear	no. samples
joy	**104**	0	0	0	0	1	105
ang.	0	**39**	0	6	3	1	49
sur.	1	0	**72**	0	1	17	91
disg.	1	12	1	**54**	1	12	81
sad.	0	6	2	2	**70**	1	81
fear	1	1	3	6	1	**13**	25

a finite set of emotional classes such as anger, joy, sadness, etc, another way is to model emotions using continuous scales such as valence (the pleasantness of the emotion) and arousal (the level of activity) of an expression [12]. In this paper, a discrete representation in six emotions was used. Finding the most relevant features is definitely the most important step in designing a recognition system. In our approach, prominent facial regions such as the eyes, including the eyebrows, the mouth and for comparison the full facial region have been considered. For these four regions orientation histograms, principal components and optical flow features have been computed. Principal components (eigenfaces approach) are very well known in face recognition [19] and orientation histograms were successfully applied for the recognition of hand gestures [10] and faces [18], both on single images. In order to extract the facial motion in these regions, optical flow[2] features from pairs of consecutive images have been computed, as suggested in [17].

4   Experiments and Results

In this section, experiments which concern both the number of the distributions and the architecture of the HMMs are presented. The results of the achieved single models will be shown and the results of the two different fusion methods will be demonstrated.

4.1   Adjusting the Number of Gaussian-Components

We started with simple models with only one state and mixture. By doing so, a model is constructed which neglects any sequential dependencies. Thus, the transition matrix was just a vector. The results of the twelve feature-region pairs with this plain model are shown in Table 2. The next step was to improve the results by optimizing the models.

Numerical test runs were conducted with 10-fold cross-validation. The number of states were evaluated from 1 to 9 and the number of mixture components

---

[2] We were using a biologically inspired optical flow estimator which was developed by the Vision and Perception Science Lab of the Institute of Neural Processing at the University of Ulm [1,2].

from 1 to 4. The transition matrices were fully connected. The results of this process are shown in Table 2. The adjustments on the number of distributions led to higher detection rates than they were in the plain model. The experiments showed that by increasing the number of states, the corresponding number of mixture components per state became smaller. We determined that a total number of 8-20 for all distributions is sufficient for this application. One negative effect of a high number of distributions was, that some emotions were recognized very poorly: since the more states and mixtures were chosen, the more parameters had to be estimated. There were a few emotions with only little data (see Table 1), so the fine-tuning could not be adjusted well enough and it led to overfitting. Additionally, there was no model which was optimal for all emotions. The emotion "joy" had the best recognition rate with PCA-features in the region face, whereas the emotion "anger" was best recognized in the region of the left eye and features with optical flow estimation. The best results for the different emotions and the corresponding models are shown in Table 3.

Table 2. Results for expression concerning the number of states, the corresponding number of mixtures per state and the detection rate for all twelve models. In case of more than one state, the transition matrix was fully connected.

no.	feature	region	no. st/mix	det. rate	no. st/mix	det. rate
1	PCA	face	1 / 1	0.727	9 / 2	0.742
2	PCA	mouth	1 / 1	0.641	8 / 2	0.671
3	PCA	right eye	1 / 1	0.421	7 / 2	0.444
4	PCA	left eye	1 / 1	0.472	3 / 2	0.479
5	Orientation histograms	face	1 / 1	0.678	4 / 2	0.710
6	Orientation histograms	mouth	1 / 1	0.678	4 / 3	0.732
7	Orientation histograms	right eye	1 / 1	0.440	4 / 2	0.473
8	Orientation histograms	left eye	1 / 1	0.417	9 / 2	0.475
9	Optical flow	face	1 / 1	0.627	8 / 2	0.638
10	Optical flow	mouth	1 / 1	0.639	9 / 2	0.646
11	Optical flow	right eye	1 / 1	0.437	7 / 3	0.471
12	Optical flow	left eye	1 / 1	0.449	8 / 4	0.491

Table 3. Best recognition rate and corresponding model for each emotion

emotion	det. rate	feature	region	no. st/mix
joy	0.981	PCA	face	9 / 2
anger	0.816	Optical flow	left eye	8 / 4
surprise	0.978	Orientation histograms	mouth	4 / 3
disgust	0.593	Optical flow	face	8 / 2
sadness	0.889	Orientation histograms	mouth	4 / 3
fear	0.200	PCA	mouth	8 / 2

4.2   Investigating Temporal Dependence

In this subsection, the temporal properties of the learned models and the underlying data is studied. To study the influence of the topology of the HMM to the classifier, fully connected models were compared with forward models (see Sect. 2.1). It was discovered that the transition matrix of the fully connected model approximated the transition matrix of the forward model during training. Since the detection rate of the forward model was higher, we concluded that the forward model can model the time dependence better in this case. By using the forward model, also computation time can be saved because not all possible entries in the transition matrix have to be adapted. Another advantage of the forward model is the avoidance of local minima. The fully connected model has a higher risk to descend toward a local minimum [8].

The expectation values of the GMMs (see Sect. 2.2) form the centers of the normal Gaussian distributions. For initialization, the k-means procedure was used to ensure that the centers cover the entire feature space. To investigate whether the centers are different from each other after the training, the pairwise distances of the expected values were calculated. If the observed probabilities in the states are equal, using HMMs does not provide any benefit. Figure 1 shows the differences of an example with four states and two mixtures per state as grayscale values. It can be observed, that both, the distances between centers in different states as well as centers within a state, differ. Thus, there is a corresponding difference between the data assigned to the respective states. This can be demonstrated more descriptive by producing average images of each state. Figure 2 shows average images of the emotion "joy" made from 103 sequences. The images were assigned to the states using the Viterbi algorithm.

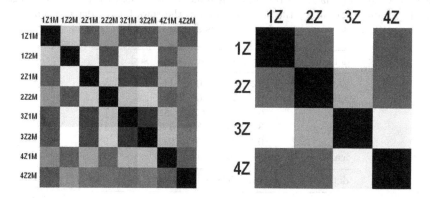

Fig. 1. Pairwise distances of the centers of the Gaussian distributions as grayscale values: the brighter the grayscale value, the greater the distance. On the diagonal are the distances to itself (no distance = black). (iZjM designates the i-th mixture M of the j-th state Z).

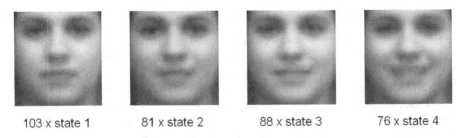

103 x state 1   81 x state 2   88 x state 3   76 x state 4

Fig. 2. Average pictures for each state. A four state model of the emotion "joy" was utilized and the paths of 103 sequences were computed.

4.3   Results of the Single Models

Based on the considerations above, the individual models were developed. Forward models with the optimized configurations for the number of states and mixture components were used (see Table 2). The best single model with a total detection rate of 76.4 % was the model with PCA-features in the region face. Table 4 shows the confusion matrix of this model. The emotions "joy", "anger" and "disgust" were often confused. This occured particularly in the eye-regions. One reason for this could be that for these three emotions either the eyes have to be pinched together or the cheek muscles goes up because of a smile. However, with "surprise", "sadness" and "fear" the eyes are open. The emotion "fear" was often confused with the emotion "surprise", in which the greatest confusion occurs in the region mouth. The open mouth of the two emotions might play a role here.

Table 4. Confusion matrix of the best single model: features with PCA of the region face. The columns stand for the labeled emotion, the rows for the detected emotions. The recognition rate of each single emotion is shown in the last column.

det.emo\emo.	joy	ang.	sur.	disg.	sad.	fear
joy	**102**	4	0	11	2	5
ang.	0	**26**	0	8	6	0
sur.	0	0	**82**	2	3	7
disg.	3	9	3	**47**	1	3
sad.	0	10	5	8	**68**	5
fear	0	0	1	5	1	**5**
det.rate	0.981	0.531	0.901	0.580	0.840	0.200

4.4   Results of the Fused Models

To improve the total detection rate even further, the results of the twelve individual models were fused. For a better overview, the individual models were coded according to the number of the rows of Table 2. For instance, model 1

Table 5. Top ten combinations with vote-fusion and probability-fusion of the developed twelve single models. The coding of the models corresponds to the first column in Table 2.

vote-fusion		probability-fusion	
no. of models	det. rate	no. of models	det. rate
1 2 5 6 9 10	0.817	1 2 5 6 7 9 10 11	0.861
1 2 5 6 9	0.815	1 2 3 5 6 9 10 12	0.859
1 2 6 8 9 10	0.815	1 2 3 5 6 9 10 12	0.859
1 2 6 9 10 11	0.815	1 2 5 6 7 9 10 12	0.859
1 2 4 5 6 9 10	0.815	1 5 6 9	0.857
1 2 6 7 9 10	0.813	2 5 6 9 10 12	0.857
1 5 6 9 10 12	0.813	1 2 4 5 6 7 9	0.857
1 2 6 9 12	0.810	1 2 4 5 6 9 11	0.857
1 2 5 6 8 9 10	0.810	1 5 6 9 10	0.854
1 2 5 6 10 11 12	0.810	1 2 4 5 6 10 12	0.854

recognized the emotion "joy" with 97.1 %, while it reached less than 25 % for the emotion "anger". On the other hand, model 12 showed different results: for emotion "anger" a detection rate of 63.3 % was obtained, but the detection rate for "joy" was lower than with model 1. To combine the advantages of these models and neutralize the disadvantages as possible, the decisions of the models were fused. In the following, the results of the two fusion approaches will be presented:

Vote-Fusion. A simple fusion approach is to vote the detected emotions of all twelve classifiers. If a test sequence is classified incorrectly from one model, but the other eleven models detect it correctly, it will be outvoted. After the training of the individual classifiers, we investigated whether the classification rate could be improved, if not all twelve models would be combined. Therefore, all 2^{12} model combinations were computed. The results of the top ten combinations are shown in Table 5. The appearance of the models 1 and 6 in all displayed combinations is originated from their good individual recognition performance. Models using the eye regions occur only sporadically in this list. This shows that these models could reduce the classification rate because of their minor accuracy. The combination of all face and mouth regions led to a detection rate of 81.7 %. This means that, compared to the best single model, this combination has the ability to recognize 23 sequences more.

Probability-Fusion. This approach, as mentioned in Sect. 2.3, combines the logarithmic probabilities which are obtained from the six HMMs of each log-likelihood estimator. In many cases, when a sample was misclassified, the probability of the incorrect emotion was just slightly higher than the one of the correct emotion. The class-wise multiplication of the posterior probabilities has the potential to overcome the limitations of the early hardening in the vote-fusion. The results are also shown in Table 5. Again, the face and mouth regions played an important role. The result, showing the highest classification rate of 86.1 % (372 out of 432), consisted of all models using the features of the regions face and mouth and the models 7 and 11. Table 6 shows the confusion matrix. This was the highest achieved recognition rate in this study. Thus, the probability-fusion could recognize 19 sequences more than the vote-fusion.

Table 6. Confusion matrix of the best result, obtained with probability-fusion: combination of all models with features of the regions face and mouth and the two models with orientation histograms and optical flow features in the right eye region. Total detection rate: 86.1 %.

det.emo\emo.	joy	ang.	sur.	disg.	sad.	fear
joy	**105**	0	0	11	2	4
ang.	0	**36**	0	3	2	0
sur.	0	1	**89**	1	0	8
disg.	0	7	1	**65**	1	10
sad.	0	5	1	1	**76**	2
fear	0	0	0	0	0	**1**
det. rate	1.000	0.735	0.978	0.803	0.938	0.040

5   Summary

The aim of this study was to investigate the behavior of HMMs in facial expression recognition and to demonstrate whether HMMs are capable of recognizing emotions in image sequences sufficiently well.

432 image sequences from the Cohn-Kanade database were used as dataset. Four different regions in each image were extracted (face, mouth, right and left eye) and three different methods for feature extraction were applied (PCA, orientation histograms and optical flow estimation). The data of each combination of region and feature was utilized to train a log-likelihood estimator which consisted of six HMMs (one for each emotion). The emotions were: "joy", "anger", "surprise", "disgust", "sadness" and "fear".

Not only the detection performance but also the architecture and the learning behavior was investigated. First, experiments with different number of states and mixtures per state were performed. It was observed that by increasing the number of states, the number of Gaussian distributions had to be reduced to achieve a good classification performance. As a result of this investigation, twelve single models for the feature-region pairs were identified. Also, experiments with the transition matrix, i.e. the structure of the HMMs, were conducted. We discovered that the forward models were slightly better than the fully connected models. The best single model with a detection rate of 76.9 % was the forward model with PCA-features in the region face (see Table 4). This model classified 332 out of 432 sequences correctly.

Furthermore, the outputs of the individual models were fused to combine the advantages of the models and possibly neutralize the shortcomings. Two fusion approaches were evaluated: vote-fusion and probability-fusion. These fusion approaches showed that the classification rates can be improved further by combining the results of the different models. The best fusion result with a classification rate of 86.1 % is shown in Sect. 4.4, including a confusion matrix in Table 6. This was the highest achieved recognition rate with 372 out of 432 detected sequences.

The experiments and results show that HMMs are very well capable of emotion recognition in image sequences. Time dependencies can simply be modeled by HMMs and the experiments also show that they play an important role.

Acknowledgments

The authors would like to thank Prof. Dr. H. Neumann, Dr. P. Bayerl and S. Ringbauer from the Vision and Perception Science Lab of the Institute of Neural Processing at the University of Ulm for generous assistance in extracting the optical flow features.

This paper is based on work done within the project SCHW623/4-3, and the "Information Fusion" subproject of the Transregional Collaborative Research Center SFB/TRR 62 "Companion-Technology for Cognitive Technical Systems", both funded by the German Research Foundation (DFG). The work of Martin Schels is supported by a scholarship of the Carl-Zeiss Foundation.

References

1. Bayerl, P., Neumann, H.: Disambiguating visual motion through contextual feedback modulation. Neural Computation 16, 2041–2066 (2004)
2. Bayerl, P., Neumann, H.: A fast biologically inspired algorithm for recurrent motion estimation. IEEE Transactions on Pattern Analysis and Machine Intelligence 29, 246–260 (2007)
3. Bishop, C.M.: Pattern Recognition and Machine Learning. Springer, New York (2007)
4. Cohen, I., Garg, A., Huang, T.S.: Emotion recognition from facial expressions using multilevel HMM. In: Neural Information Processing Systems (2000)
5. Cohn, J.F., Kanade, T., Tian, Y.: Comprehensive database for facial expression analysis. In: Proceedings of the 4th IEEE International Conference on Automatic Face and Gesture Recognition, pp. 46–53 (2000)
6. Darwin, C.: The Expression of the Emotions in Man and Animals, 1st edn. Oxford University Press Inc., New York (1872)
7. Lin, D.T., Chen, J.: Facial expressions classification with hierarchical radial basis function networks. In: Proceedings of the 6th International Conference on Neural Information Processing, ICONIP, pp. 1202–1207 (1999)
8. Durbin, R., Eddy, S., Krogh, A., Mitchison, G.: Biological Sequence Analysis: Probabilistic Models of Proteins and Nucleic Acids. Cambridge University Press, Cambridge (1998)

9. Ekman, P., Friesen, W.V.: Facial Action Coding System: A Technique for the Measurement of Facial Movement. Consulting Psychologists Press, Palo Alto (1978)
10. Freeman, W.T., Roth, M.: Orientation histograms for hand gesture recognition. In: International Workshop on Automatic Face and Gesture Recognition, pp. 296–301 (1994)
11. Kuncheva, L.I., Whitaker, C.J.: Measures of diversity in classifier ensembles and their relationship with the ensemble accuracy. Mach. Learn. 51(2), 181–207 (2003)
12. Lang, P.J.: The emotion probe. studies of motivation and attention. The American psychologist 50(5), 372–385 (1995)
13. Lien, J.J.J., Kanade, T., Cohn, J., Li, C.: A multi-method approach for discriminating between similar facial expressions, including expression intensity estimation. In: Proceedings of the IEEE Conference on Computer Vision and Pattern Recognition (CVPR 1998) (June 1998)
14. Lisetti, C.L., Rumelhart, D.E.: Facial expression recognition using a neural network. In: Proceedings of the Eleventh International Florida Artificial Intelligence Research Society Conference, pp. 328–332. AAAI Press, Menlo Park (1998)
15. Rabiner, L., Juang, B.H.: Fundamentals of Speech Recognition. Prentice Hall PTR, Englewood Cliffs (1993)
16. Rabiner, L.R.: A tutorial on hidden markov models and selected applications in speech recognition. In: Proceedings of the IEEE, pp. 257–286 (1989)
17. Rosenblum, M., Yacoob, Y., Davis, L.: Human expression recognition from motion using a radial basis function network architecture. IEEE Transactions on Neural Networks 7(5), 1121–1138 (1996)
18. Schwenker, F., Sachs, A., Palm, G., Kestler, H.A.: Orientation histograms for face recognition. In: Schwenker, F., Marinai, S. (eds.) ANNPR 2006. LNCS (LNAI), vol. 4087, pp. 253–259. Springer, Heidelberg (2006)
19. Turk, M., Pentland, A.: Eigenfaces for recognition. Journal of Cognitive Neuroscience 3(1), 71–86 (1991)
20. Yeasin, M., Bullot, B., Sharma, R.: From facial expression to level of interest: A spatio-temporal approach. In: IEEE Computer Society Conference on Computer Vision and Pattern Recognition, vol. 2, pp. 922–927 (2004)

Analysis, Interpretation, and Recognition of Facial Action Units and Expressions Using Neuro-Fuzzy Modeling

Mahmoud Khademi[1], Mohammad Hadi Kiapour[2], Mohammad T. Manzuri-Shalmani[1], and Ali A. Kiaei[1]

[1] DSP Lab, Sharif University of Technology, Tehran, Iran
[2] Institute for Studies in Fundamental Sciences (IPM), Tehran, Iran
{khademi@ce.,kiapour@ee.,manzuri@,kiaei@ce.}sharif.edu

Abstract. In this paper an accurate real-time sequence-based system for representation, recognition, interpretation, and analysis of the facial action units (AUs) and expressions is presented. Our system has the following characteristics: 1) employing adaptive-network-based fuzzy inference systems (ANFIS) and temporal information, we developed a classification scheme based on neuro-fuzzy modeling of the AU intensity, which is robust to intensity variations, 2) using both geometric and appearance-based features, and applying efficient dimension reduction techniques, our system is robust to illumination changes and it can represent the subtle changes as well as temporal information involved in formation of the facial expressions, and 3) by continuous values of intensity and employing top-down hierarchical rule-based classifiers, we can develop accurate human-interpretable AU-to-expression converters. Extensive experiments on Cohn-Kanade database show the superiority of the proposed method, in comparison with support vector machines, hidden Markov models, and neural network classifiers.

Keywords: biased discriminant analysis (BDA), classifier design and evaluation, facial action units (AUs), hybrid learning, neuro-fuzzy modeling.

1 Introduction

Human face-to-face communication is a standard of perfection for developing a natural, robust, effective and flexible multi modal/media human-computer interface due to multimodality and multiplicity of its communication channels. In this type of communication, the facial expressions constitute the main modality [1]. In this regard, automatic facial expression analysis can use the facial signals as a new modality and it causes the interaction between human and computer more robust and flexible. Moreover, automatic facial expression analysis can be used in other areas such as lie detection, neurology, intelligent environments and clinical psychology.

Facial expression analysis includes both measurement of facial motion (e.g. mouth stretch or outer brow raiser) and recognition of expression (e.g. surprise or anger). Real-time fully automatic facial expression analysis is a challenging complex topic in computer vision due to pose variations, illumination variations, different age, gender,

F. Schwenker and N. El Gayar (Eds.): ANNPR 2010, LNAI 5998, pp. 161–172, 2010.
© Springer-Verlag Berlin Heidelberg 2010

ethnicity, facial hair, occlusion, head motions, and lower intensity of expressions. Two survey papers summarized the work of facial expression analysis before year 1999 [2, 3]. Regardless of the face detection stage, a typical automatic facial expression analysis system consists of facial expression data extraction and facial expression classification stages. Facial feature processing may happen either holistically, where the face is processed as a whole, or locally. Holistic feature extraction methods are good at determining prevalent facial expressions, whereas local methods are able to detect subtle changes in small areas.

There are mainly two approaches for facial data extraction: geometric-based methods and appearance-based methods. The geometric facial features present the shape and locations of facial components. With appearance-based methods, image filters, e.g. Gabor wavelets, are applied to either the whole face or specific regions in a face image to extract a feature vector [4].

The sequence-based recognition method uses the temporal information of the sequences (typically from natural face towards the frame with maximum intensity) to recognize the expressions. To use the temporal information, the techniques such as hidden Markov models (HMMs) [5], recurrent neural networks [6] and rule-based classifier [7] were applied.

The facial action coding system (FACS) is a system developed by Ekman and Friesen [8] to detect subtle changes in facial features. The FACS is composed of 44 facial action units (AUs). 30 AUs of them are related to movement of a specific set of facial muscles: 12 for upper face (e.g. AU 1 inner brow raiser, AU 2 outer brow raiser, AU 4 brow lowerer, AU 5 upper lid raiser, AU 6 cheek raiser, AU 7 lid tightener) and 18 for lower face (e.g. AU 9 nose wrinkle, AU 10 upper lip raiser, AU 12 lip corner puller, AU 15 lip corner depressor, AU 17 chin raiser, AU 20 lip stretcher, AU 23 lip tightener, AU 24 lip pressor, AU 25 lips part, AU 26 jaw drop, AU 27 mouth stretch). Facial action units can occur in combinations and vary in intensity. Although the number of single action units is relatively small, more than 7000 different AU combinations have been observed. To capture such subtlety of human emotion paralinguistic communication, automated recognition of fine-grained changes in facial expression is required (for more details see [8, 9]).

The main goal of this paper is developing an accurate real-time sequence-based system for representation, recognition, interpretation, and analysis of the facial action units (AUs) and expressions. We summarize the advantages of our system as follows:

1) The facial action unit intensity is intrinsically fuzzy. We developed a classification scheme based on neuro-fuzzy modeling of the AU intensity, which is robust to intensity variations. Applying this accurate method, we can recognize lower intensity and combinations of AUs.
2) Recent work suggests that spontaneous and deliberate facial expressions may be discriminated in term of timing parameters. Employing temporal information instead of using only the last frame, we can represent these parameters properly.
3) By using both geometric and appearance features, we can increase the recognition rate and also make the system robust against illumination changes.
4) By employing top-down hierarchical rule-based classifiers such as J48, we can automatically extract human interpretable classification rules to interpret each expression.
5) Due to the relatively low computational cost, the proposed system is suitable for real-time applications.

The rest of the paper has been organized as follows: In section 2, we describe the approach which is used for facial data extraction and representation using both geometric and appearance features. Then, we discuss the proposed scheme for recognition of facial action units and expressions in section 3 and section 4 respectively. Section 5 reports our experimental results, and section 6 presents conclusions and a discussion.

2 Facial Data Extraction and Representation

2.1 Biased Learning

Biased learning is a learning problem in which there are an unknown number of classes but we are only interested in one class. This class is called "positive" class. Other samples are considered as "negative" samples. In fact, these samples can come from an uncertain number of classes. Suppose $\{x_i | i = 1, ..., N_x\}$ and $\{y_i | i = 1, ..., N_y\}$ are the set of positive and negative d-dimensional samples (feature vectors) respectively. Consider the problem of finding d × r transformation matrix w (r ≪ d), such that separates projected positive samples from projected negatives in the new subspace. The dimension reduction methods like fisher discriminant analysis (FDA) and multiple discriminant analysis have addressed this problem simply as a two-class classification problem with symmetric treatment on positive and negative examples. For example in FDA, the goal is to find a subspace in which the ratio of between-class scatter over within-class scatter matrices is maximized. However, it is part of the objective function that negative samples shall cluster in the discriminative subspace. This is an unnecessary and potentially damaging requirement because very likely the negative samples belong to multiple classes. In fact, any constraint put on negative samples other than stay away from the positives is unnecessary and misleading. With asymmetric treatment toward the positive samples, Zhou and Huang [10] proposed the following objective function:

$$w_{opt} = \arg\max_w \frac{\text{trace}(w^T S_y w)}{\text{trace}(w^T S_x w)} \tag{1}$$

where S_y and S_x are within-class scatter matrices of negative and positive samples with respect to positive centroid, respectively. The goal is to find w that clusters only positive samples while keeping negatives away. The problem of finding optimal w, becomes finding the generalized eigenvectors α's associated with the largest eigenvalues λ's in the below generalized eigenanalysis problem:

$$S_y \alpha = \lambda S_x \alpha \tag{2}$$

Our goal is developing a facial action unit recognition system that can detect whether the AUs occur or not. The input of the system is a sequence of frames from natural face towards one of the facial expressions with maximum intensity. Suppose we have extracted a feature matrix or a feature vector from each frame. In order to embed facial features in a low-dimensionality space and deal with curse of dimensionality dilemma, we should use a dimension reduction method. For recognition of each AU, we are facing an asymmetric two-class classification problem. For example when the goal is detecting whether AU 27 (mouth stretch) occur or not, the positive class includes all of sequences in the train set that represent stretching of the mouth; other sequences are considered as negative samples. These samples can come from an uncertain number of classes. They can represent any single AU or AU combinations except AU 27. In fact, our problem is a biased learning problem.

2.2 Appearance-Based Facial Feature Extraction Using Gabor Wavelets

In order to extract the appearance-based facial features from each frame, we use a set of Gabor wavelets. They allow detecting line endings and edge borders of each frame over multiple scales and with different orientations. Gabor wavelets remove also most of the variability in images that occur due to lighting changes [4]. Each frame is convolved with p wavelets to form the Gabor representation of the t frames (Fig. 1).

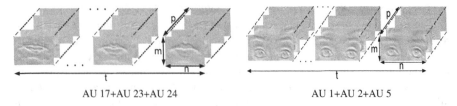

AU 17+AU 23+AU 24   AU 1+AU 2+AU 5

Fig. 1. Examples of the image sequences and their representation using Gabor wavelets

However, for applying the Zhou and Huang's method, which is called biased discriminant analysis (BDA), to the facial action unit recognition problem we should first transform the feature matrices of the sequence into a one-dimensional vector that ignores the underlying data structure (temporal and local information) and leads to the curse of dimensionality dilemma and the small sample size problem. Thus, we use two-dimension version of BDA algorithm by simply replacing the image vector with image matrix in computing the corresponding variance matrices to reduce the dimensionality of each feature matrix in two directions [11]. Then, we apply BDA algorithm to the vectorized representation of the reduced feature matrices. Also, In order to deal with singularity in the matrices we use 2D and 1D principle component analysis (PCA) algorithms [12], before applying 2DBDA and BDA respectively.

2.3 Geometric-Based Facial Feature Extraction Using Optical Flow

In order to extract geometric features we use a facial feature extraction method presented in [13]. The points of a 113-point grid, which is called Wincanide-3, are placed on the first frame manually. Automatic registering of the grid with the face has been addressed in many literatures (e.g. see [14]). For upper face and lower face action units a particular set of points are selected. The pyramidal optical flow tracker [15] is employed to track the points of the model in the successive frames towards the last frame (see Fig. 2). The loss of the tracked points is handled through a model deformation procedure (for detail see [13]). For each frame, the displacements of the points in two directions with respect to the first frame are calculated and placed in the columns of a matrix. Then, we apply 2DBDA algorithm [11] to the matrix in two directions. The vectorized representation of the reduced feature matrix is used as geometric feature vector.

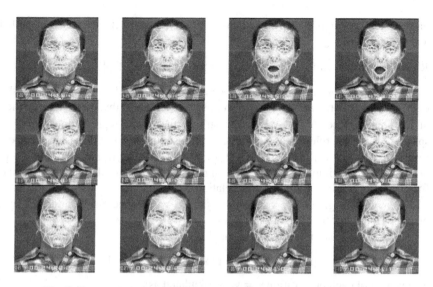

Fig. 2. Geometric-based facial feature extraction using grid tracking

3   Facial Action Unit Recognition Using Neuro-Fuzzy Modeling

3.1   Takagi-Sugeno Fuzzy Inference System and Training Data Set

The flow diagram of the proposed system is shown in Fig. 3. In order to recognize each single AU we construct a fuzzy rule-based system. The reduced feature vector is used as the input of the system.

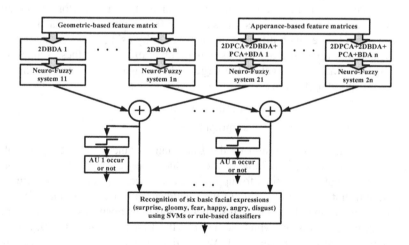

Fig. 3. Block diagram of the proposed system (n is number of the facial action units)

Each system is composed of n Takagi-Sugeno type fuzzy if-then rules of the below format:

R^i: if x_1 is A_1^i and ... and x_k is A_k^i then $y = p_0^i + p_1^i x_1 + \cdots + p_k^i x_k$ $(i = 1, ..., n)$

Here, y is variable of the consequence whose value is the AU intensity and we should infer it. $x_1, ..., x_k$ are variables of the premise, i.e. features, that appear also in the part of the consequence. $A_1^i, ..., A_k^i (i = 1, ..., n)$ are fuzzy sets representing a fuzzy subspace in which the rule R^i can be applied for reasoning (we use Gaussian membership function with two parameters), and $p_0^i, p_1^i ..., p_k^i$ $(i = 1, ..., n)$ are consequence parameters. The fuzzy implication is based on a fuzzy partition of the input space. In each fuzzy subspace, a linear input-output relation is formed.

When we are given $(x_1 = x_1^0, x_2 = x_2^0, ..., x_k = x_k^0)$, the fuzzy inference system produces output of the system as follows:

For each implication R^i, y_i is calculated as:

$$y_i = p_0^i + p_1^i x_1^0 + \cdots + p_k^i x_k^0 \qquad i = 1, ..., n \qquad (3)$$

The weight of each proposition $y = y_i$ is calculated as:

$$w_i = A_1^i(x_1^0) \times ... \times A_k^i(x_k^0) \qquad i = 1, ..., n \qquad (4)$$

Then, the final output y inferred from n rules is given as the average of all y_i ($i = 1, ..., n$) with the weights w_i, i.e. $y = (\sum_{i=1}^n y_i w_i)/(\sum_{i=1}^n w_i)$.

For modeling each system, i.e. each single AU, we should extract the rules using training data. Depending on the AU that we want to model it, the sequences in the training set are divided into two subsets: positive set and negative set. The positive set includes all of sequences that the AU occurs on them; other sequences are placed in the negative set. Then, by applying the method discussed in section 2, we extract the feature vector for each sequence. Assuming the last frames of the sequences in the training set have maximum intensity, the target value of feature vectors which are in the positive set and negative set is labeled 1 and 0 respectively. We also use some sequences several times with different intensities, i.e. by using intermediate frames as the last frame and removing the frames which come after it. For these sequences, if the original sequence was negative the target value is again 0. Otherwise, the target value of corresponding feature vector for produced sequence is calculated by:

$$t = \frac{\text{sumdistances}_1}{\text{sumdistances}_2} \qquad (5)$$

where sumdistances$_1$ is the sum of the Euclidian distances between point of the Wincandide-3 grid in the last frame of the produced sequence and their positions in the first frame (a subset of points for upper face and lower face action units are used). Similarly, sumdistances$_2$ is the sum of the Euclidean distances between points of the model in the last frame of the original sequence and their positions in the first frame; e.g. if we remove all frames which come after the first frame then $t = 0$, and if we remove none of the frames then $t = 1$. We model each single AU two times, using geometric and appearance features separately. In test phase, the outputs are added and the result is passed through a threshold. When several outputs were on, it signals that a combination of AUs has been occurred.

Modeling of each system, i.e. each single AU, is composed of two parts: structure identification and parameter identification. The structure identification relate to partition of the input space, i.e. number of rules. In parameter identification process the premise parameters and consequence parameters are determined.

3.2 Structure Identification Algorithm

We use some of the training samples as the validation set. This set is used for avoiding the problem of overfitting data in the modeling process. Suppose $x_1, x_2, ..., x_k$ are the inputs of the system. Moreover, suppose d_i is the number of divided fuzzy subspace for x_i. The initial value of d_i ($i = 1, ..., k$) is 1, because at first the range of each variable is undivided. Also, let V, i.e. the value of mean squares of errors of the model on validation set be a big number. The algorithm of modeling is as follows:

1) The range of x_1 is divided into one more fuzzy subspace (e.g. "big" and "small" if $d_1 = 1$ or "big", "medium" and "small" if $d_1 = 2$) and the range of the other variables $x_2, x_3, ..., x_k$ are not more divided. This model is called model 1; e.g. in the first iteration, model 1 consisting of two rules of the below format:

$$R^1: \text{if } x_1 \text{ is big then } y = p_0^1 + p_1^1 x_1$$

$$R^2: \text{if } x_1 \text{ is samll then } y = p_0^2 + p_1^2 x_1$$

Similarly the model in which the range of x_2 is divided in to one more subspace and the rage of other variables $x_1, x_3, ..., x_k$ are not more divided, is called mode 2. In this way we have k models.

2) For each model the optimum premise parameters (mean and variance of the membership functions) and consequence parameters are found by the parameter identification algorithm described in the next subsection. This algorithm applies hybrid learning, to determine the premise and consequence parameters.

3) For each model, the mean squares of errors (MSE) using training data is calculated:

$$\text{MSE} = \frac{\sum_{j=1}^{P}(y^j - t^j)^2}{P} \tag{6}$$

Here, P is number of the training data, $y^j (j = 1, ..., P)$ is the final output inferred from rules of the model for j'th feature vector in the training set. $t^j (j = 1, ..., P)$ is target value for j'th input vector in training set, which is a number between 0 to 1. Then, the model with least mean squares of errors is selected. This model called stable state model. Let T be the MSE of the stable state model using validation set.

4) If $T \geq V$ stop otherwise, let $d_s = d_s + 1$. where s is index of the stable state model; let $V = T$ and go to step 1.

After each iteration of the modeling algorithm, the range of a variable is divided to one more fuzzy subspace. In each fuzzy subspace, a linear input-output relation in consequence part of the corresponding rule is used to approximate the intensity of AU. Consequently, a highly non-linear system can be approximated efficiently by this method. Applying this accurate approach, we can recognize the lower intensity and combinations of AUs.

3.3 Parameter Identification Using Hybrid Learning

The goal of this section is determining the optimum premise parameters (mean and variance of the membership functions), and consequent parameters of the model, assuming fixed structure. We use an adaptive-network-based fuzzy inference system (ANFIS) to determine the parameters (for more details see [16]). This architecture represents the fuzzy inference described in subsection 4.1. Given the values of premise parameters, the overall output can be express as a linear combinations of consequence parameters. In forward pass of the hybrid learning algorithm, functional signals go forward till layer 4 of the ANFIS and the consequence parameters are identified by the least squares estimate. In the backward pass, the error rates propagate backward and the premise parameters are updated by gradient descent procedure.

Alternatively, we could apply the gradient descent procedure to identify all parameters. But this method is generally slow and likely to become trapped in local minima. By using the hybrid algorithm, we can decrease the dimension of search space and cut down the convergence time.

4 Facial Expression Recognition and Interpretation

Although we can use a SVMs for classification of six basic facial expressions (by feature vectors directly or AU intensity values), employing rule-based classifiers such as J48 [17], we can automatically extract human interpretable classification rules to interpret each expression. Thus, novel accurate AU-to-expression converters by continues values of the AU intensities can be created. These converters would be useful in animation, cognitive science, and behavioral science areas.

5 Experimental Results

To evaluate the performance of the proposed system and other methods like support vector machines (SVMs), hidden Markov models (HMMs), and neural network (NN) classifiers, we test them on Cohn-Kanade database [18]. The database includes 490 frontal view image sequences from over 97 subjects. The final frame of each image sequence has been coded using Facial Action Coding System which describes subject's expression in terms of action units. For action units that vary in intensity, a 5-point ordinal scale has been used to measure the degree of muscle contraction. In order to test the algorithm in lower intensity situation, we used each sequence five times with different intensities, i.e. by using intermediate frames as the last frame. Of theses, 1500 sequences were used as the training set. Also, for upper face and lower face AUs, 240 and 280 sequences were used as the test set respectively. None of the test subjects appeared in training data set. Some of the sequences contained limited head motion.

Image sequences from neutral to the frame with maximum intensity, were cropped into 57×102 and 52×157 pixel arrays for lower face and upper face action units respectively. To extract appearance features we applied 16 Gabor kernels to each frame. We used the same dimension reduction method in the proposed and SVMs methods. Depending on the single AU that we want to model it, the geometric and appearance feature vectors were of dimension 4 to 8 after applying the dimension reduction techniques. In training phase we allowed the target value of feature vector for multiple systems (single AUs) to set 1, when the input consists of AU combinations. The value of the threshold is set to 1 (see Fig. 3).

Table 1 and Table 2 show the upper face and lower face action unit recognition results respectively. In the proposed method, an average recognition rate of 88.8 and 95.4 percent were achieved for upper face and lower face action units respectively. Also, an average false alarm rate of 7.1 and 2.9 percent were achieved for upper face and lower face action units respectively. In SVMs method, we first concatenated the reduced geometric and appearance feature vectors for each single AU. Then, we classify them using a two-class SMVs classifier with Gaussian kernel. Due to use of crisp value for targets, this method suffers from intensity variations. In HMMs method, the best performance was obtained by three Gaussians and five states. The Gabor coefficients were reduced to 100 dimensions per image sequence using PCA and 2DPCA (like [19]). The geometric features were reduced to 8 dimension using PCA. Then, we concatenated the geometric and appearance feature vectors. For each single AU and also each AU combination, a hidden Markov model was trained, i.e. in this method we consider each AU as a class. We used the same dimension reduction method in the NN and HMMs methods.

Table 1. Upper face action unit recognition results (R=recognition rate, F=false alarm)

AUs	Sequences	Recognized AUs (Proposed method)			AUs	Sequences	Recognized AUs (HMMs)		
		True	Missing or extra	False			True	Missing or extra	False
1	20	17	2(1+2+4), 1(1+2)	0	1	20	15	2(1+2+4), 2(1+2)	1(2)
2	10	7	1(1+2+4), 2(1+2)	0	2	10	6	2(1+2+4), 1(1+2)	1(1)
4	20	19	1(1+2+4)	0	4	20	18	1(1+2+4)	1(2)
5	20	20	0	0	5	20	20	0	0
6	20	19	0	1(7)	6	20	18	1(1+6)	1(7)
7	10	9	0	1(6)	7	10	7	2(6+7)	1(6)
1+2	40	37	2(2), 1(1+2+4)	0	1+2	40	38	1(1+2+4)	1(4)
1+2+4	20	18	1(1), 1(2)	0	1+2+4	20	16	2(2), 2(1+2)	0
1+2+5	10	8	2(1+2)	0	1+2+5	10	7	3(1+2)	0
1+4	10	7	3(1+2+4)	0	1+4	10	5	3(1+2+4)	2(5)
1+6	10	8	1(1+6+7)	1(7)	1+6	10	6	2(1+6+7)	2(7)
4+5	20	17	2(4), 1(5)	0	4+5	20	14	3(4), 1(5)	2(2)
6+7	30	27	2(1+6+7), 1(7)	0	6+7	30	25	2(1+6+7), 3(7)	0
Total	240	213	24	3	Total	240	195	33	12
R		88.8%			R		81.3%		
F		7.1%			F		12.9%		

AUs	Sequences	Recognized AUs (SVMs)			AUs	Sequences	Recognized AUs (NN)		
		True	Missing or Extra	False			True	Missing or Extra	False
1	20	15	2(1+2+4), 1(1+2)	2(2)	1	20	14	3(1+2+4)	3(2)
2	10	6	2(1+2+4)	2(1)	2	10	5	4(1+2+4)	1(1)
4	20	18	1(1+2+4)	1(2)	4	20	18	1(1+2+4)	1(2)
5	20	20	0	0	5	20	18	1(4+5)	1(5)
6	20	19	1(1+6)	0	6	20	18	2(1+6)	0
7	10	7	0	3(6)	7	10	6	2(6+7)	2(6)
1+2	40	35	1(2), 2(1+2+4)	2(4)	1+2	40	36	2(2), 2(1+2+4)	0
1+2+4	20	15	2(1), 2(2)	1(5)	1+2+4	20	16	2(1), 2(2)	0
1+2+5	10	6	2(1+5)	2(4)	1+2+5	10	7	2(1+2)	1(4)
1+4	10	4	3(1+2+4)	3(5)	1+4	10	5	4(1+2+4)	1(5)
1+6	10	6	3(1+6+7)	1(7)	1+6	10	6	2(1+6+7)	2(7)
4+5	20	15	2(1+2+5)	3(2)	4+5	20	16	3(4)	1(1)
6+7	30	24	2(1+6+7), 2(7)	2(1)	6+7	30	24	3(1+6+7), 3(7)	0
Total	240	190	28	22	Total	240	189	38	13
R		79.2%			R		78.8%		
F		17.1%			F		15.4%		

Although this method can deal with AU dynamics properly, it needs the probability density function for each state. Moreover, the number of AU combinations is too big and the density estimation methods may lead to poor result especially when the number of training samples is low. Finally, in NN methods we trained a neural network with an output unit for each single AU and by allowing multiple output units to fire when the input sequence consists of AU combinations (like [20]).

Table 2. Lower facial action unit recognition results (R=recognition rate, F=false alarm)

Proposed method				
AUs	Sequences	Recognized AUs		
		True	Missing or extra	False
9	8	8	0	0
10	12	12	0	0
12	12	12	0	0
15	8	8	0	0
17	16	16	0	0
20	12	12	0	0
25	48	48	0	0
26	24	18	4(25+26)	2(25)
27	24	24	0	0
9+17	24	24	0	0
9+17+23+24	4	3	1(19+17+24)	0
9+25	4	4	0	0
10+17	8	5	2(17), 1(10)	0
10+15+17	4	3	1(15+17)	0
10+25	8	8	0	0
12+25	16	16	0	0
12+26	8	6	2(12+25)	0
15+17	16	16	0	0
17+23+24	8	8	0	0
20+25	16	16	0	0
Total	280	267	11	2
R	95.4%			
F	2.9%			

HMMs				
AUs	Sequences	Recognized AUs		
		True	Missing or extra	False
9	8	8	0	0
10	12	11	1(10+17)	0
12	12	12	0	0
15	8	6	1(15+17)	1(17)
17	16	16	0	0
20	12	12	0	0
25	48	45	2(25+26)	1(26)
26	24	19	3(25+26)	2(25)
27	24	24	0	0
9+17	24	22	2(9)	0
9+17+23+24	4	2	2(19+17+24)	0
9+25	4	4	0	0
10+17	8	3	3(10+12)	2(12)
10+15+17	4	2	2(15+17)	0
10+25	8	7	1(25)	0
12+25	16	16	0	0
12+26	8	5	2(12+25)	1(25)
15+17	16	16	0	0
17+23+24	8	6	1(17+23)	1(10)
20+25	16	12	2(20+26)	2(26)
Total	280	248	22	10
R	88.6%			
F	8.9%			

SVMs				
AUs	Sequences	Recognized AUs		
		True	Missing or extra	False
9	8	8	0	0
10	12	8	2(10+7)	2(17)
12	12	12	0	0
15	8	6	2(15+17)	0
17	16	14	2(10+17)	0
20	12	12	0	0
25	48	43	2(25+26)	3(26)
26	24	18	3(25+26)	3(25)
27	24	24	0	0
9+17	24	22	2(9)	0
9+17+23+24	4	1	3(9+17+24)	0
9+25	4	4	0	0
10+17	8	2	4(10+12)	2(12)
10+15+17	4	2	2(15+17)	0
10+25	8	7	1(25)	0
12+25	16	16	0	0
12+26	8	3	3(12+25)	2(25)
15+17	16	16	0	0
17+23+24	8	6	2(17+24)	0
20+25	16	11	3(20+26)	2(26)
Total	280	235	31	14
R	83.9%			
F	12.5%			

NN				
AUs	Sequences	Recognized AUs		
		True	Missing or extra	False
9	8	7	1(9+17)	0
10	12	8	2(10+7)	2(17)
12	12	11	1(12+25)	0
15	8	6	2(15+17)	0
17	16	13	2(10+17)	1(10)
20	12	12	0	0
25	48	42	3(25+26)	3(26)
26	24	19	2(25+26)	3(25)
27	24	23	1(27+25)	0
9+17	24	22	2(9)	0
9+17+23+24	4	1	3(9+17+24)	0
9+25	4	4	0	0
10+17	8	4	2(10+12)	2(12)
10+15+17	4	2	2(15+17)	0
10+25	8	7	1(25)	0
12+25	16	16	0	0
12+26	8	3	3(12+25)	2(25)
15+17	16	16	0	0
17+23+24	8	6	2(17+24)	0
20+25	16	12	3(20+26)	1(26)
Total	280	234	32	14
R	83.6%			
F	12.9%			

The best performance was obtained by one hidden layer. Although this method can deal with intensity variations by using continues values for target of feature vectors, it suffer from trapping in local minima. Also unlike the proposed method, in NN classifier there is no any systematic approach to determine the structure of the network, i.e. number of hidden layer and hidden units. Table 3 shows the facial expression recognition results using J48 [17] classifier. As discussed in section 4, by applying each rule-based classifier we can develop an AU-to-expression converter.

Table 3. Facial expression recognition results using J48 [17] classifier

Confusion matrix for J48 classifier (total number of samples=2916, correctly classified samples=2710 (92.935%), incorrectly classified samples=206 (7.065%) :

Classified as →	Surprise	Gloomy	Fear	Happy	Angry	Disgust
Surprise	579	7	7	0	7	0
Gloomy	6	467	0	0	13	0
Fear	23	0	402	0	49	0
Happy	0	0	0	618	0	0
Angry	29	11	0	0	404	18
Disgust	6	0	0	0	30	24

Detailed accuracy by class for J48 classifier:

True positive rate	False positive rate	Precision	ROC area	Class
0.965	0.028	0.900	0.992	Surprise
0.961	0.007	0.963	0.996	Gloomy
0.848	0.003	0.983	0.987	Fear
1.000	0.000	1.000	1.000	Happy
0.874	0.040	0.803	0.973	Angry
0.870	0.007	0.930	0.987	Disgust

The resulted tree for converting the AU intensities to expressions using J48. Each path from root to leaf represents a rule (S=surprise, G=gloomy, F=fear, H=happy, A=angry, D=disgust, the value of each AU is between 0 and 1):

```
AU12 <= 0
|   AU20 <= 0
|   |   AU9 <= 0
|   |   |   AU15 <= 0
|   |   |   |   AU27 <= 0
|   |   |   |   |   AU26 <= 0.083173
|   |   |   |   |   |   AU1 <= 0.268941
|   |   |   |   |   |   |   AU24 <= 0
|   |   |   |   |   |   |   |   AU7 <= 0
|   |   |   |   |   |   |   |   |   AU4 <= 0.152609: A
|   |   |   |   |   |   |   |   |   AU4 > 0.152609: G
|   |   |   |   |   |   |   |   AU7 > 0: A
|   |   |   |   |   |   |   AU24 > 0: A
|   |   |   |   |   |   AU1 > 0.268941
|   |   |   |   |   |   |   AU5 <= 0.360907
|   |   |   |   |   |   |   |   AU16 <= 0: G
|   |   |   |   |   |   |   |   AU16 > 0: F
|   |   |   |   |   |   |   AU5 > 0.360907: S
|   |   |   |   |   AU26 > 0.083173: S
|   |   |   |   AU27 > 0: S
|   |   |   AU15 > 0
|   |   |   |   AU2 <= 0.838493
|   |   |   |   |   AU24 <= 0.935031: G
|   |   |   |   |   AU24 > 0.935031
|   |   |   |   |   |   AU4 <= 0.390682: G
|   |   |   |   |   |   AU4 > 0.390682: A
|   |   |   |   AU2 > 0.838493
|   |   |   |   |   AU5 <= 0.360907: G
|   |   |   |   |   AU5 > 0.360907: S
|   |   AU9 > 0
|   |   |   AU24 <= 0: D
|   |   |   AU24 > 0: A
|   AU20 > 0
|   |   AU27 <= 0.360907: F
|   |   AU27 > 0.360907: S
AU12 > 0
|   AU2 <= 0.390682
|   |   AU4 <= 0.268941: H
|   |   AU4 > 0.268941: F
|   AU2 > 0.390682: S
```

6 Discussion and Conclusions

We proposed an efficient system for representation, recognition, interpretation, and analysis of the facial action units (AUs) and expressions. As an accurate tool, this system can be applied to many areas such as recognition of spontaneous and deliberate facial expressions, multi modal/media human computer interaction and lie detection efforts. In our neuro-fuzzy classification scheme each fuzzy rule applies a linear approximation to estimate the AU intensity in a specific fuzzy subspace. In addition combining geometric and appearance features increases the recognition rate.

Although the computational cost of the proposed method can be high in the training phase, when the fuzzy inference systems were created, it needs only some matrix products to reduce the dimensionality of the geometric and appearance features in the test phase. Employing a 3× 3 Gabor kernel and a grid with low number of vertices, we can construct the Gabor representation of the input image sequence and also track the grid in less than two seconds with moderate computing power. As a result, the proposed system is suitable for real-time applications. Future research direction is to consider variations on face pose in the tracking algorithm.

Acknowledgment. The authors would like to thank the Robotic Institute of Carnegie Mellon University for allowing us to use their database.

References

1. Mehrabian, A.: Communication without words. Psychology Today 2(4), 53–56 (1968)
2. Patnic, M., Rothkrantz, J.: Automatic analysis of facial expressions: the state of art. IEEE Transactions on PAMI 22(12) (2000)
3. Fasel, B., Luettin, J.: Automatic facial expression analysis: a survey. Pattern Recognition 36(1), 259–275 (2003)
4. Lyons, M., Akamatsu, S., Kamachi, M., Gyoba, J.: Coding facial expressions with Gabor wavelets. In: 3rd IEEE Int. Conf. on Automatic Face and Gesture Recognition, pp. 200–205 (1998)
5. Cohen, I., Sebe, N., Cozman, F., Cirelo, M., Huang, T.: Coding, analysis, interpretation, and recognition of facial expressions. Journal of Computer Vision and Image Understanding Special Issue on Face Recognition (2003)
6. Rosenblum, M., Yacoob, Y., Davis, L.: Human expression recognition from motion using a radial basis function network architecture. IEEE Transactions on Neural Network 7(5), 1121–1138 (1996)
7. Cohn, J., Kanade, T., Moriyama, T., Ambadar, Z., Xiao, J., Gao, J., Imamura, H.: A comparative study of alternative faces coding algorithms, Technical Report CMU-RI-TR-02-06, Robotics Institute, Carnegie Mellon University, Pittsburgh (2001)
8. Ekman, P., Friesen, W.: The facial action coding system: A technique for the measurment of facial movement. Consulting Psychologist Press, San Francisco (1978)
9. Tian, Y., Kanade, T., Cohn, F.: Recognizing action units for facial expression analysis. IEEE Transactions on PAMI 23(2) (2001)
10. Sean, Z., Huang, T.: Small sample learning during multimedia retrieval using bias map. In: IEEE Int. Conf. on Computer Vision and Pattern Recognition, Hawaii (2001)
11. Lu, Y., Yu, J., Sebe, N., Tian, Q.: Two-dimensional adaptive discriminant analysis. In: IEEE International Conf. on Acoustics, Speech and Signal Processing, vol. 1, pp. 985–988 (2007)
12. Yang, D., Frangi, A., Yang, J.: Two-dimensional PCA: A new approach to appearance-based face representation and recognition. IEEE Transactions on PAMI 26(1), 131–137 (2004)
13. Kotsia, I., Pitas, I.: Facial expression recognition in image sequences using geometric deformation features and support vector machines. IEEE Transactions on Image Processing 16(1) (2007)
14. Wiskott, L., Fellous, K.N., Malsburg, C.: Face recognition by elastic bunch graph matching. IEEE Transactions on PAMI 19(7), 775–779 (1997)
15. Bouguet, J.: Pyramidal implementation of the Lucas Kanade feature tracker description of the algorithm, Technical Report, Intel Corporation, Microprocessor Research Labs (1999)
16. Jang, R.: ANFIS: Adaptive-network-based fuzzy inference systems. IEEE Transactions on Systems, Man and Cybernetics 23(3), 665–685 (1993)
17. Quinlan, R.: C4.5: Programs for machine learning. Morgan Kaufmann Publishers, San Mateo (1993)
18. Kanade, T., Tian, Y.: Comprehensive database for facial expression analysis. In: IEEE In. Conf. on Face and Gesture Recognition, pp. 46–53 (2000)
19. Bartlett, M., Braathen, B., Littlewort-Ford, G., Hershey, J., Fasel, I., Marks, T., Smith, E., Sejnowski, T.: Movellan. J.:Automatic analysis of spontaneous facial behavior: A final project report, Technical Report INC-MPLab-TR-2001.08, UCSD (2001)
20. Tian, Y., Kanade, T., Cohn., J.: Evaluation of gabor-wavelet-based facial action unit recognition in image sequences of increasing complexity. In: IEEE Int. Conf. on Automatic Face and Gesture Recognition (2002)
21. Takagi, T., Sugeno, M.: Fuzzy identification of systems and its applications to modeling and control. IEEE Transactions on systems, Man and Cybernetics 15(1) (1985)

Content-Based Retrieval and Classification of Ultrasound Medical Images of Ovarian Cysts

Abu Sayeed Md. Sohail[1], Prabir Bhattacharya[2], Sudhir P. Mudur[1],
Srinivasan Krishnamurthy[3], and Lucy Gilbert[3]

[1] Dept. of Computer Science and Software Engineering, Concordia University, Canada
{a_sohai,mudur}@encs.concordia.ca
[2] Dept. of Computer Science, University of Cincinnati, Ohio, USA
bhattapr@ucmail.uc.edu
[3] Dept. of Obstetrics and Gynecology, Royal Victoria Hospital, Montreal, Canada
{srinivasan.krishnamurthy,lucy.gilbert}@muhc.mcgill.ca

Abstract. This paper presents a combined method of content-based retrieval and classification of ultrasound medical images representing three types of ovarian cysts: *Simple Cyst, Endometrioma,* and *Teratoma.* Combination of histogram moments and Gray Level Co-Occurrence Matrix (GLCM) based statistical texture descriptors has been proposed as the features for retrieving and classifying ultrasound images. To retrieve images, relevance between the query image and the target images has been measured using a similarity model based on Gower's similarity coefficient. Image classification has been performed applying Fuzzy k-Nearest Neighbour (k-NN) classification technique. A database of 478 ultrasound ovarian images has been used to verify the retrieval and classification accuracy of the proposed system. In retrieving ultrasound images, the proposed method has demonstrated above 79% and 75% of average precision considering the first 20 and 40 retrieved images respectively. Further, 88.12% of average classification accuracy has been achieved in classifying ultrasound images using the proposed method.

Keywords: Ultrasound Medical Image Retrieval, Ovarian Cyst Classification, Texture Feature, Histogram Moments, Fuzzy k-NN.

1 Introduction

With advancements in image processing and pattern classification techniques, Content-Based Image Retrieval (CBIR) has become one of the most active research topics of computer vision during the last 15-years. As medical images are produced in large number everyday during regular clinical practice, they have been frequently used in developing and analyzing the performance of such retrieval systems. This has eventually lead medical domain to be cited as one of the principal application domains of CBIR technologies in terms of potential for high impact. A large number of propositions have already been made for content-based retrieval of medical images including radiology images, X-ray images, CT images of lung, dermatology images, MRI images

F. Schwenker and N. El Gayar (Eds.): ANNPR 2010, LNAI 5998, pp. 173–184, 2010.
© Springer-Verlag Berlin Heidelberg 2010

of heart and brain, ultrasound images of kidney and breast [1, 2]. However, to the best of our knowledge, no published research has yet reported the application of CBIR and image classification techniques over ultrasound images of ovarian abnormalities.

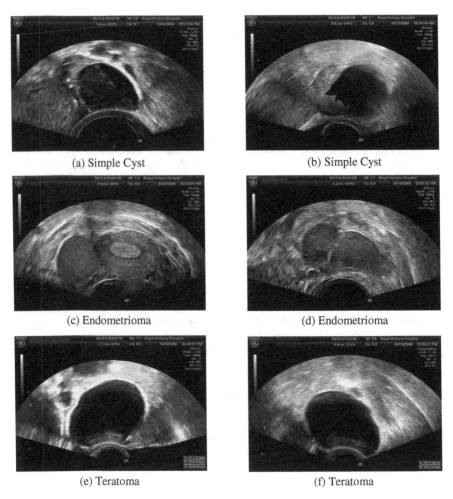

(a) Simple Cyst

(b) Simple Cyst

(c) Endometrioma

(d) Endometrioma

(e) Teratoma

(f) Teratoma

Fig. 1. Sample ultrasound images representing three types of ovarian cysts: *Simple Cyst, Endometrioma*, and *Teratoma*

Ovarian cysts are particularly heterogeneous in nature (Fig. 1) and profiling them accurately is very important in order to arrive at a diagnosis. During the past decade, visual ultrasound examination has been accepted as the optimal diagnostic modality for the non-invasive assessment of ovarian cysts and other types of ovarian abnormalities [3]. Although, several ultrasound-based algorithms have been proposed for this purpose, recognition of inherent patterns through visual observation of ultrasound images remains the best way for assessing their nature and category. However, this method largely depends on the accumulation of practical experience in identifying the morphology and characteristics of various types of ovarian cysts present in their corresponding ultrasound images. Therefore, inexperienced ultrasonographers and gynecologists always face difficulties in differentiating among different types of cysts resulting in lower rate of correct diagnosis. Since incorrect diagnosis can either result in unnecessary biopsies/surgery, or worse, missed cases, there is a huge need for inexperienced operators to be given supporting tools to help increase their diagnostic accuracy. The proposed system for retrieval and classification of ultrasound images could serve the purpose of such a decision support tool in the diagnosis of ovarian abnormalities.

The subsequent discussions of the paper have been organized as follows: methods for calculating and combining histogram moments and GLCM based texture feature have been explained in Section 2. Section 3 provides a brief theoretical overview covering the similarity matching and image classification techniques applied. Section 4 demonstrates the experimental results achieved applying the proposed method in retrieving and classifying ultrasound images of ovarian cysts, and finally, Section 5 concludes the paper.

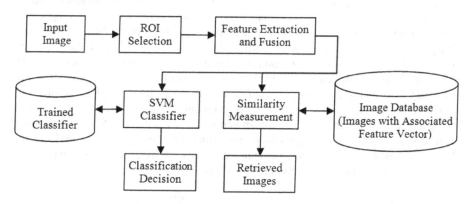

Fig. 2. Flowchart of the proposed system for retrieval and classification of ultrasound ovarian images

2 Feature Extraction

The proposed method of ultrasound ovarian image retrieval and classification works in three main stages: (i) feature extraction and fusion, (ii) image retrieval by similarity measurement, and (iii) image classification using Fuzzy k-NN. Since automated segmentation of ultrasound images is still an open problem, the proposed method requires manual intervention in selecting the Region of Interest (ROI) for feature extraction from an ultrasound image. Once the region has been specified using a number of points on the image surface, the rest of the process of feature extraction, feature fusion, image retrieval and classification is done automatically without further intervention from the user. The flowchart of the proposed system is given in Fig. 2.

In this work, the following 2-types of features have been extracted from ultrasound images for the purpose of their retrieval and classification:

(a) Histogram Moments
(b) GLCM Based Statistical Texture Descriptors

2.1 Histogram Moments

Histogram based feature is one of the most widely used visual features in image retrieval and classification. The histogram of an image f_n is an N-dimensional vector $H(f_n, i)$; $i = 0,1,2, \cdots, N - 1$, where N is the number of gray levels and $H(f_n, i)$ is the number of pixels with gray level value i. Histogram feature is relatively robust to background complication. Besides, it is also insensitive to changes such as image size, rotation and slight transition, each of which has little or no impact on the distribution of the gray levels in an image. However, the disadvantage of using histogram feature is that different images can have similar overall histogram in a large image database. Consequently, this can lead to poor performance in terms of retrieval and classification accuracy. Another drawback of using histogram features for ultrasound image classification is that an ultrasound image can have histogram with many empty bins due to the quantization process involved in the imaging system. As a result, slight changes in illumination may cause a shift in the histogram pdf, which can introduce huge changes between the set of features obtained from two similar images. The moment-based approach can minimize these problems since it smoothes the gram pdf. The histogram distribution of an image can be interpreted as a probability distribution and can be characterized by its moments. Stricker and Orengo [4] introduced a method using the central moment approach. Mandal et al. [5] used orthogonal Legendre moments for histogram indexing, which resulted in a better performance than regular or central moments. Legendre moments are based on orthogonal Legendre polynomials. A Legendre polynomial is defined as follows:

$$P_k(x) = \frac{1}{k! \, 2^k} \frac{d^k}{dx^k} (x^2 - 1)^k = \sum_{j=0}^{k} a_{kj} x^j \tag{1}$$

The value of a_{kj} can be expressed as:

$$a_{kj} = \sum_{\substack{j=0 \\ j,k \text{ odd} \\ j,k \text{ even}}}^{k} (-1)^{(k-j)/2} \frac{(k+j)!}{2^k \left(\frac{k-j}{2}\right)! \left(\frac{k+j}{2}\right)! j!}$$

The k-th Legendre moment of a function $f(x)$ is defined by

$$\lambda_k = \frac{2k+1}{2} \int_{-\infty}^{\infty} P_k(x) f(x) dx \qquad k \geq 0, k \in \mathbb{Z}$$

Replacing the value of $P_k(x)$ from Eq. (1), and applying the definition of centrals moments, λ_k can be expressed as:

$$\lambda_k = \frac{2k + 1}{2} \sum_{j=0}^{k} a_{kj} M_j$$

It can be observed that the Legendre moment of any order depends only on the regular moments of the same order and lower and can be calculated easily using the above equation.

Generally, the image grey levels range from 0 to 255. Since, the Legendre polynomials are orthogonal only in the interval $[-1, 1]$, the dynamic range of the *pdf* has to be mapped onto this interval. The histogram *pdf* function $f(x)$ can then be written as an infinite series expansion in terms of the Legendre polynomials as:

$$f(x) = \sum_{k=0}^{\infty} \lambda_k P_k(x)$$

From this equation, the image *pdf* can be reconstructed using first $(N + 1)$ moments as follows:

$$f'(x) = \sum_{k=0}^{N} \lambda_k P_k(x)$$

The reconstructed *pdf* function $f'(x)$ is then free from any quantization effect and has no empty bins. The optimum number of moments required for reconstructing the image *pdf* accurately is an important concern. Usually, 10-16 moments can give a good representation of an image *pdf* when it does not contain any sharp peak [5]. Ultrasound ovarian images usually contain sharp peaks with empty bins and therefore cannot be represented well using a small number of moments. To quantify the reconstruction efficiency as well as the no. of moments required to better represent ultrasound images, we have calculated the Signal to Error Ratio (*SER*) as follows [5]:

$$SER = \frac{\int f^2(x)dx}{\int [f(x) - f'(x)]^2 dx}$$

Where, $f(x)$ and $f'(x)$ are the original histogram and the moment reconstructed histogram respectively. Table 1 shows the percentage of images having a *SER* greater than a certain threshold value using 8, 16, 32, 64, 128 moments.

Table 1. *SER* of the re-constructed image histogram *pdf* with finite number of moments. Using 128 moments, 98% of the reconstructed histograms have *SER* > 8 dB.

	No. of Moments				
SER	8	16	32	64	128
> 8 dB	52%	79%	91%	97%	98%
> 9 dB	41%	72%	86%	97%	97%
> 10 dB	35%	67%	80%	93%	95%
> 11 dB	31%	63%	77%	91%	94%
> 12 dB	25%	57%	72%	88%	92%

As can be observed from Table 1, use of 64 and 128 moments demonstrates the best re-construction capability. However, calculation of 128 moments is computationally at least twice as expensive as calculating 64 moments. In addition, use of 64 moments instead of 128 has a significant impact on reducing the dimension of the extracted feature vector. Therefore, we opted for using the first 64 moments as image histogram based features.

2.2 GLCM Based Texture Feature

The co-occurrence probabilities of GLCM provide a second-order method for generating texture features [6]. These probabilities represent the conditional joint probabilities of all pair wise combinations of grey levels in the spatial window of interest with respect to two parameters: inter-pixel distance (d) and orientation (θ). The probability measure can be defined as: $P(x) = \{C_{ij}|(d,\theta)\}$, where C_{ij}, the co-occurrence probability between grey level i and j, is defined as:

$$C_{ij} = \frac{P_{i,j}}{\sum_{i,j=1}^{G} P_{i,j}}$$

Here $P_{i,j}$ represents the number of occurrences of grey level i and j within a specific window, given a certain pair of (d,θ); and G is the quantized number of grey levels. The sum in the denominator thus represents the total number of grey level pairs (i,j) within the window. For extracting GLCM based texture features from ultrasound ovarian images, we obtained four co-occurrence matrices from each image using $\theta = \{0, 45, 90, 135\}$ degree and $d = 1$ pixel. After that, 19-statistical texture descriptors have been calculated from each of these co-occurrence matrices as proposed in [6] and [7]. These descriptors are: angular second moment (energy), contrast, correlation, sum of squares, inverse difference moment, sum average, sum variance, sum entropy, entropy, difference variance, difference entropy, two information measures of correlation, maximal correlation coefficient, autocorrelation, dissimilarity, cluster shade, cluster performance, and maximum probability. Then, by taking the range statistics (maximum and minimum), and average of these texture descriptors calculated using the four GLCMs, a total of $19 \times 3 = 57$ texture features were extracted from each of the images and used for their retrieval and classification.

2.3 Feature Fusion and Normalization

After extracting the histogram moments and GLCM based texture features from an image, they are organized into a single feature vector. Each feature vector x_k, consisting of 121 features $(64 + 57)$, is then normalized as:

$$\hat{x}_k = \frac{x_k - \mu_k}{\sigma_k}$$

where, μ_k and σ_k are the mean and standard deviation of feature vector x_k.

3 Image Retrieval and Classification

3.1 Similarity Model for Image Retrieval

For retrieving ultrasound images, Gower's similarity coefficient [8] based similarity model has been used in which combination of features to constitute a global similarity is done as an average of each of the individual similarities on each feature. The model is defined as follows [9]:

$$GS_{ij} = \frac{\sum_{k=1}^{n} S_{ij}^{(k)}}{\sum_{k=1}^{n} \delta_{ij}^{(k)}} \tag{2}$$

Here, $S_{ij}^{(k)}$ is the result of comparing image i and j on their feature k, and $\delta_{ij}^{(k)}$ represents the possibility of comparing image i and j on their feature k. In Eq. (2), $\delta_{ij}^{(k)} = 1$ if image i and j can be compared on feature k, otherwise, $\delta_{ij}^{(k)} = 0$. If the image i and image j can be compared across all the considered features, $\sum_{k=1}^{n} \delta_{ij}^{(k)} = N$, which is the dimension of the feature vector. So, global similarity GS_{ij} between images i and j is defined as an average of the similarities on each feature between image and i and j. The quantity $S_{ij}^{(k)}$ can be defined as follows:

$$S_{ij}^{(k)} = 1 - \frac{|x_{ik} - x_{jk}|}{R_k} \tag{3}$$

Where R_k represents a normalization factor and is calculated as: $R_k = Max(x_{ik}) - Min(x_{ik})$ where x_{ik}; $i = 1, 2, \cdots, n$ is the set of values taken by each of the image i of the sample considered for the feature k. $S_{ij}^{(k)} = 1$ if image i and j are identical and $S_{ij}^{(k)} = 0$ if they are completely different. $S_{ij}^{(k)}$ can take a positive value between 0 and 1 if the two images have a certain degree of similarity according to feature k. Using Eq. (3) and considering that all features can be compared, global similarity GS_{ij} between two images i and j, as defined in Eq. (2), can be re-written as:

$$GS_{ij} = \frac{1}{N} \sum_{k=1}^{n} \left(1 - \frac{|x_{ik} - x_{jk}|}{R_k}\right) \tag{4}$$

3.2 Image Classification Using Fuzzy k-NN

Let $X \subseteq \mathfrak{R}^N$ be the set of all possible input patterns, and $\chi = \{x_1, x_2, \cdots, x_n\} \subseteq X$ be a set of input training patterns for which the corresponding class labels are already known. In the conventional k-NN algorithm, the Euclidean distances between the test pattern and all training patterns are calculated, and the test pattern is assigned the class label that most of the k-closest training patterns have [10]. Let the Euclidean distance between the test pattern $y \in X \subseteq \mathfrak{R}^N$ and the training pattern $x_j \in X$ be denoted by $\|y - x_j\| = \sqrt{\sum_{i=1}^{N}(y_i - x_{ji})^2}$. Since the number of training patterns is n, a total of n such distances are calculated, and the closest k-training patterns are identified as neighbors. The output of the conventional k-NN algorithm attains a richer semantic when the output is interpreted as a posteriori probability [11]. Hence, instead of labeling the output class label equal to the class label that most of the neighbors have, the following class confidence values are assigned to the test pattern y:

$$o_c(y) = \frac{1}{k}(\text{no. of neighbors with class label } c); \forall_c$$

$$= \frac{1}{k}\sum_{j=1}^{k} \delta_c(x_j); \forall_c \tag{5}$$

where $\delta_c(x_j)$ is the characteristic function corresponding to the j-th neighbor, i.e., $\delta_c(x_j) = 1$ if x_j has the class label c, and $\delta_c(x_j) = 0$ otherwise. Here, $o_c(y)$ is the *posteriori* probability that y belongs to the class c. With this formulation, we can still derive the hard decision by assigning the class label j to the test pattern y where $o_j(y) = \max_{1,2,\ldots,c}\{o_c(y)\}$ and C is the total number of classes.

In the conventional k-NN algorithm, all k neighbors receive equal importance, and the class label of each training pattern is considered crisp. The fuzzy k-NN algorithm refines the conventional k-NN algorithm-

1. By weighing the contribution of each of the k neighbors based on its distance to the test pattern. Evidently, the closest neighbor should receive the highest weight. Hence, the i-th closest neighbor is weighted based on its relative distance with respect to all k-closest neighbors. Thus, the relative weight of the i-th neighbor $i \in \{1, 2, \ldots, k\}$ is $\frac{1/\|y-x_i\|^{2/(q-1)}}{\sum_{j=1}^{k}\left(1/\|y-x_j\|^{2/(q-1)}\right)}$; where $\|y - x_i\|$ is the Euclidean distance between y and x_i; and q determines how strongly the distance is weighted while calculating each neighbor's contribution to the membership value. Here, the denominator is used for normalization such that the sum of the weights = 1.

2. By considering that each neighbor may belong to more than one class. It typically happens where the classes overlap. Hence, the crisp class membership of each training pattern, i.e., δ, is modified to the fuzzy membership function μ.

To consider the above two refinements, Eq. (5) is modified to the following [12]:

$$o_c(y) = \sum_{i=1}^{k}\left(\frac{\dfrac{1}{\|y-x_i\|^{2/(q-1)}}}{\sum_{j=1}^{k}\dfrac{1}{\|y-x_j\|^{2/(q-1)}}}\right)\mu_{c_c}(x_i)$$

Where $\mu_{c_c}(x_i)$ is the fuzzy membership of the i-th neighbor in the c-th class. In this case $o_c(y)$ is interpreted as the *fuzzy membership function* [13].

In the fuzzy k-NN algorithm, the initial membership on each training pattern can be assigned in the following two ways [12, 14]:

1. *Crisp Membership:* Each training pattern can have complete membership in their known class and non-memberships in all other classes.
2. *Constrained Fuzzy Membership:* The k-nearest neighbors of each training pattern (say x_i) are found, and the membership of x_i in each class is assigned as:

$$\mu_{c_c}(x_i) = \begin{cases} 0.51 + \dfrac{0.49n_j}{k} & \text{if } j = c, \\ \dfrac{0.49n_j}{k} & \text{If } j \neq c \end{cases} \tag{6}$$

The value n_j is the number of the neighbors found that belongs to the j-th class. This initialization technique fuzzifies the memberships of the labeled samples that are in the region where classes are overlapping. Moreover, the patterns that are well away from the overlapping area are assigned with the complete membership in the known class. Consequently, a test pattern lying in the overlapping region will be influenced to a lesser extent by the labelled samples that are also in the overlapping area [14].

4 Performance Analysis

Performance of the proposed method of ultrasound ovarian image retrieval and classification has been tested using 478 ultrasound images of ovarian cysts collected during regular clinical practice at the Department of Obstetrics and Gynecology, Royal Victoria Hospital, Montreal. The collected images were classified into three types of ovarian cyst: *Simple Cyst* (187 images), *Endometrioma* (154 images) and *Teratoma* (137 images). This categorization was performed by at least one expert and the categorization decision was verified by consulting the proven pathological diagnosis associated with the respective ultrasound ovarian images.

4.1 Retrieval Performance

To evaluate the retrieval performance, randomly selected 50 images of each category have been used as the query images. We adopted "Query by Example" for submitting the query to the retrieval system where the query is specified by providing an example image to the system. A retrieved image is considered a match if it belongs to the same category as that of the query image. For quantitative evaluation, retrieval performances of each category (simple cyst, endometrioma, and teratoma) were compared by calculating "Precision" values for $N = \{10, 20, 30, 40, 50, 60, 70, 80, 90, 100, 120, 140\}$ retrieved results as:

$$Precision = \frac{True\ Positive}{True\ Positive + False\ Positive}$$

Fig 3. demonstrates the precision curves drawn by calculating the average precision values from the retrieved images of each category. As can be observed from this graph, the best overall retrieval performance has been achieved in retrieving the ultrasound images of simple cyst. The average precision value lies above 79% for the first 20 retrieved images and above 75% for the first 40 retrieved images, which indicates very satisfactory and consistent retrieval performance.

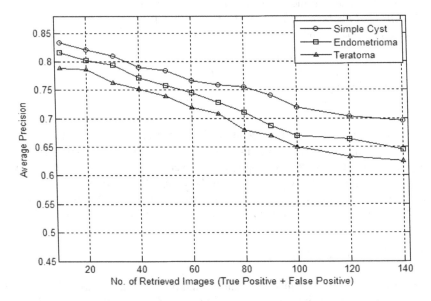

Fig. 3. Performance of the proposed method in retrieving ultrasound images of ovarian cysts

4.2 Classification Performance

Due to its capability of providing high classification accuracy over small training sets as well as comparatively better generalization performance on data that belongs to a limited number of classes, fuzzy k-Nearest Neighbor (fuzzy k-NN) classification technique has been chosen in this work for classifying ultrasound images into 3-categories: simple cyst, endometrioma, and teratoma. The initial membership on each training pattern was assigned using *Constrained Fuzzy Membership* as defined by Eq. (6). The choice of the number of nearest neighbors (k) is the most important customizations that can be made while adjusting fuzzy k-NN classifier to a particular application domain. By performing experiments with fuzzy k-NN applying different values of k starting with $k = 3$, we found that a stable average classification accuracy of 88.12% was achieved using $k = 21$ (Fig . 4). Feature extracted from 200 images of the databases has been used to train the classifier applying "K-Fold Cross Validation" technique with $K = 5$. Performance of the proposed ultrasound image classification method has also been compared with other popular classification techniques namely, SVM (with RBF, Sigmoid and Polynomial kernels), ordinary k-Nearest Neighbor (k-NN) and Neural Network (NN). Results regarding these comparisons have been summarized in Table 2.

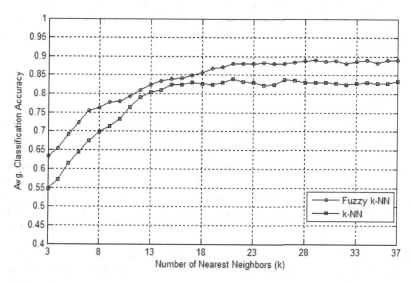

Fig. 4. Selection of a suitable value of k (no. of nearest neighbors) for Fuzzy k-NN and ordinary k-NN classifier

Table 2. Comparison of classification accuracy (%) achieved using different classifiers in classifying ultrasound ovarian images

	Simple Cyst	Endometrioma	Teratoma	Average
Fuzzy k-NN	91.44	88.96	83.94	88.12
k-NN	86.10	80.52	81.02	82.55
SVM (RBF Kernel)	88.77	86.36	84.67	86.60
SVM (Polynomial Kernel)	86.63	83.12	80.29	83.35
SVM (Sigmoid Kernel)	88.24	85.71	78.10	84.02
Neural Network	78.07	74.68	73.72	75.49

5 Conclusions

We have presented a combined method for retrieval and classification of ultrasound ovarian images using combination of histogram moments and gray level co-occurrence matrix based texture descriptors. With 88.12% of average classification accuracy and above 79% and 75% precision for the first 20 and 40 retrieved images respectively, the proposed method has been able to demonstrate significant potential in support of combining histogram moments and GLCM based texture feature for retrieving and classifying ultrasound images. In future, we plan to investigate the classification and retrieval performance of the proposed method using ultrasound images of other types of ovarian cysts. Our ultimate objective is to incorporate the proposed method in developing a Computer-Aided Diagnosis (CAD) system for providing decision support in the diagnosis of ovarian abnormalities. It is expected that by querying such a CAD system with new images and consulting the retrieved results along with their proven pathological diagnosis, the physician would gain more confidence in his/her decision or even sometimes see the scope of considering other possibilities towards improving their overall accuracy in the diagnosis of ovarian abnormalities.

References

1. Müller, H., Michoux, N., Bandon, D., Geissbuhler, A.: A Review of Content-Based Image Retrieval Systems in Medical Applications-Clinical Benefits and Future Directions. Journal of Medical Informatics 73(1), 1–23 (2004)
2. Lehman, T.M., Güld, M.O., Thies, C., Fischer, B., Spitzer, K., Keysers, D., Ney, H., Kohnen, M., Schubert, H., Wein, B.B.: Content-Based Image Retrieval in Medical Applications. Methods of Information in Medicine 43(4), 354–361 (2004)
3. Van Nagell, J.R., Depriest, P.D., Donaldson, E.S., Gallion, H.H., Pavlik, E.J., Kryscio, R.J.: Ovarian Cancer Screening in Asymptomatic Postmenopausal Women by Transvaginal Sonography. Cancer 68(3), 458–462 (2006)
4. Stricker, M., Orengo, M.: Similarity of Color Images. In: SPIE: Storage and Retrieval for Image and Video Databases III, vol. 2420, pp. 381–392. SPIE Publications, Bellingham (1995)
5. Mandal, M.K., Aboulnsar, T., Panchanathan, S.: Image Indexing Using Moments and Wavelet. IEEE Trans. Consumer Electronics 41, 557–565 (1996)
6. Haralick, R.M., Shanmugan, K., Dinstein, I.: Textural Features for Image Classification. IEEE Trans. Systems, Man and Cybernetics. 3(6), 610–621 (1973)
7. Soh, L.-K., Tsatsoulis, C.: Texture Analysis of SAR Sea Ice Imagery Using Gray Level Co-Occurrence Matrices. IEEE Trans. Geoscience Remote Sensing 37(2), 780–795 (1999)
8. Gower, J.C.: A General Coefficient of Similarity and Some of Its Properties. Biometrics 27(4), 857–871 (1971)
9. Abbadeni, N.: Content Representation and Similarity Matching for Texture-Based Image Retrieval. In: 5th ACM SIGMM international Workshop on Multimedia information Retrieval (MIR 2003), pp. 63–70. ACM Publications, New York (2007)
10. Mitchell, T.M.: Machine Learning. McGraw-Hill, New York (1997)
11. Duda, R., Hart, P.: Pattern Classification and Scene Analysis. Wiley, New York (1973)
12. Keller, J.M., Gray, M.R., Givens, J.A.: A Fuzzy k-Nearest Neighbor Algorithm. IEEE Trans. Systems Man Cybernetics 15(4), 580–585 (1985)
13. Klir, G.S., Yuan, B.: Fuzzy Sets and Fuzzy Logic Theory and Applications. Prentice-Hall, Englewood Cliffs (1995)
14. Sarkar, M.: Fuzzy-Rough Nearest Neighbor Algorithm in Classification. Fuzzy Sets and Systems 158, 2134–2152 (2007)

A Novel Word Spotting Algorithm Using Bidirectional Long Short-Term Memory Neural Networks

Volkmar Frinken, Andreas Fischer, and Horst Bunke

Institute of Computer Science and Applied Mathematics, University of Bern, Neubrückstrasse 10, CH-3012 Bern, Switzerland
{frinken,afischer,bunke}@iam.unibe.ch

Abstract. Keyword spotting refers to the process of retrieving all instances of a given key word in a document. In the present paper, a novel keyword spotting system for handwritten documents is described. It is derived from a neural network based system for unconstrained handwriting recognition. As such it performs template-free spotting, i.e. it is not necessary for a keyword to appear in the training set. The keyword spotting is done using a modification of the CTC Token Passing algorithm. We demonstrate that such a system has the potential for high performance. For example, a precision of 95% at 50% recall is reached for the 4,000 most frequent words on the IAM offline handwriting database.

1   Introduction

The automatic recognition of handwritten text – such as letters, manuscripts or entire books – has been a focus of intensive research for several decades [1,2]. Yet the problem is far from being solved. Particularly in the field of unconstrained handwriting recognition where the writing styles of various writers must be dealt with, severe difficulties are encountered.

Making handwritten texts available for searching and browsing is of tremendous value. For example, one might be interested in finding all occurrences of the word "complain" in the letters a company receives. As another example, libraries all over the world store huge numbers of handwritten books that are of crucial importance for preserving the world's cultural heritage. Making these books available for searching and browsing would greatly help researchers and the public alike. Finally, it is worth mentioning that Google and Yahoo have announced to make handwritten books accessible through their search engines as well [3].

Transcribing the entire text of a handwritten document for searching is not only inefficient as far as computational costs are concerned, but it may also result in poor performance, since misrecognized words cannot be found. Therefore, techniques especially designed for the task of keyword spotting have been developed.

Current approaches to word spotting can be split into two categories, viz. query-by-example (QBE) and query-by string (QBS). With the former approach,

F. Schwenker and N. El Gayar (Eds.): ANNPR 2010, LNAI 5998, pp. 185–196, 2010.
© Springer-Verlag Berlin Heidelberg 2010

all instances of the search word in the training set are compared with all word images in the test set. Among the most popular approaches in this category are dynamic time warping (DTW) [4,5,6] and classification using global features [7,8]. Word shape methods using Gradient, Structural and Concavity features (GSC) have been shown to outperform DTW in [9,10]. Algorithms based on QBE suffer from the drawback that they can only find words appearing in the training set. The latter approach of QBS models the key words according to single characters in the training set and searches for sequences of these characters in the test set. The approach proposed in [11,12] requires a character-position based ground truth for the training set. Consequently, not only bounding boxes around each word are required, but around each single character. In addition to expensive manual preprocessing, this imposes a problem since in cursive handwriting it is often not clear how to segment a word into individual characters.

The approach proposed in this paper uses a neural network based handwriting recognition system which has several advantages compared to the above mentioned approaches. First, by treating an entire text line at a time, it is not necessary to split the text lines of the test set into separate words. Secondly, being derived from a general neural network based handwritten text recognition system, any arbitrary string can be searched for, not just the words appearing in the training set. Thirdly, it is not required to have the bounding box of each word or character included in the training set. The ASCII transcription of the text lines in the training set is sufficient to train the neural network.

The rest of this paper is organized as follows. In Section 2, document preprocessing procedures and a neural network for handwritten text recognition are introduced. In Section 3, we describe how the proposed system can be adopted to the task of word spotting. An experimental evaluation of this system is presented in Section 4, and conclusions are drawn in Section 5.

2   Neural Network Based Handwritten Text Recognizer

2.1   Preprocessing

We follow the common approach to offline handwriting recognition by first automatically segmenting an input document into individual text lines. From each line, a sequence of feature vectors is extracted, which is then submitted to the neural network.

The words used in the experiments described in this paper come from the IAM database [13]. They are extracted from pages of handwritten texts, which were scanned and separated into individual text lines. After binarizing an image with a suitable threshold on the gray scale value, the slant and skew of each textline is corrected and the width and height of the handwriting are normalized [14].

Given the image of a single word, a horizontally sliding window with a width of one pixel is used to extract nine geometric features at each position from left to right, three global and six local ones. The global features are the 0th, 1st and 2nd moment of the black pixels' distribution within the window. The local features are the position of the top-most and bottom-most black pixel, the inclination of

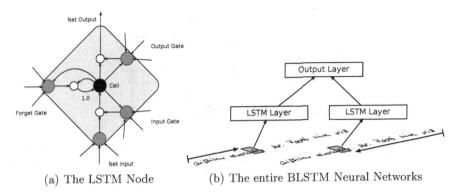

(a) The LSTM Node   (b) The entire BLSTM Neural Networks

Fig. 1. The schematics of the BLSTM Neural Network. LSTM memory cells (a) in two distinct recurrent layers process the text line from different directions.

the top and bottom contour of the word at that position, the number of vertical black/white transitions, and the average gray scale value between the top-most and bottom-most black pixel. For details on these steps, we refer to [14].

2.2 BLSTM Neural Network

The recognizer used in this paper is a recently developed recurrent neural network, termed *bidirectional long short-term memory* (BLSTM) neural network [15]. In general, recurrent NN offer a natural way for neural networks to process sequential data by reading the sequence one step at a time. Due to recurrent connections within the hidden layer, information from previous times steps can be accessed. Unfortunately, recurrent neural networks suffer from the *vanishing gradient problem*, which describes the exponential increase or decay of values as they cycle through recurrent network layers.

A way to circumvent this problem is the introduction of so-called *long short-term memory* blocks. In Fig. 1(a) such a LSTM node is displayed. At the core of the node, a simple cell, which is connected to itself with a recurrence multiplication factor of 1.0 stores the information. New information via the *Net Input* enters only if the *Input Gate* opens and leaves the cell into the network when the *Output Gate* is open. The activation of the *Forget Gate* resets the cell's value to 0. The gates and the *Net Input* are conventional nodes using an arctan activation function. This architecture admits changes to the cell's memory only when one of the gates is open and is therefore able to carry information across arbitrarily long sequence positions. Thus, at any point in the sequence, the usage of contextual information is not restricted to the direct neighborhood.

The input layer contains one node for each of the nine geometrical features and is connected with two distinct recurrent hidden layers. The hidden layers are both connected to the output layer. The network is *bidirectional*, i.e. a sequence of feature vectors is fed into the network in both forward and backward modes. One hidden layer deals with the forward sequence, and the other layer with the

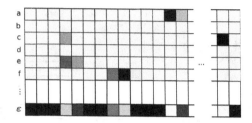

Fig. 2. The activation level for all nodes in the output layer. The activation is close to 0 most of the time for normal letters and peaks only at distinct position. In contrast, the activation level of the ε node is nearly constantly 1.

backward sequence. An illustration of the network can be seen in Fig. 1(b). At each position k of the input sequence of length t, the output layer sums up the values coming from the hidden layer that has processed positions 1 to k and the hidden layer that has processed positions t down to k. The output layer contains one node for each possible character in the sequence plus a special ε node, to indicate "no character". At each position, the output activations of the nodes are normalized so that they sum up to 1, and are treated as probabilities that the node's corresponding character occurs at this position. A visualization of the output activations along a text line can be seen in Fig. 2. For more details about BLSTM networks, we refer to [15,16].

The neural network produces a sequence of probabilities for each letter and each position in the text line. This sequence can be efficiently used for word and text line recognition as well as for word spotting as shown in the present paper, where the Connectionist Temporal Classification (CTC) Token Passing algorithm is utilized for the latter task.

3   Word Spotting Using BLSTM

The neural network described in the previous section produces a sequence of probabilities for each letter and each position in the text line. This sequence can be efficiently used for word and text line recognition [15] as well as for word spotting as shown in the present paper, where the Connectionist Temporal Classification (CTC) Token Passing algorithm is adapted to the latter task. To the knowledge of the authors, this is the first attempt to spot keywords using the CTC algorithm in conjunction with BLSTM neural networks.

3.1   CTC Token Passing Algorithm

The CTC Token Passing algorithm for single words expects a sequence of letter probabilities of length t as output by the neural network, together with a word w as a sequence of ASCII characters, and returns a matching score, i.e. the probability that the input to the neural network was indeed the given word (Algorithm 1).

Let $n(l, k)$ denote the probability of the letter l occurring at position k according to the neural network output[1] and let $w = l_1 l_2 \ldots l_n$ denote the word to be matched. The algorithm first expands w into a sequence

$$w' = \varepsilon l_1 \varepsilon l_2 \ldots \varepsilon l_n \varepsilon = c_1 c_2 c_3 \ldots c_{2n+1}$$

and creates for every character c_i ($i = 1, \ldots, 2n + 1$) and every position $j = 1, \ldots, t$ in the text line a token $\vartheta(i, j)$ to store the probability that character c_i is present at position j together with the probability of the best path from the beginning to position j. The tokens are initialized so that their probability is 0 except for the first ε (c_1) and the first letter (c_2), which are initialized to hold the value of ε and the probability value of c_1 at the first position of the sequence, respectively (Lines 3–5).

During the following loop over all input sequence positions j, the tokens $\vartheta(\cdot, j)$ are updated, so that a) the token's corresponding letter l occurs at position j, b) in the best path, all letters of the word occur in the given order, c) between two subsequent letters of the word, only ε-node activations are considered and d) if two subsequent letters of a given word are the same (e.g. positions 3 and 4 in "Hello"), at least one ε node must lie between them. To compute the value of the token $\vartheta(i, j)$, a set $\mathbb{T}_{best}$ is created in which all valid tokens are stored that can act as predecessor to the token $\vartheta(i, j)$ according to the above mentioned constraints. If at sequence position j the letter c_i is considered (which might be a real letter or ε), the token corresponding to the same letter c_i at sequence position $j - 1$ is valid (Line 9). The token corresponding to the letter c_{i-1} (ε if c_i is a real letter and a real letter if $c_i = \varepsilon$) at sequence position $j - 1$ is valid for each but the first letter (Line 10 and 11). Since two different letters might follow each other without an ε-node activation, the token corresponding to the letter c_{i-2} is valid for these cases, too (Line 12 and 13). Afterwards, the probability of the best token in $\mathbb{T}_{best}$ is multiplied with $n(i, j)$ to give the probability of $\vartheta(i, j)$. Algorithm 1 is a slightly simplified version of the one given in [15] and only suitable for single word recognition, but it is sufficient for our task of keyword spotting.

The main contribution of this paper is the modification of Algorithm 1 to search for any given word in a text line s of arbitrary length. First, we add a virtual node to the output nodes, called the *any-* or *-node and set $\forall j \, n(*, j) = 1$. Then, assuming that the considered word actually occurs in the given text line, we distinguish three different cases: a) the word to be searched occurs at the beginning of the sequence, b) the word occurs at the end of the sequence, and c) the word occurs in the given text line, but neither at the end nor at the beginning. This distinction is important in view of whitespace characters $'\_'$ following or preceding the word to be searched.

Consider again the given word $w = l_1 l_2 \ldots l_n$. To cope with case a), we append a whitespace character to the word, followed by the *any*-character. For case b)

---

[1] Due to our normalization procedure, the following statement holds $\forall k : \sum_l n(l, k) = 1$. Therefore, the output values of the neural network can be indeed considered as probabilities.

Algorithm 1. The CTC Token Passing Algorithm for single word recognition

---

Require: input word $w = l_1 l_2 \ldots l_n$
Require: sequence of letter probabilities, accessible via $n(\cdot, \cdot)$
1: **Initialization:**
2: expand w to $w' = \varepsilon l_1 \varepsilon l_2 \varepsilon \ldots \varepsilon l_n \varepsilon = c_1 c_2 \ldots c_{2n+1}$
3: $\vartheta(1,1) = n(\varepsilon, 1)$
4: $\vartheta(2,1) = n(l_1, 1)$
5:
6: **Main Loop:**
7: **for all** sequence positions $2 \le j \le t$ **do**
8:     **for all** positions i of the extended word $1 \le i \le 2n+1$ **do**
9:         $\mathbb{T}_{best} = \{\vartheta(i, j-1)\}$
10:         **if** $i > 1$ **then**
11:             $\mathbb{T}_{best} = \mathbb{T}_{best} \cup \vartheta(i-1, j-1)$
12:             **if** $c_i \ne \varepsilon$ and $c_i \ne c_{i-2}$ **then**
13:                 $\mathbb{T}_{best} = \mathbb{T}_{best} \cup \vartheta(i-2, j-1)$
14:             **end if**
15:         **end if**
16:         $\vartheta(i, j) = \max(\mathbb{T}_{best}) \cdot n(i, j)$ ▷ multiply the best token's probability with the letter probability
17:     **end for**
18: **end for**
      return $\max\{\vartheta(2n+1, t), \vartheta(2n, t)\}$ ▷ The word can either end on the last ε (c_{2n+1}) or on the last regular letter (c_{2n})

---

we prefix the word with the *any*-character and a whitespace. Finally, for case c), we prefix and append a whitespace and an *any*-character:

$$w_a = l_1 l_2 \ldots l_n \_ *$$
$$w_b = * \_ l_1 l_2 \ldots l_n$$
$$w_c = * \_ l_1 l_2 \ldots l_n \_ *$$

If we now use the CTC-Algorithm for single word recognition to compute the probability of the word being w_a, we compute in fact the probability that the text line starts with the first letter of the word w, followed by the second letter, and so on until the word's last letter, followed by a whitespace and then by *any* character. Obviously, the size and content of the text following the whitespace after word w is irrelevant, since $n(\text{`*'}, j) = 1$. Similarly, if we run the CTC-Algorithm with the word w_b, we compute the probability that the textline ends with word w. If the CTC Token Passing algorithm is run with w_c, we get the probability that the word w occurs somewhere in the middle. We can now easily combine the output of the three runs of the algorithm with w_a, w_b and w_c by using the maximum

$$p_{CTC}(w|s) = \max\{p_{CTC}(w_a|s), \ p_{CTC}(w_b|s), \ p_{CTC}(w_c|s)\} \ .$$

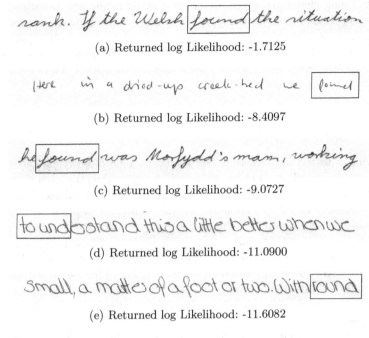

(a) Returned log Likelihood: -1.7125

(b) Returned log Likelihood: -8.4097

(c) Returned log Likelihood: -9.0727

(d) Returned log Likelihood: -11.0900

(e) Returned log Likelihood: -11.6082

Fig. 3. Search results for the word "found"

Of course, the returned probability of a word still depends upon the word's length. To receive a normalized value which can then be thresholded, we divide $p_{CTC}(w|s)$ by the search word's length[2]

$$p'_{CTC}(w|s) = \frac{p_{CTC}(w|s)}{|w|} \; .$$

How the results of such a search may look like can be seen in Fig. 3. Note that the system just returns a likelihood of the word being found. Afterwards, this likelihood can be compared to a threshold to decide whether or not this is a true match.

4   Experimental Evaluation

4.1   Setup

For testing the proposed keyword spotting method, we used the IAM offline database[3]. This database consists of 1,539 pages of handwritten English text, written by 657 writers. From this database, we used 6,161 lines as a training set, 920 lines as a writer independent validation set, and an additional 920 lines as

---

[2] We define the length according to the number of letters.

[3] http://www.iam.unibe.ch/fki/databases/iam-handwriting-database

a test set. Using the training set, we trained seven randomly initialized neural networks and used the validation set to stop the back propagation iterations in the training process. See [15] for details on the neural network training algorithm. Then we selected the 4,000 most frequent words from all three sets and performed keyword spotting using these words. Note that by far not all the keywords used in the test occur in every set.

4.2   Results

Every word w tested on each text line s returns a probability $p'_{CTC}(w|s)$. The word spotting algorithm compares this probability against a global threshold to decide whether or not it is a match. We used all returned values $p'_{CTC}(w|s)$ as a global threshold in oder to make the results as precise as possible. For each of these thresholds, we computed the number *true positives* (TP), *true negatives* (TN), *false positives* (FP), and *false negatives* (FN). These number were then used to plot a *precision-recall* curve for each neural network (see Fig. 4(a)). *Precision* is defined as number of relevant objects found by the algorithm divided by the number of all objects found $\frac{TP}{TP+FP}$, while *recall* is defined as the number of relevant objects found divided by the number of all relevant objects in the test set $\frac{TP}{TP+FN}$.

A precision-recall curve therefore gives us an idea about the noise in the returned results, given the percentage of how many true elements are found. It can be seen that the performance of the different networks varies greatly. The network that performed best on the validation set achieves an average precision rate of 87.6%. At a recall rate of 50%, the precision is 97.3%.

Based on the validation set, it is possible to pick good networks from the ensemble of all networks. The network that performed best (Network 6) was further analyzed. Its *precision-recall* curve on the independent test set is shown in Fig. 4(b). The average precision rate is 82.8% and its precision at a recall rate of 50% is 95.5%. This performance rivals those of the best existing systems, e.g. [9,8], although on a different, but not harder data set.

In Fig. 5, a *rank plot* is shown. A rank plot visualizes the quality of the search results. Each row corresponds to one keyword and dots correspond to positions of the right occurrences of keywords in the rank. Keywords are sorted according to the number of occurrences in the test set. Consequently, an optimal rank plot would have all the black dots on the left hand side of the plot, more black dots in the lower left hand corner than in the upper left hand corner, while the rest of the plot should be white.

The rank plot using all search words actually occurring in the test set can be seen in Fig 5. Although there are few black dots spread over the entire plot, it can be seen that it comes quite close to the ideal form. Since rank plots cannot easily be converted into a single number, it is not straight forward to compare different rank plots. However, they give a good impression about a system's performance. The keyword spotting ability of the proposed approach seems to be independent of the frequency of search word. Both common and uncommon

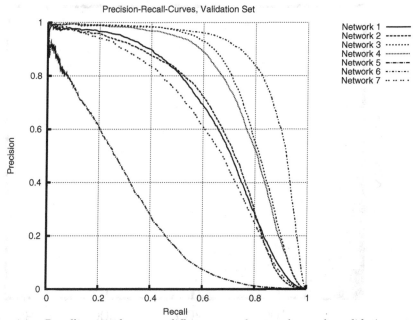

(a) Precision-Recall curves for seven different neural networks on the validation set.

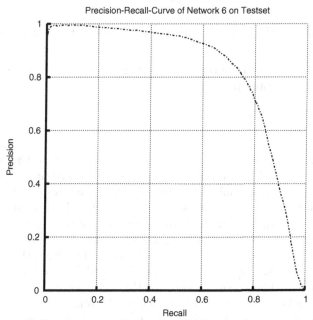

(b) Precision-Recall curves on the test set for network 6.

Fig. 4. The Precision-Recall curves of the neural networks

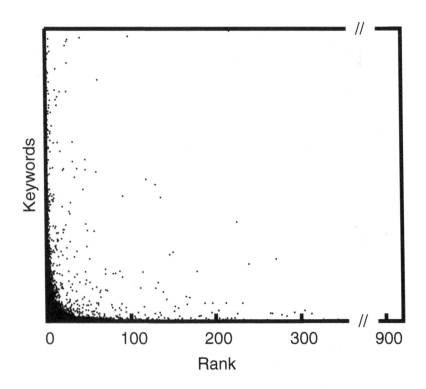

Fig. 5. The best network's rank plot for all search words that appear in the test set

words can be spotted since the black dots are aligned on the left side in the lower and upper part of the rank plot.

5   Conclusion

I this paper, we developed a word spotting algorithm that is derived from a recurrent neural network based handwriting recognition system. We were able to demonstrate that it not only performs very well with respect to the word spotting task, but it also overcomes some of the drawbacks of existing system. Not only words that appear in the training set can be searched for as in the query-by-example approach, but any character string. Secondly, to train the neural network, it is not necessary to have a bounding box for each character in the training set; just the correct transcription is needed. The performance varies greatly from one individual neural network to the other; however, an independent validation set can be used to identify the best performing network.

A future line of research will evaluate the proposed form of keyword spotting on historical documents. Desirable is also a direct experimental comparison to other available methods. Combining different neural networks to build one single improved system as well as modifying different handwriting recognition systems, such as HMM based recognizers for the task of keyword spotting, are further aspects worth to be considered in future research.

Acknowledgments

This work has been supported by the Swiss National Center of Competence in Research (NCCR) on Interactive Multimodal Information Management (IM2) and the the Swiss National Science Foundation (Project CRSI22_125220/1). We thank Alex Graves for kindly providing us with the BLSTM Neural Network source code.

References

1. Vinciarelli, A.: A Survey On Off-Line Cursive Word Recognition. Pattern Recognition 35(7), 1433–1446 (2002)
2. Plamondon, R., Srihari, S.N.: On-Line and Off-Line Handwriting Recognition: A Comprehensive Survey. IEEE Transaction on Pattern Analysis and Machine Intelligence 22(1), 63–84 (2000)
3. Levy, S.: Google's two revolutions, Newsweek (December 27/January 3, 2004)
4. Kołcz, A., Alspector, J., Augusteijn, M.F., Carlson, R., Popescu, G.V.: A Line-Oriented Approach to Word Spotting in Handwritten Documents. Pattern Analysis and Applications 3, 153–168 (2000)
5. Manmatha, R., Rath, T.M.: Indexing of Handwritten Historical Documents - Recent Progress. In: Symposium on Document Image Understanding Technology, pp. 77–85 (2003)
6. Rath, T.M., Manmatha, R.: Word Image Matching Using Dynamic Time Warping. Computer Vision and Pattern Recognition 2, 521–527 (2003)
7. Ataer, E., Duygulu, P.: Matching Ottoman Words: An Image Retrieval Approach to Historical Document Indexing. In: 6th Int'l. Conf. on Image and Video Retrieval, pp. 341–347 (2007)
8. Leydier, Y., Lebourgeois, F., Emptoz, H.: Text Search for Medieval Manuscript Images. Pattern Recognition 40, 3552–3567 (2007)
9. Srihari, S.N., Srinivasan, H., Huang, C., Shetty, S.: Spotting Words in Latin, Devanagari and Arabic Scripts. Indian Journal of Artificial Intelligence 16(3), 2–9 (2006)
10. Zhang, B., Srihari, S.N., Huang, C.: Word Image Retrieval Using Binary Features. In: Proceedings of the SPIE, vol. 5296, pp. 45–53 (2004)
11. Edwards, J., Whye, Y., David, T., Roger, F., Maire, B.M., Vesom, G.: Making Latin Manuscripts Searchable using gHMM's. In: Advances in Neural Information Processing Systems (NIPS), vol. 17, pp. 385–392. MIT Press, Cambridge (2004)
12. Cao, H., Govindaraju, V.: Template-free Word Spotting in Low-Quality Manuscripts. In: 6th Int'l. Conf. on Advances in Pattern Recognition (2007)
13. Marti, U.V., Bunke, H.: The IAM-Database: An English Sentence Database for Offline Handwriting Recognition. Int'l. Journal on Document Analysis and Recognition 5, 39–46 (2002)
14. Marti, U.V., Bunke, H.: Using a Statistical Language Model to Improve the Performance of an HMM-Based Cursive Handwriting Recognition System. Int'l. Journal of Pattern Recognition and Artificial Intelligence 15, 65–90 (2001)

15. Graves, A., Liwicki, M., Fernández, S., Bertolami, R., Bunke, H., Schmidhuber, J.:
    A Novel Connectionist System for Unconstrained Handwriting Recognition. IEEE
    Transaction on Pattern Analysis and Machine Intelligence 31(5), 855–868 (2009)
16. Graves, A., Fernández, S., Gomez, F., Schmidhuber, J.: Connectionist Temporal
    Classification: Labelling Unsegmented Sequential Data with Recurrent Neural Net-
    works. In: 23rd Int'l. Conf. on Machine Learning, pp. 369–376 (2006)

Swarm Based Fuzzy Discriminant Analysis for Multifunction Prosthesis Control

Rami N. Khushaba, Ahmed Al-Ani, and Adel Al-Jumaily

Faculty of Engineering, University of Technology, Sydney, Australia
{rkhushab,ahmed,adel}@eng.uts.edu.au

Abstract. In order to interface the amputee's with the real world, the myoelectric signal (MES) from human muscles is usually utilized within a pattern recognition scheme as an input to the controller of a prosthetic device. Since the MES is recorded using multi channels, the feature vector size can become very large. In order to reduce the computational cost and enhance the generalization capability of the classifier, a dimensionality reduction method is needed to identify an informative moderate size feature set. This paper proposes a new fuzzy version of the well known Fisher's Linear Discriminant Analysis (LDA) feature projection technique. Furthermore, based on the fact that certain muscles might contribute more to the discrimination process, a novel feature weighting scheme is also presented by employing Particle Swarm Optimization (PSO) for the weights calculation. The new method, called PSOFLDA, is tested on real MES datasets and compared with other techniques to prove its superiority.

Keywords: Discriminant Analysis, Myoelectric Control, Particle Swarm Optimization.

1 Introduction

The myoelectric signal (MES), also known as the Electromyogram (EMG), is a one dimensional non-stationary signal that carries the distinct signature of the voluntary intent of the central nervous system. It is usually recorded in a noninvasive scheme utilizing a set of surface electrodes mounted on the human forearm. One of the most important application of the MES is its use in controlling prosthetic devices functioning as artificial alternatives to missing limbs. Advances in myoelectric signals studies revealed that the MES exhibits different temporal structure for different kinds of the arm movements. This in turn facilitated the use of a pattern recognition based myoelectric control strategies for prosthetics control. To this end, a wide set of pattern recognition methods were proposed in the literature to produce a computationally efficient and accurate MES recognition systems [1].

In order to capture the complete muscle activity, a multi channel approach is usually utilized when measuring the MES signal to capture novel motor information from different muscles. However, this will increase the number of extracted

F. Schwenker and N. El Gayar (Eds.): ANNPR 2010, LNAI 5998, pp. 197–206, 2010.
© Springer-Verlag Berlin Heidelberg 2010

features (variables that describe these movements) and hence it will increase the learning parameters of the classifier and may degrade its performance. A straight forward solution to these problems is to project the data onto low-dimensional subspaces to extract the most significant features. Many feature projections techniques were used in myoelectric control with the aim to produce a statistically uncorrelated or independent feature set, a desirable goal in any pattern recognition system. Various approaches in dimensionality reduction were utilized in myoelectric control. These include, principal component analysis (PCA), a well-known feature projection method which, according to [2], achieved good performance in the myoelectric control problem , a combination of PCA and a self organizing feature map (SOFM), proving better results than PCA alone, according to [3], linear discriminant analysis (LDA) based feature projection, proving better results than PCA with SOFM, according to [4].

Although LDA is a well known projection technique, but there are many limitations with the classical LDA [5]. The first is that it requires the scatter matrices to be nonsingular, while in real world problems they can be singular. The second limitation with LDA is that it treats all the data points equivalently where as in the real world problems each sample may belong to each of the different classes with to certain degree. Finally, classical LDA pays no attention to the decorrelation of the data, which is a desirable property in many applications. One possible approach to overcome the first and the third problems was utilized in myoelectric control, this is based on the use of the uncorrelated linear discriminant analysis (ULDA) that requires the reduced features to be statistically uncorrelated with one another [6].

As a variation to the ULDA approach which is based on Singular Value Decomposition (SVD) that is known to be expensive in terms of time and memory requirements for large datasets, this paper proposes a new mixture of fuzzy logic and discriminant analysis as a novel dimensionality reduction technique. The proposed method aims to reduce the dimensionality of the extracted feature set and cluster features, such that the classification accuracy is improved. Due to the fact that most of the biosignals generated by the human body tend to produce patterns that are fuzzy in nature (i.e., belongs to different classes with certain degrees), then the incorporation of the concept of fuzzy memberships is required to reduce the effect of overlapping and outliers points. The new method, called PSOFLDA, unlike the current available variations to Fisher's linear discriminant analysis (LDA), accounts for the different contribution of different muscles into the discrimination process. Thus it assumes that certain features are more important than others. In order to reflect this importance, a novel feature weighting scheme is introduced employing Particle Swarm Optimization (PSO) technique for the weights calculation. Also in order to overcome the singularity problem, a regularization parameter is included within each particle (i.e., member of the population).

This paper is structured as follows: Section 2 explains the proposed methodology. The experimental results are given in section 3. Finally the conclusion is given in section 4.

2   Methodology

In this section, first the theory behind the proposed fuzzy discriminant analysis is introduced. Secondly, an introduction to the Particle Swarm Optimization (PSO) technique is presented as one possible way to reach the near optimal solution.

2.1   Weighted Fuzzy Discriminant Analysis

Consider a classification problem with c classes, in which the data set of labeled training samples is given as:

$$S = \{(x_1, y_1), (x_2, y_2), ..., (x_l, y_l)\} \subseteq (X, Y)^l \tag{1}$$

Where X is the input space and Y is the output space. $X \subseteq \Re^n$, and l is the number of samples. Each training point x_i originally belongs to one of the c classes and is given a label $y_i \in \{1, 2, 3, ..., c\}$ for $i = \{1, 2, 3, ..., l\}$. The goal is to find an optimal hyper-plane using the training samples that can recognize the test points, i.e., the classifier will have a good generalization capability. In PSOFLDA each point, x_i, belongs to each of the c classes with a certain degree of membership. The fuzzy within class scatter matrix S_W, fuzzy between class scatter matrix S_B, and the fuzzy total class scatter matrix S_T are given as follows:

$$S_W = \sum_{i=1}^{c} \sum_{k=1}^{l} u_{ik}^m (x_k - v_i)(x_k - v_i)^T \odot (ww^T) \tag{2}$$

$$S_B = \sum_{i=1}^{c} \sum_{k=1}^{l} u_{ik}^m (v_i - \overline{x})(v_i - \overline{x})^T \odot (ww^T) \tag{3}$$

$$S_T = \sum_{i=1}^{c} \sum_{k=1}^{l} u_{ik}^m (x_k - \overline{x})(x_k - \overline{x})^T \odot (ww^T) \tag{4}$$

where u_{ik} is the membership of pattern k in class i, m (given that $m > 1$) is the fuzzification parameter, x_{kj} is the value of the k'th sample across the j'th dimension, v_i is the mean of the patterns belonging to class i, and v_{ij} is its value across the j'th dimension. $\odot$ refers to the *Hadamard product* operation, w is the weight vector associated with all features, i.e., $w = \{w_1, w_2, ..., w_f\}$, where f is the total number of features. $\overline{x}$ is the mean of the training samples which is given in Eq.(5) below.

$$\overline{x} = \frac{1}{l} \sum_{k=1}^{l} x_k \tag{5}$$

In this paper, the value of the membership u_{ik} is calculated using a possibilistic fuzzy clustering approach. The cost function of the possibilistic clustering approach is adopted from [7], as given in Eq.(6) below.

$$J(\theta, U) = \sum_{k=1}^{l} \sum_{i=1}^{c} u_{ik}^m (x_k - \theta_i)^2 + \sum_{i=1}^{c} \eta_i \sum_{k=1}^{l} (1 - u_{ik})^m \tag{6}$$

where θ_i is the i'th cluster center, η_i are positive constants that are suitably chosen. The first term in Eq. (6) is the same objective function used in probabilistic clustering approach, while the second term is added to reduce the effect of outliers. In order to find the membership values from the above equation, then the values of the clusters centers are needed. A direct way would be to differentiate Eq. (6) with respect to θ_i, but this in turn would cancel the second term leaving only the first term. A general look at the first term of Eq. (6) reveals that it represents the classical within class scatter matrix S_W given in Eq. (2) if the weight is removed. Thus applying the values of the clusters means ensures that the objective function given by Eq. (6) would settle at a global optimum value. Then in order to compute the membership values, a differentiation of the resultant function with respect to u_{ik} needs to be done as follows.

$$\frac{\partial J(\theta, U)}{\partial u_{ik}} = m u_{ik}^{m-1}(x_k - v_i)^2 - m\eta_i(1 - u_{ik})^{m-1} = 0 \qquad (7)$$

This would in turn result in the following function

$$u_{ik} = \frac{1}{1 + \left(\frac{(x_k - v_i)^2}{\eta_i}\right)^{\frac{1}{m-1}}} \qquad (8)$$

The values of η_i, where $i = \{1, 2, 3, \ldots, c\}$ were chosen to be equal to the maximum distance between the samples belonging to that class and the class center.

After computing all the variables, PSOFLDA finds the vector G that would maximize the ratio of the between class scatter matrix to the within class scatter matrix by solving the following equation:

$$G = arg \max_{G} \; trace\left((G^T S_W G)^{-1} G^T S_B G\right) \qquad (9)$$

The solution can be readily computed by applying an Eigen-decomposition on $S_W^{-1} S_B$, provided that the within class scatter matrix S_W is nonsingular. In this paper, we are using a regularized version of S_W given by $S_W = S_W + zI$, for some $z > 0$ that is included in the particle representation of the weights, where I is an identity matrix. In this way the scatter matrix is guaranteed to be nonsingular. Since the rank of the between class scatter matrix is bounded from above by $c - 1$, there are at most $c - 1$ discriminant vectors by PSOFLDA.

2.2   PSO Based Weight Optimization

One possible solution for finding the best values of the weights is to employ evolutionary algorithms, or EAs. Powerful EA algorithms include genetic algorithm (GA) and Particle Swarm Optimization (PSO). PSO is an effective continuous function optimizer as it encodes the parameters as floating-point numbers and manipulate them with arithmetic operators. By contrast, GAs are often better suited for combinatorial optimization because they encode the parameters as bit strings and modify them with logical operators. There are many variants to both approaches, but because PSO is primarily a numerical optimizer, the PSO is considered in this paper.

Particle swarm optimization, is a population based stochastic optimization technique developed by Eberhart and Kennedy in 1995 [8]. It represents an example of a modern search heuristics belonging to the category of *Swarm Intelligence* methods. PSO mimics the behavior of a swarm of birds or a school of fish. The swarm behavior is modeled by particles in multidimensional space that have two characteristics: position (p) and velocity (s). These particles wander around the hyper space and remember the best position that they have discovered. A particle's position in the multi-dimensional problem space represents one solution for the problem. They exchange information about good positions to each other and adjust their own position and velocity with certain probabilities based on these good positions. The original formula developed by Kennedy and Eberhart was improved by Shi and Eberhart with the introduction of an inertia weight ϖ that decreases over time, (typically from 0.9 to 0.4), to narrow the search that would induce a shift from an exploratory to an exploitative mode. Though the maximum velocity of a particle (s_{max}) was no longer necessary for controlling the explosion of the particles, Shi and Eberhart continued to use it, often setting $s_{max} = p_{max}$ that is the maximum velocity is equaled to the maximum value along the specific dimension, in order to keep the system within the relevant part of the search space. This was found to be a good idea that significantly improves the PSO performance and at the same time it costs very little computationally. During iterations each particle adjusts its own trajectory in the space in order to move towards its best position and the global best according to the following equations:

$$s_{ij}(t+1) = \varpi s_{ij}(t) + c_1 r_1 (pbest_{ij} - p_{ij}) + c_2 r_2 (gbest_{ij} - p_{ij}) \qquad (10)$$

$$p_{ij}(t+1) = p_{ij}(t) + s_{ij}(t+1) \qquad (11)$$

Where
   i: is the particle index
   j: is the current dimension under consideration
   p_i: is the current position,
   s_i: is the current velocity
   ϖ: is the inertia weight
   t: is the current time step
   r_1 and r_2 are two random numbers uniformly distributed in the range (0,1), c_1 and c_2 are cognitive and social parameters respectively, $pbest_i$ is the local best position, the one associated with the best fitness value the particle has achieved so far, and $gbest_i$ is the global best position, the one associated with the best fitness value found among all of the particles.

The personal best of each particle is updated according to the following equation:

$$pbest_i(t+1) = \begin{cases} pbest_i(t) & \text{if } (pbest_i(t)) \leq f(p_i) \\ p_i(t) & \text{if } (pbest_i(t)) > f(p_i) \end{cases} \qquad (12)$$

Finally, the global best of the swarm is updated using the following equation:

$$gbest(t+1) = arg \min_{pbest_i} f(pbest_i(t+1)) \tag{13}$$

Where $f(.)$ is a function that evaluates the fitness value for a given position. This model is referred to as the *gbest* (global best) model. In this paper, each particle will represent a vector whose elements are the weights assigned to each feature plus the regularization parameter z. The idea here is to generate a new weight vector by utilizing a set of particles that wander through the solution space searching for the best possible representation achieving the minimum error rates. In such a system, the fitness function was chosen to be the error rates achieved by a suitable classifier. The details of the classifiers chosen will be given in the experiments section.

3   Experiments and Results

In order to present a fair comparison with the available techniques, we include many of them in the experiments. The details of the experiments carried on are listed below:

- **Comparison with other methods:** The PSOFLDA will be compared against two groups of other techniques: the first has already been applied into myoelectric control like ULDA [5], and PCA [2]. The second group include technique that were not used within the myoelectric control problems, like Orthogonal Linear Discriminant Analysis (OLDA) [9], and Fuzzy Discriminant Analysis (FLDA) [10]. These were included because they represent new variations to Fisher's LDA.
- **Testing method employed:** The general testing scheme employed is a three way data split. The dataset utilized is divided into three sets: training, validation, and testing. An initial projection matrix is calculated based on the training set. Then a validation set is used in order to optimize the weights to produce the optimum projection matrix that can minimize the mean of the training and validation errors. Finally a completely unseen testing set is utilized to measure the generalization capability of the proposed system.
- **Parameters of PSO:** Specifically the following parameters values were used: maximum number of generations, 100; maximum velocity s_{max}: 20% of the range of the corresponding variable; maximum value along a specific dimension $p_{max} = 1$ and minimum $p_{min} = 0$; w decreases linearly from 0.9 to 0.4; and acceleration constants c_1, c_2 are set to 2.0.

The MES dataset utilized in this research is the same one that was originally collected and used by Chan et al [6]. Eight channels of surface MES were collected from the right arm of thirty normally limbed subjects (twelve males and eighteen females). Each session consisted of six trials. Seven distinct limb motions were used, hand open (HO), hand close (HC), supination (S), pronation (P), wrist flexion (WF), wrist extension (WE), and rest state (R). Data from the first two

Table 1. Classification Results Achieved by Different Methods

Feature Set	Divisions	PSOFLDA	FLDA	ULDA	OLDA	PCA
WT	Validate	96.26	93.28	93.35	93.35	89.90
	Test	94.60	92.32	92.44	92.44	88.21
TDAR	Validate	95.02	92.25	92.38	92.38	83.51
	Test	93.68	91.67	91.79	91.79	81.93

trials were used as training set and data from the remaining four trials were divided equally into two trials for validation (trails 1 & 2) and two trials for testing (trails 3 and 4).

As a first part of the MES pattern recognition system, two sets of features were extracted from the original dataset in order to test the performance of the proposed method with different feature extraction techniques. The first set of extracted features included a combination of the first four autoregressive (AR) coefficients and the root mean square value (Time-Domain (TD) feature) as the feature vector (dimensionality is $40 = 8$ channels 5 features/channel). This feature set was referred to as the **TDAR** feature set. The second feature set extracted included the mean of the square values of the wavelet coefficients using a Symmlet wavelet family with five levels of decomposition (dimensionality is $48 = 8$ channels 6 features/channel). This feature set was referred to as **WT** feature set. The analysis window size was 256 msec. Data that were 256 msec before or after a change in limb motion were removed from the training set to avoid transitional data. As a dimensionality reduction part, all of the following five methods: PSOFLDA, ULDA, OLDA, FLDA, and PCA were utilized to compare their performance with different feature sets. The final step of the MES recognition system involves a suitable classifier that can be chosen at the disposal of the designer. In the current experiments a Linear Discriminant Analysis (LDA) classifier was chosen. The advantage of this classifier is that it does not require iterative training, avoiding the potential for under- or over-training [6].

The classification results averaged across thirty subjects (with one standard deviation) using both the TDAR and the WT feature sets reduced in dimensionality with PSOFLDA, ULDA, OLDA, FLDA, and PCA are shown in Fig.1. The number of extracted features from all methods was set to $c - 1$, where c is the number of classes, since the discriminant analysis based techniques usually ends up with $c - 1$ features. The results shown for both the validation and the testing sets were given first without post processing (referred to as Initial), then with a majority vote (MV) as a post processing step, followed by the transitional data between classes removed (NT), and finally with both majority vote and the removal of the transitional data (MV+NT). The results for both the validation and testing sets are given in the Table-1.

It is clear from the results that the PSOFLDA was able to outperform all other methods. This is due to the fact the PSOFLDA is assigning higher importance to good features compared to those that are less useful. At the same time, the using of the classification accuracy as a judgment criterion on the weights values moved

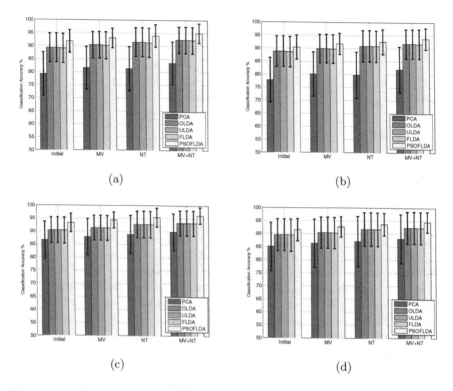

Fig. 1. Classification accuracies using different feature sets averaged across 30 subjects with different dimensionality reduction techniques (a) Using the TDAR validation set and (b) Using the TDAR testing set (c) Using the WT validation set and (d) Using the WT testing set

the projection matrix closer toward the optimal projection matrix than all other techniques. Also the PSOFLDA assigns lower fuzzy membership values to the outliers points, thus reducing their effect. Another issue to be mentioned here is that with all of the feature projection techniques, the WT features achieved higher accuracies than that achieved using the simple TDAR features. But from the computational cost point of view the performance of the system with the TDAR features is still highly accepted.

In order to provide a rigorous validation or comparison with existing techniques for dimensionality reduction, the confusion matrix for all the subjects was also computed for the different feature sets. A plot of the diagonal values of the confusion matrices (class wise classification accuracy) with both the TDAR and the WT feature are presented in Fig.2 respectively, each with the validation and testing sets results. All the results indicate that there were more significant enhancements when applying the PSOFLDA method than that of the other techniques.

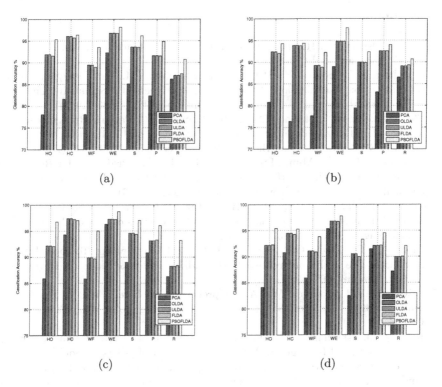

Fig. 2. Diagonal values of the confusion matrix averaged across 30 subjects using the proposed PSOFLDA in comparison with FLDA, ULDA, OLDA, and PCA (a) Using the TDAR validation set and (b) Using the TDAR testing set (a) Using the WT validation set and (b) Using the WT testing set

4   Conclusion

In this paper, a novel feature projection technique based on a mixture of fuzzy logic and Fisher's LDA was developed. Unlike the typical variations to LDA, The new technique assigned higher importance to good features compared with others. The importance was based on a weighting scheme that was optimized with PSO technique. This in turn caused the PSOFLDA's projection matrix to be closer to the optimal. The proposed PSOFLDA technique was fairly compared with other techniques like FLDA, ULDA, OLDA, and PCA proving to present better results on real MES datasets. This in turn proves the ability of the proposed technique in enhancing the performance of the multifunction myoelectric hand control system.

Acknowledgments. The Authors would like to acknowledge the support of Dr. Adrian D. C. Chan from Carleton University for providing us with the MES datasets.

References

1. Asghari Oskoei, M., Hu, H.: Myoelectric Control Systems–A Survey. Biomedical Signal Processing and Control 2, 275–294 (2007)
2. Englehart, K., Hudgins, B., Parker, P.A., Stevenson, M.: Time-Frequency Representation for Classification of the Ttransient Myoelectric Signal. In: The 20th EMBS Annual International Conference, China, pp. 2627–2630 (1998)
3. Chu, J.U., Moon, I., Mun, M.S.: A Real-Time EMG Pattern Recognition System based on Linear-Nonlinear Feature Projection for a Multifunction Myoelectric Hand. IEEE Transactions on Biomedical Engineering 53, 2232–2239 (2006)
4. Chu, J.U., Moon, I., Mun, M.S.: A Supervised Feature Projection for Real-Time Multifunction Myoelectric Hand Control. In: The 28th IEEE EMBS Annual International Conference, New York City, pp. 2417–2420 (2006)
5. Ye, J., Janardan, R., Li, Q., Park, H.: Feature Reduction via Generalized Uncorrelated Linear Discriminant Analysis. IEEE Transactions on Knowledge and Data Engineering 18(10), 1312–1322 (2006)
6. Chan, A.D.C., Green, G.C.: Myoelectric Control Development Toolbox. In: Proceedings of The 30'th Conference of the Canadian Medical & Biological Engineering Society, Toronto, ON (2007)
7. Oliveira, J.V.d., Pedrycz, W.: A Comprehensive, Coherent, and in Depth Presentation of the State of the Art in Fuzzy Clustering. John Wiley & Sons Ltd., Chichester (2007)
8. Kennedy, J., Eberhart, R.C., Shi, Y.: Swarm Intelligence. The Morgan Kaufmann Series in Artificial Intelligence. Morgan Kaufmann Publishers, London (2001)
9. Ye, J.: Characterization of a Family of Algorithms for Generalized Discriminant Analysis on Undersampled Problems. Journal of Machine Learning Research 6, 483–502 (2005)
10. Chen, Z.P., Jiang, J.H., Li, Y., Liang, Y.Z., Yu, R.Q.: Fuzzy Linear Discriminant Analysis for Chemical Data Sets. Chemometrics and Intelligent Laboratory Systems 45(1-2), 295–302 (1999)

Bayesian Learning of Generalized Gaussian Mixture Models on Biomedical Images

Tarek Elguebaly and Nizar Bouguila

CIISE, Faculty of Engineering and Computer Science, Concordia University,
Montreal, Qc, Canada H3G 2W1
t_elgue@encs.concordia.ca, bouguila@ciise.concordia.ca

Abstract. In the context of biomedical image processing and bioinformatics, an important problem is the development of accurate models for image segmentation and DNA spot detection. In this paper we propose a highly efficient unsupervised Bayesian algorithm for biomedical image segmentation and spot detection of cDNA microarray images, based on generalized Gaussian mixture models. Our work is motivated by the fact that biomedical and cDNA microarray images both contain non-Gaussian characteristics, impossible to model using rigid distributions like the Gaussian. Generalized Gaussian mixture models are robust in the presence of noise and outliers and are more flexible to adapt the shape of data.

1 Introduction

In recent years a lot of different algorithms were developed in the aim of automatically learning to recognize complex patterns, and to make intelligent decisions based on observed data. Machine learning, a branch of artificial intelligence, offers a principled approach for developing and studying automatic techniques capable of learning models and their parameters based on training data. Recent advances in machine learning have fascinated researchers from biology/bioinformatics community because they offer promise for the development of novel supervised and unsupervised methods that can help in specifying, detecting, and diagnosing different diseases, while at the same time increasing objectivity of the decision-making process. The relation history between biology and the field of machine learning is long and complex. The flexibility of machine learning techniques is expected to improve the efficiency of discovery and understanding in the mounting volume and complexity of biological data. Machine learning techniques have been used, for instance, in [1] for microarray analysis and classification, in [2] for DNA microarray image spot detection, in [3] for biomedical image analysis, and in [4] for multiple limb motion classification.

Mixture models are one of the machine learning techniques receiving considerable attention in different applications. Mixture models are normally used to model complex datasets. In most of biomedical applications, the Gaussian density is applied for data modeling [4,5]. However, data are generally non-Gaussian [6]. Many studies have demonstrated that the generalized Gaussian distribution

F. Schwenker and N. El Gayar (Eds.): ANNPR 2010, LNAI 5998, pp. 207–218, 2010.
© Springer-Verlag Berlin Heidelberg 2010

(GGD) can be a good alternative to the Gaussian thanks to its shape flexibility which allows the modeling of a large number of non-Gaussian signals [7,8]. The GGD for a variable $x \in \mathbb{R}$ is defined as follows:

$$P(x|\mu, \alpha, \beta) = \frac{\beta\alpha}{2\Gamma(1/\beta)}e^{-(\alpha|x-\mu|)^{\beta}} \tag{1}$$

where $\alpha = \frac{1}{\sigma}\sqrt{\frac{\Gamma(3/\beta)}{\Gamma(1/\beta)}}$, $-\infty < \mu < \infty$, $\beta > 0$, and $\alpha > 0$, and $\Gamma(.)$ is the Gamma function given by: $\Gamma(x) = \int_0^{\infty} t^{x-1}e^{-t}dt$, $x > 0$. μ, α and β denote the distribution mean, the inverse scale parameter, and the shape parameter, respectively. The GGD is flexible thanks to its shape parameter β that controls the decay rate of the density function. In other words, β allows the GGD to take different shapes depending on the data. Fig. 1 shows us two main reasons to use GGD. First, the parameter β controls the shape of the pdf. The larger the value, the flatter the pdf; and the smaller the value, the more picked the pdf. Second, when $\beta = 2$ and $\beta = 1$, the GGD is reduced to the Gaussian and Laplacian distributions, respectively. In the past few years, several approaches have been applied for GGDs parameters estimation such as moment estimation [9], entropy matching estimation [10,11], and maximum likelihood estimation [12,13]. It is noteworthy that these approaches consider a single distribution. Concerning finite mixture models parameters estimation, approaches can be arranged into two categories: deterministic, and Bayesian methods. In deterministic approaches, parameters are taken as fixed and unknown, and inference is founded on the likelihood of the data. In the recent past, some deterministic approaches have been proposed for the estimation of finite generalized Gaussian mixture (GGM) models parameters (see, for instance, [14,15]). Despite the fact that deterministic approaches have controlled mixture models estimation due to their small computational time, many works have demonstrated that these methods have severe problems such as convergence to local maxima, and their tendency to overfitt the data [16] especially when data are sparse or noisy. With computational tools evolution, researchers were encouraged to implement and use Bayesian MCMC methods and techniques as an alternative approach. Bayesian methods consider parameters to be random, and to follow different probability distributions (prior distributions). These distributions are used to describe our

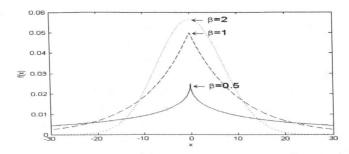

Fig. 1. Generalized Gaussian Distributions with different values of the shape parameter

knowledge before considering the data, as for updating our prior beliefs the like-lihood is used. Please refer to [16] for interesting and in depth discussions about the general Bayesian theory. In this paper, we describe a Bayesian algorithm for GGM learning and provide two examples of their applications in Biomed-ical/Bioinformatics fields. Biomedical image segmentation is chosen to be our first application since medical images are highly corrupted by noise, and con-tain non-gaussian characteristics. For the second application, we are interested in developing an algorithm capable of automatically detecting the spots in DNA microarray images.

The rest of this paper is organized as follows. Section 2 describes the Bayesian estimation algorithm. In section 3, we demonstrate the efficacy of the model on two applications. Our last section is devoted to the conclusion.

2   Bayesian Estimation of the GGM Model

A general Gaussian mixture with M components is defined as:

$$P(x|\Theta) = \sum_{j=1}^{M} P(x|\mu_j, \alpha_j, \beta_j)p_j \tag{2}$$

where p_j are the mixing proportions which are constrained to be non-negative and sum to one, and $p(x|\mu_j, \alpha_j, \beta_j)$ is the GGD describing component j. The symbol $\Theta = (\xi, p)$ refers to the whole set of parameters to be estimated, know-ing that $\xi = (\mu_1, \alpha_1, \beta_1, ..., \mu_M, \alpha_M, \beta_M)$, and $p = (p_1, ..., p_M)$. The two main problems in finite mixture models are the estimation of the parameters vector Θ and the number of components M. Supposing that the number of classes M is known then for N observations , $\mathcal{X} = (x_1, ..., x_N)$, the likelihood corresponding to this case is:

$$P(\mathcal{X}|\Theta) = \prod_{i=1}^{N} \sum_{j=1}^{M} P(x_i|\xi_j)p_j \tag{3}$$

where $\xi_j = (\mu_j, \alpha_j, \beta_j)$. For each variable x_i, let Z_i, the unobserved or missing vector, be an M-dimensional vector that indicates to which component x_i be-longs. In other words, Z_{ij} equals 1 if x_i belongs to class j and 0, otherwise. The complete-data likelihood for this case is then:

$$P(\mathcal{X}, Z|\Theta) = \prod_{i=1}^{N} \sum_{j=1}^{M} (P(x_i|\xi_j)p_j)^{Z_{ij}} \tag{4}$$

where $Z = \{Z_1, Z_2, ..., Z_N\}$. Then, the estimation of each Z_{ij}, defined as the posterior probability that the ith observation arises from the jth component of the mixture is:

$$\widehat{Z}_{ij}^{(t)} = \frac{P^{(t-1)}(x_i|\xi_j^{(t-1)})p_j^{t-1}}{\sum_{j=1}^{M} P^{(t-1)}(x_i|\xi_j^{(t-1)})p_j^{t-1}} \tag{5}$$

where t denotes the current iteration step and $\xi_j^{(t)}$ and $p_j^{(t)}$ are the current evaluations of the parameters.

Bayesian MCMC simulation methods are applied to get the posterior distribution $\pi(\Theta|\mathcal{X}, Z)$. Generally MCMC methods are found on the Bayesian theory, which means that they allow for probability statements to be made directly about the unknown parameters of a mixture model, while taking into consideration prior or expert opinion. In order to get the posterior distribution using MCMC, we need to combine the prior information about the parameters, $\pi(\Theta)$, with the observed value or realization of the complete data $P(\mathcal{X}, Z|\Theta)$. This can be reached from Bayes formula:

$$\pi(\Theta|\mathcal{X}, Z) = \frac{\pi(\Theta)P(\mathcal{X}, Z|\Theta)}{\int \pi(\Theta)p(\mathcal{X}, Z|\Theta)} \propto \pi(\Theta)P(\mathcal{X}, Z|\Theta) \tag{6}$$

where $(\mathcal{X}, Z)$ is the complete data. With the joint distribution, $\pi(\Theta)P(\mathcal{X}, Z|\Theta)$, in hand we can deduce the posterior distribution (Eq. 6). Having $\pi(\Theta|\mathcal{X}, Z)$ we can simulate our model parameters Θ, rather than computing them. Taking advantage of the missing data Z, we simulate Z according to the posterior probability $\pi(Z|\Theta, \mathcal{X})$. This is done by associating with each observation x_i a missing Multinomial variable $Z_i \sim \mathcal{M}(1; \widehat{Z}_{i1}; ...; \widehat{Z}_{iM})$. This choice is based on two reasons, first, we know that each Z_i is a vector of zero-one indicator variables. Second, the probability that the ith observation, x_i, arises from the jth component of the mixture is given by $\widehat{Z}_{ij}$.

For p simulation we need to get $\pi(p|Z^{(t)})$, using Bayes rule: $\pi(p|Z)=\frac{\pi(Z|p)\pi(p)}{\int \pi(Z|p)\pi(p)}$ $\propto \pi(Z|p)\pi(p)$. This indicates that we need to determine $\pi(Z|p)$, and $\pi(p)$. Moreover, we know that the vector P is defined on the simplex ($0 \leq p_j \leq 1$ and $\sum_{j=1}^{M} p_j = 1$), then the typical choice, as a prior, for this vector is a Dirichlet distribution with parameters $\eta = (\eta_1, ..., \eta_M)$

$$\pi(p) = \frac{\Gamma(\sum_{j=1}^{M} \eta_j)}{\prod_{j=1}^{M} \Gamma(\eta_j)} \prod_{j=1}^{M} p_j^{\eta_j - 1} \tag{7}$$

As for $\pi(Z|p)$ we have:

$$\pi(Z|p) = \prod_{j=1}^{M} \pi(Z_i|p) = \prod_{i=1}^{N} \prod_{j=1}^{M} p_j^{Z_{ij}} = \prod_{j=1}^{M} p_j^{n_j} \tag{8}$$

Where $n_j = \sum_{i=1}^{N} \mathbf{I}_{Z_{ij}=1}$, then we can conclude that:

$$\pi(p|Z) = \pi(Z|P)\pi(p) = \frac{\Gamma(\sum_{j=1}^{M} \eta_j)}{\prod_{j=1}^{M} \Gamma(\eta_j)} \prod_{j=1}^{M} p_j^{\eta_j - 1} \prod_{j=1}^{M} p_j^{n_j} = \frac{\Gamma(\sum_{j=1}^{M} \eta_j)}{\prod_{j=1}^{M} \Gamma(\eta_j)} \prod_{j=1}^{M} p_j^{\eta_j + n_j - 1} \tag{9}$$

$$\propto \mathcal{D}(\eta_1 + n_1, ..., \eta_M + n_M)$$

$\mathcal{D}$ denotes the Dirichlet distribution with parameters $(\eta_1 + n_1, ..., \eta_M + n_M)$. From (Eq. 9) we can deduce that the Dirichlet distribution is a conjugate prior for the mixture proportions, which means that the prior and the posterior have the same form. Let us now define ξ priors, which are supposed to be drawn independently. For the parameters ξ, we assigned independent Normal priors for the distributions means, and Gamma priors for the inverse scale and shape parameters [17,18]:

$$\mu_j \sim \mathcal{N}(\mu_0, \sigma_0^2) \ , \ \beta_j \sim \mathcal{G}(\alpha_\beta, \beta_\beta) \ , \ \alpha_j \sim \mathcal{G}(\alpha_\alpha, \beta_\alpha)$$

Where $\mathcal{N}(\mu_0, \sigma_0^2)$ is the normal distribution with mean μ_0 and variance σ_0^2, $\mathcal{G}(\alpha_\beta, \beta_\beta)$ is the gamma distribution with shape parameter α_β and rate parameter β_β. μ_0, σ_0^2, α_β, β_β, α_α, β_α are called the hyperparameters of the model. With this priors, we can deduce the posterior distributions for μ, α, and β to be:

$$\pi(\mu_j|Z, \mathcal{X}) \propto e^{\frac{(\mu_j - \mu_0)^2}{2\sigma_0^2} + \sum z_{ij=1}(-\alpha_j|x_i - \mu_j|)^{\beta_j}} \tag{10}$$

$$\pi(\alpha_j|Z, \mathcal{X}) \propto \alpha_j^{\alpha_\alpha - 1} e^{-\beta_\alpha \alpha_j} (\alpha_j)^{n_j} e^{\sum z_{ij=1}(-\alpha_j|x_i - \mu_j|)^{\beta_j}} \tag{11}$$

$$\pi(\beta_j|Z, \mathcal{X}) \propto \beta_j^{\alpha_\beta - 1} e^{-\beta_\beta \beta_j} \left(\frac{\beta_j}{\Gamma(1/\beta_j)}\right)^{n_j} e^{\sum z_{ij=1}(-\alpha_j|x_i - \mu_j|)^{\beta_j}} \tag{12}$$

It is quite easy to notice that we cannot simulate directly from these posterior distributions because they are not in well known forms. To solve this problem we applied the well known Metropolis-Hastings (M-H) algorithm given in [19]. The major problem in the M-H algorithm is the choice of the proposal distribution. Random Walk M-H given in [19] is used here to solve this problem, then the proposals are considered to be: $\widetilde{\mu_j} \sim \mathcal{N}(\mu_j^{(t-1)}, \zeta^2)$, $\widetilde{\alpha_j} \sim \mathcal{LN}(\log(\alpha_j^{(t-1)}), \zeta^2)$, $\widetilde{\beta_j} \sim \mathcal{LN}(\log(\beta_j^{(t-1)}), \zeta^2)$, where $\mathcal{LN}$ is the log-normal distribution, since, we know that $\widetilde{\alpha_j} > 0$ and $\widetilde{\beta_j} > 0$. ζ^2 is the scale of the random walk.

In fact, choosing a relevant model consists both of choosing its form and the number of components M. The integrated or marginal likelihood using the Laplace-Metropolis estimator [19] is applied in order to rate the ability of the tested models to fit the data or to determine the number of clusters M. The integrated likelihood is defined by [19]

$$p(\mathcal{X}|M) = \int \pi(\Theta|\mathcal{X}, M)d\Theta = \int p(\mathcal{X}|\Theta, M)\pi(\Theta|M)d\Theta \tag{13}$$

where Θ is the vector of parameters of a finite mixture model, $\pi(\Theta|M)$ is its prior density, and $p(\mathcal{X}|\Theta, M)$ is the likelihood function taking into account that the number of clusters is M.

3   Experimental Results

In this section, we apply the Bayesian estimation of the GGM in biomedical image segmentation, and microarray image spot detection. We validate the algorithm by comparing it to different state of the art algorithms. In the following applications, we used 5000 iteration for our Metropolis-within-Gibbs sampler (we discarded the first 800 iterations as "burn-in" and kept the rest), and our specific choices for the hypeparameters were $(\mu_0, \sigma_0^2, \alpha_\alpha, \beta_\alpha, \alpha_\beta, \beta_\beta) = (0, 1, 0.2, 2, 0.2, 2)$. To increase the sensitivity of the random walk sampler, the scale of the random walk was chosen to be $\zeta^2 = 0.01$.

3.1   Biomedical Image Segmentation

Image segmentation is one of the major challenges in image analysis, since image analysis tasks highly depend on how well previous segmentation is accomplished. Image segmentation is the procedure of dividing an image into different groups with each group enjoying similar properties such as texture, color, boundary, and intensity. Despite, the existence of different segmentation methods, many of them fail to provide satisfactory results when applied on biomedical images. Reasons behind this failure are numerous. First, image segmentation is strongly influenced by the quality of data and biomedical images contain different noises such as speckle, shadows which may cause the boundaries of structures to be indistinct and disconnected. Second, most of image segmentations algorithms are founded on the assumption that the data are Gaussian which is not the case for biomedical images. Further complications arise as the contrast between areas of interest in biomedical images is low, which make the extraction of the desired regions impossible as they are statistically indistinguishable. Last but not least, most of existed segmentation methods do not integrate uncertain prior knowledge.

In this section, we develop a new segmentation methodology, using the Bayesian MCMC algorithm developed in section 2. We can divide our method into two main steps: histogram adjustment, and identification of object of interest using the Bayesian GGM with the integrated likelihood. We validate our algorithm by comparing it to a state of the art segmentation algorithm [20]. This method is divided into two stages: preprocessing, and object segmentation. Preprocessing stage contains histogram adjustment, noise reduction, and layer of interest extraction using K-means algorithm. For the object segmentation a marker-controlled watershed technique is used.

The first image used is the image of a rat spleen tissue pulps (Fig. 2(a)). For visual differentiation of cellular components, the tissue section was stained with haematoxylin and eosin ($H\&E$). Under a microscope, nuclei are usually dark blue, red blood cells orange/red, and muscle fibers deep pink/red. The feature used to differentiate red and white is the density of the lymphocytes. The white pulp has lymphocytes and macrophages surrounding central arterioles. The distribution of the lymphocytes in red pulp is much looser than those in white pulp. Evaluating the severity of infection requires identifying the white pulps. We started by transforming the color image to a gray level image (Fig. 3(a)) in order to simplify the processing procedure. For grayscale image nuclei are dark objects within a gray background. Then histogram adjustment [21] is applied on the image to increase image contrast (Fig. 3(b)). At this point, we applied our Bayesian GGM to identify the object of interest in the image (Fig. 3(c)).

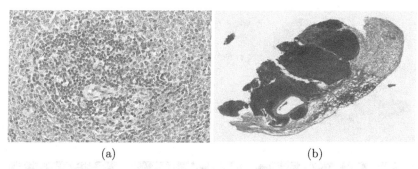

Fig. 2. Microscopic images used, (a) The rat spleen tissue pulps (Courtesy of Dr. Jinglu Tan), (b) Lung Carcinoid tumor (Courtesy of Dr. Robert Cardiff)

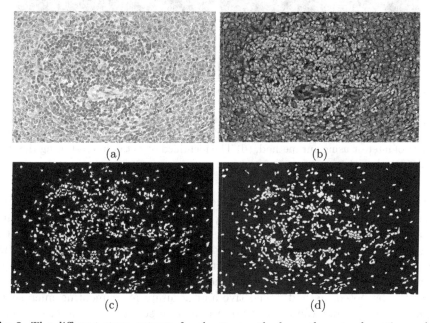

Fig. 3. The different stage outputs for the two methods on the rat spleen tissue, (a) The gray scale image, (b) The image after histogram adjustment, (c) The identified object of interest using our method, (d) The identified object of interest using the state of the art algorithm

Comparing the output from our method to the one from the watershed method (Fig. 3(d)), we can find that we were able to reach a higher identification for the infected regions. Also, the proposed method is less complex due to the fact that we did not need to use neither noise reduction, nor marker-controlled watershed techniques.

The second image is an image of a carcinoid tumor seen in the lung of eighty one years old female (Fig. 2(b)). To be able to differentiate visually the cellular

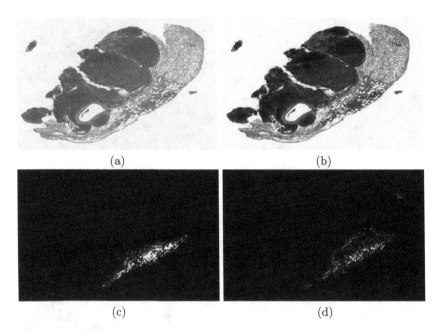

Fig. 4. The different stage outputs for the two methods on the Lung Carcinoid tumor, (a) The gray scale image, (b) The image after histogram adjustment, (c) The identified object of interest using our method, (d) The identified object of interest using the state of the art algorithm

components, the tissue section was stained with haematoxylin and eosin ($H\&E$). The size of the tumor is of 2.5 cm long as shown in the image. First we transformed the image into gray scale image (Fig. 4(a)). Next, we applied the histogram adjustment on the image (Fig. 4(b)), and last we applied our algorithm on it to reach the object of interest (Fig. 4(c)). Also, it is quite clear here that our algorithm outperformed the watershed algorithm. Experimental results show that the proposed method is effective and accurate in segmenting microscopic images even without the need of noise reduction stages and marker-controlled watershed techniques to separate the touching objects.

3.2   Spot Detection and Image Segmentation in DNA Microarray Data

In this section, we propose an optimized clustering-based method for microarray image segmentation using GGM. Our algorithm is based on the fact that GGM is flexible to model the shape of data, and have high immunity to noise. To assess the performance of our method, we compare it to two well known algorithms: k-means clustering microarray image segmentation (SKMIS) [22], and optimized k-means microarray image (OKMIS) [23]. We evaluate the segmentation performance of the three methods on the spot images from ApoA1 Data [24].

DNA microarray technology is a high throughput technique allowing the comprehensive measurement of the expression level of thousands of genes simultaneously in the studies of genomics for biology and medicine [25]. Complementary single stranded DNA (cDNA) microarrays consist of thousands of individual DNA sequences printed in a high density array. Nowadays, microarray experiments are used to compare gene expression from two samples: target or experimental, and control. The mRNA of both biological tissues (normal and tumor) is extracted, then reversed transcribed into complementary DNA (cDNA) copy, followed by a labeling procedure using two fluorescence dyes, Cyanine Cy3 (green channel) and Cy5 (red channel). After labeling, the two samples are mixed and hybridized with the arrayed DNA sequences. Afterwards, fluorescence measurements are made for each dye separately, and the digital image scanner records the intensity level at each microarray location producing two grayscale images [26].

Image analysis is a highly important aspect of cDNA microarray experiments, as it is responsible for reducing an image of spots into a table with a measure of the intensity for each spot. Efficient, accurate and automatic analysis of cDNA spot images is necessary in order to apply this technology in different biological experiments. cDNA microarray gene expression data analysis involves three main stages: spot localization or gridding, background separation or image segmentation, and intensity estimation. Spot localization or gridding is used to identify blocks and to position rows and columns of spots within each block. Background separation or image segmentation is used to segment the image into background or foreground, and the intensity estimation step gets the red and green intensities and assigns the log ratio after background correction in order to represent the log relative abundance of each spot. These stages are quite important, since the accuracy of the resulting data is essential in posterior analysis.

In cDNA microarray experiments, noise is a challenging problem as it can be produced by laser light reflection, dust on the glass slide, and photon and electronic noise. These noises force microarray images to vary in intensity, in the spot sizes and positions. For this reason, we decided to apply the Bayesian GGM on this problem for its immunity to noise. Over the past few years, many approaches have been proposed for microarray image segmentation. Fixed circle segmentation is the first applied technique on microarray images, its idea is to assign the same circle size to all the spots. Another proposed method in order to avoid the drawback of the fixed circle segmentation is the adaptive circle segmentation technique. This algorithm fits a circle with adaptive size around each spot, in order to characterize the pixels in the circle as signal pixels and the pixels out of the circle as background pixels (i.e. foreground or background). Another technique that has been efficiently used in microarray image segmentation is clustering, since it is not restricted to a particular shape and size for the spots. Single k-means clustering microarray image segmentation (SKMIS) attempts to cluster the pixels into two groups, one for foreground, and the other for background. Therefore in SKMIS, feature vector is reduced to a single variable in the Euclidean one-dimensional space. Optimized k-means microarray image segmentation (OKMIS) not only consider the intensity of the pixel but also the shape

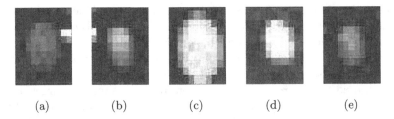

(a)   (b)   (c)   (d)   (e)

Fig. 5. Five noisy spots obtained from the 1230c1G/R microarray image

SKMIS

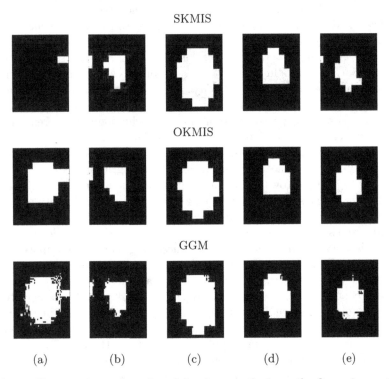

OKMIS

GGM

(a)   (b)   (c)   (d)   (e)

Fig. 6. The experimental results of the three methods on the five noisy spots

of the spot based on the fact that the position of the pixel could also influence the result of the clustering. Our algorithm is very simple as we only apply a two component GGM to classify the data to either foreground or background.

In order to compare GGM, OKMIS and SKMIS, we applied the three methods on the 1230c1G/R microarray image obtained from the ApoA1 data. Fig. 5 shows some examples for the noisy spots in our microarray image. From Fig. 6 it is clear that our method was able to retrieve the true foreground from the background. We also observe that the GGM outperformed the SKMIS and OKMIS in identifying noisy pixels from foreground. Note that, the GGM was able to take the data form. Hence, the GGM is more suitable when dealing with cDNA microarray image segmentation.

4 Conclusion

In this paper, we have presented a new Bayesian algorithm for biomedical image segmentation and multi-class DNA classification. Our method is based on GGM models which chief advantage is their flexibility and immunity to noise. We have used the Monte Carlo simulation technique of Gibbs sampling mixed with a Metropolis-Hasting step for parameters estimation. The Bayesian estimation of the model parameters incorporates uncertainty which disease diagnosis, for instance, are in need. For the selection of number of clusters we have used the integrated Likelihood. The experimental results show the effectiveness of the proposed method in two interesting applications.

Acknowledgment

The completion of this research was made possible thanks to the Natural Sciences and Engineering Research Council of Canada (NSERC), a NATEQ Nouveaux Chercheurs Grant, and a start-up grant from Concordia University.

References

1. Cho, S.-B., Won, H.-H.: Machine Learning in DNA Microarray Analysis for Cancer Classification. In: Proc. of the First Asia-Pacific Bioinformatics Conference, pp. 189–198 (2003)
2. Katzer, M., Kummert, F., Sagerer, G.: Methods for Automatic Microarray Image Segmentation. IEEE Transactions on NanoBioscience 2(4), 202–214 (2003)
3. Pappas, T.N.: An Adaptive Clustering Algorithm for Image Segmentation. IEEE Transactions on Signal Processing 40(4), 901–914 (1992)
4. Yonghong, H., Englehart, K.B., Hudgins, B., Chan, A.D.C.: A Gaussian Mixture Model Based Classification Scheme for Myoelectric Control of Powered Upper Limb Prostheses. IEEE Transactions on Biomedical Engineering 52(11), 1801–1811 (2005)
5. Rocke, D.M., Durbin, B.: A Model for Measurement Error for Gene Expression Arrays. Journal of Computational Biology 8(6), 557–569 (2004)
6. Bouguila, N., Ziou, D., Monga, E.: Practical Bayesian Estimation of a Finite Beta Mixture Through Gibbs Sampling and its Applications. Statistics and Computing 16(2), 215–225 (2006)
7. Gao, Z., Belzer, B., Villasenor, J.: A Comparison of the Z, E_8, and Leech Lattices for Quantization of Low Shape-Parameter Generalized Gaussian Sources. IEEE Signal Processing Letters 2(10), 197–199 (1995)
8. Meignen, S., Meignen, H.: On the Modeling of Small Sample Distributions with Generalized Gaussian Density in a Maximum Likelihood Framework. IEEE Transactions on Image Processing 15(6), 1647–1652 (2006)
9. Sharifi, K., Leon-Garcia, A.: Estimation of Shape Parameter for Generalized Gaussian Distributions in Subband Decomposition of Video. IEEE Transactions on Circuits and Systems for Video Technology 5(1), 52–56 (1995)
10. Aiazzi, B., Alpaone, L., Baronti, S.: Estimation Based on Entropy Matching for Generalized Gaussian PDF Modeling. IEEE Signal Processing Letters 6(6), 138–140 (1999)

11. Kokkinakis, K., Nandi, A.K.: Exponent Parameter Estimation for Generalized Gaussian Probability Density Functions with Application to Speech Modeling. Signal Processing 85(9), 1852–1858 (2005)
12. Varanasi, M.K., Aazhang, B.: Parametric Generalized Gaussian Density Estimation. The Journal of the Acoustical Society of America 86(4), 1404–1415 (1989)
13. Pi, M.: Improve Maximum Likelihood Estimation for Subband GGD Parameters. Pattern Recognition Letters 27(14), 1710–1713 (2006)
14. Allili, M.S., Bouguila, N., Ziou, D.: Finite General Gaussian Mixture Modeling and Application to Image and Video Foreground Segmentation. Journal of Electronic Imaging 17(1), 1–13 (2008)
15. Fan, S.-K.S., Lin, Y.: A Fast Estimation Method for the Generalized Gaussian Mixture Distribution on Complex Images. Computer Vision and Image Understanding 113(7), 839–853 (2009)
16. Robert, C.P.: The Bayesian Choice From Decision-Theoretic Foundations to Computational Implementation, 2nd edn. Springer, Heidelberg (2007)
17. Robert, C.P., Casella, G.: Monte Carlo Statistical Methods, 2nd edn. Springer, Heidelberg (2004)
18. Gentle, J.E., Härdle, W.: Handbook of Computational Statistics. In: Concepts and Fundamentals, vol. 1, Springer, Heidelberg (2004)
19. Lewis, S.M., Raftery, A.E.: Estimating Bayes Factors via Posterior Simulation with the Laplace-Metropolis Estimator. Journal of the American Statistical Association 90, 648–655 (1997)
20. Yu, J., Tan, J.: Object Density-Based Image Segmentation and its Applications in Biomedical Image Analysis. Computer Methods and Programs in Biomedicine 96(3), 193–204 (2009)
21. Larson, G.W., Rushmeier, H., Piatko, C.: A Visibility Matching Tone Reproduction Operator for High Dynamic Range Scenes. IEEE Transactions on Visualization and Computer Graphics 3(4), 291–306 (1997)
22. Wu, S., Yan, H.: Microarray Image Processing Based on Clustering and Morphological Analysis. In: Proc. of the First Asia Pacific Bioinformatics Conference, pp. 111–118 (2003)
23. Rueda, L., Qin, L.: An Improved Clustering-based Approach for DNA Microarray Image Segmentation. In: Campilho, A.C., Kamel, M.S. (eds.) ICIAR 2004. LNCS, vol. 3212, pp. 644–652. Springer, Heidelberg (2004)
24. Callow, M.J., Dudoit, S., Gong, E.L., Speed, T.P., Rubin, E.M.: Microarray Expression Profiling Identifies Genes with Altered Expression in HDL Deficient Mice. Genome Research 10(12), 2022–2029 (2000)
25. Brown, P., Botstein, D.: Exploring the new world of the genome with DNA microarrays. Nature Genetics, 33–37 (1999)
26. Qin, L., Rueda, L., Ali, A., Ngom, A.: Spot Detection and Image Segmentation in DNA Microarray Data. Applied Bioinformatics 4(1), 1–11 (2005)

Defective Areas Identification in Aircraft Components by Bivariate EMD Analysis of Ultrasound Signals

Marco Leo[1], David Looney[2], Tiziana D'Orazio[1], and Danilo P. Mandic[2]

[1] Institute of Intelligent Systems for Automation,
Italian National Research Council, Bari, Italy
{leo,dorazio}@ba.issia.cnr.it
[2] Department of Electrical and Electronic Engineering,
Imperial College of Science, Technology and Medicine, London, UK
{david.looney06,d.mandic}@imperial.ac.uk

Abstract. In recent years many alternative methodologies and techniques have been proposed to perform non-destructive inspection and maintenance operations of moving structures. In particular, ultrasonic techniques have shown to be very promising for automatic inspection systems. From the literature, it is evident that the neural paradigms are considered, by now, the best choice to automatically classify ultrasound data. At the same time the most appropriate pre-processing technique is still undecided. The aim of this paper is to propose a new and innovative data pre-processing technique that allows the analysis of the ultrasonic data by a complex extension of the Empirical Mode Decomposition (EMD). Experimental tests aiming to detect defective areas in aircraft components are reported and a comparison with classical approaches based on data normalization or wavelet decomposition is also provided.

1 Introduction

The challenge of guaranteeing reliable and efficient safety checks for engineering structures has received much attention in recent years in many industrial contexts. This is of crucial importance in the case of moving structures (such as with transportation vehicles or rotating machinery) and in the case of aircraft inspection and maintenance. Traditionally inspection and maintenance operation are performed by trained human operators but, unfortunately, this approach does not ensure an adequate reliability level and, at the same time, it requires prohibitive amounts of time and high costs. In addition, humans cannot detect cracks or any other irregularities in the structure components which are not visible to the naked eye. To face the above problems, many alternative methodologies and techniques have been proposed to perform non-destructive inspection and maintenance operations. These are based on the analysis of different signals such as ultrasonics, acoustic emissions, thermography, laser ultrasonics, X-radiography, eddy currents, shearography, and low-frequency methods [1].

F. Schwenker and N. El Gayar (Eds.): ANNPR 2010, LNAI 5998, pp. 219–230, 2010.
© Springer-Verlag Berlin Heidelberg 2010

In particular, in the last decade, ultrasonic techniques have shown to be very promising for non-destructive inspection and control, becoming an effective alternative to such traditional and well studied approaches such as thermography, eddy current and shearography.

Major works of literature describing ultrasound based techniques for inspection and evaluation purposes can be conveniently clustered into two categories. The first category concentrates on the study of the data acquisition and manipulation processes in order to prove the relationship between data and structural defects or composition of the material. The second category concentrates instead on the a posteriori analysis of the ultrasound data in order to (fully or partially) refer to some computational algorithm the automatic recognition of material composition, operative conditions, presence of defects and so on. Most of the works in the literature belong to the first category [12] [4] [10].

Works belonging to the second category are the least developed of the two and, moreover, their level of inspection reliability, is still inadequate, especially for those sectors (namely transportation) where an error can have serious health and safety consequences. Almost all the works in this category proposed a well-known framework based on some pre-processing technique followed by a classifier able to recognize the patterns in the data. For example in [7], the wavelet transform was used in conjunction with an artificial neural network to distinguish the ultrasonic flaw echoes from those scattered by micro-structures. A quite similar approach was introduced in [11] which addressed the problem of pipe inspection by ultrasonic guided waves.

The automatic detection of internal defects in composite materials with non-destructive techniques based on ultrasonic techniques was addressed also in [14].

In [8] the authors addressed the flaw detection problem by using a radial basis function neural network and they tried to demonstrate that a neural based approach overcame the classical threshold based for flaw detection problems. The author's idea in [13] was, instead, to cluster the signals in the similarity space (using the Kohonen Self organizing Algorithm to cluster datasets in an unsupervised manner) and then to use this result in order to distinguish between signals corresponding respectively to non defect, flat defects (cracks), and volumetric defects. A high resolution pursuit based signal processing method (an enhanced version of the matching pursuit algorithm that provides high time-resolution time-frequency representations and resolves closely time-spaced features) was proposed in [5] for detecting flaws close to the surface of strongly scattering materials, such as steel and composites, in NDT applications. In [6] an approach to non-destructive pipeline testing using ultrasonic imaging was proposed.

Finally, in [9], an evaluation of various types and configurations of neural networks developed for the purpose of assisting in accurate flaw detection in steel plates was illustrated.

Therefore, from the literature, it is evident that the neural paradigms are considered, by now, the best choice to classify ultrasound data in an automatic inspection system. At the same time the most appropriate pre-processing technique is still undecided. The wavelet based approaches seem to be the most promising ones but, considering that in this area an error could waste time, money and even endanger someone's life, further efforts have to be done in order to increase reliability.

The aim of this paper is then to propose a new and innovative data pre-processing technique to explore the pattern embedded in the data. A comparison with classical approaches based on data normalization or wavelet decomposition is also provided. In the proposed pre-processing procedure the complex valued Fourier coefficients of the ultrasound signals are decomposed by using Empirical Mode Decomposition (EMD) [16] into a set of oscillation modes or IMFs (intrinsic mode functions). In this way, the phase information is defined locally for both the real and imaginary components of the decomposition. This facilitates the detection of the temporal locking of phase information between the components, known as phase synchrony. The existence of phase synchrony between a pair of components has been used to characterise shared signal dynamics in a variety of applications.

It is this methodology of converting a real-valued data source into a complex signal in order to obtain a set of synchrony features that is novel to our work. The new data representation is then applied as an input to a supervised neural classifier trained to recognize the defective areas from the non-defective ones. To demonstrate the effectiveness of the proposed approach it has been applied to detect and classify internal defects in aircraft composite materials and in particular in a Honeycomb structure containing different inserts placed to simulate some of the most common defective situations in aircraft materials.

2 The Proposed Approach

Ultrasonic Inspection uses sound signals at frequencies beyond human hearing (more than 20KHz), to estimate some properties of the irradiated material by analysing either the reflected (reflection working modality) or transmitted (transmission working modality) signals. A typical ultrasonic based inspection system consists of several functional units, such as pulser, receiver, transducer, and display devices. A pulser is an electronic device that can produce a high voltage electrical pulse. Driven by the pulser, the transducer generates a high frequency ultrasonic energy. The sound energy is introduced into and propagates through the materials in the form of waves. In the transmission modality the receiver is placed on the opposite side of the material from the pulser whereas in the reflection modality the pulser and the receiver are placed on the same side of the material. Inspection devices can be or not be in contact with the material. In the latter case a liquid (couplant) is used to facilitate the transmission of ultrasonic vibrations from the transducer to the test surface. Ultrasonic data can be collected and displayed in a number of different formats. The three most common formats are known in the Non Destructive Testing (NDT) world as A-scan, B-scan and C-scan presentations. Each presentation mode provides a different way of looking at and evaluating the region of material being inspected. In this paper the analysis of ultrasonic data acquired from the reflection working modality and A-Scan representation is reported. This means that for each

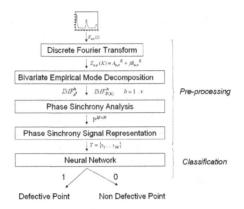

Fig. 1. Scheme of the procedure involved in the proposed approach

point of the inspecting material we have a continuous signal that represents the amount of received ultrasonic energy as a function of time.

The temporal evolution of the ultrasound signal $x_{u,v}(t)$ relative to the position (u, v) onto the plane associated to the upper surface of the material to be inspected is the input to the core of proposed approach that consists of two main steps: the pre-processing of the data in order to emphasize the differences between signals relative to the same class and its subsequent neural classification. The proposed ***pre-processing*** step includes multiple operative phases that are schematized in figure 1. In the pre-processing phase, each ultrasound signal $x_{u,v}(t)$ is, firstly, represented in term of its spectral components by using the *discrete Fourier transform (DFT)*. In this way the new signal representation $Z_{u,v}(k) = A_{u,v}(k) + jB_{u,v}(k)$ lies in the field of complex numbers and it is well suited for the next part of the analysis. The signal $Z_{u,v}(k)$ is decomposed into a set of complex IMFs, $\gamma_i(z)$ where $i = 1, \dots, M$, by a applying a complex extension of EMD, bivariate EMD [18]. The real and imaginary parts of the decomposition, $\Re\{\gamma_i(k)\}$ and $\Im\{\gamma_i(k)\}$, denote the IMFs for $A_{u,v}(k)$ and $B_{u,v}(k)$ respectively. By construction, the phase information for each IMF is well defined at each instant k and facilitates a highly localised comparison between the phase information of $A_{u,v}(k)$ and $B_{u,v}(k)$ [2]. The degree of phase synchrony, the temporal locking of phase information, between $\Re\{\gamma_i(k)\}$ and $\Im\{\gamma_i(k)\}$ is then determined to characterise the dynamics of $Z_{u,v}(k)$. The information is represented by a matrix $\rho_{u,v}(f, k)$ which denotes the synchrony at index k and frequency f.

The feature vector $Y_{u,v}(f) = y_1, y_2, \dots y_M$ relative to the considered $x_{u,v}(t)$ ultrasound signal is finally built by considering the integral of the phase synchrony values for each frequency bin, that is $Y_{u,v}(f) = \int_0^K \rho_{u,v}(f, k)dk$. The feature vector Y is, finally, normalized (zero mean and unit standard deviation) and it is given as input to a ***neural classifier*** trained with a *back propagation algorithm* to recognize defective from non defective areas.

The following sub-sections will describe the bivariate empirical mode decomposition strategy and the following phase synchrony methodology.

The EMD Algorithm. Empirical mode decomposition [16] is a data driven time-frequency technique which adaptively decomposes a signal, by means of a process called the sifting algorithm, into a finite set of AM/FM modulated components. These components, called "intrinsic mode functions" (IMFs), represent the oscillation modes embedded in the data. By definition, an IMF is a function for which the number of extrema and the number of zero crossings differ by at most one, and the mean of the two envelopes associated with the local maxima and local minima is approximately zero. The EMD algorithm decomposes the signal $x(t)$ as

$$x(t) = \sum_{i=1}^{M} C_i(t) + r(t) \tag{1}$$

where $C_i(t)$, $i = 1, \ldots, M$, are the IMFs and $r(t)$ is the residual. The first IMF is obtained as follows [16].

1. Let $\tilde{x}(t) = x(t)$;
2. Identify all local maxima and minima of $\tilde{x}(t)$;
3. Find an "envelope," $e_{min}(t)$ (resp. $e_{max}(t)$) that interpolates all local minima (resp. maxima);
4. Extract the "detail," $d(t) = \tilde{x}(t) - (1/2)(e_{min}(t) + e_{max}(t))$;
5. Let $\tilde{x}(t) = d(t)$ and go to step 2); repeat until $d(t)$ becomes an IMF.

Once the first IMF is obtained, the procedure is applied iteratively to the residual $r(t) = x(t) - d(t)$ to obtain all the IMFs. The extracted components satisfy so called monocomponent criteria and the Hilbert transform can be applied to each IMF separately. This way, it is possible to generate analytic signals, having an IMF as the real part and its Hilbert transform as the imaginary part, that is $x + j\mathcal{H}(x)$ where $\mathcal{H}(\cdot)$ is the Hilbert transform operator. Equation (1) can therefore be augmented to its analytic form given by

$$X(t) = \sum_{i=1}^{M} a_i(t) \cdot e^{j\theta_i(t)} \tag{2}$$

where the trend $r(t)$ is purposely omitted, due to its overwhelming power and lack of oscillatory behavior. Observe from (2), that now the time dependent amplitude $a_i(t)$ and phase function $\theta_i(t)$ can be extracted. By plotting the amplitude $a_i(t)$ versus time t and instantaneous frequency $f_i(t) = \frac{d\theta_i}{dk}$ [21], a time-frequency-amplitude representation of the entire signal is obtained, the so called Hilbert–Huang spectrum (HHS).

Complex Extensions of EMD. In order to obtain a set of M complex/bivariate IMFs, $\gamma_i(k)$, $i = 1, \ldots, M$, from a complex signal $z(k)$ using bivariate EMD, the following procedure is adopted [18]:

1. Let $\tilde{z}(k) = z(k)$;
2. To obtain Q signal projections, given by $\{p_{\theta_q}\}_{q=1}^Q$, project the complex signal $\tilde{z}(k)$, by using a unit complex number $e^{-J\theta_q}$, in the direction of θ_k, as

$$p_{\theta_q} = \Re\left(e^{-J\theta_q}\tilde{z}(k)\right), \quad q = 1, \dots, Q \tag{3}$$

   where $\Re(\cdot)$ denotes the real part of a complex number, and $\theta_q = 2q\pi/Q$;
3. Find the locations $\{t_j^q\}_{q=1}^Q$ corresponding to the maxima of $\{p_{\theta_q}\}_{q=1}^Q$;
4. Interpolate (using spline interpolation) between the maxima points $[t_j^q, \tilde{z}(t_j^q)]$, to obtain the envelope curves $\{e_{\theta_q}\}_{q=1}^Q$;
5. Obtain the mean of all the envelope curves, $m(k)$, and subtract from the input signal, that is, $d(k) = \tilde{z}(k) - m(k)$. Let $\tilde{z}(k) = d(k)$ and go to step 2); repeat until $d(k)$ becomes an IMF.

Similarly to real-valued EMD, once the first IMF is obtained, $\gamma_1(k)$, the procedure is applied iteratively to the residual $r(k) = z(k) - d(k)$ to obtain all the IMFs.

Once the IMFs have been obtained, the real and imaginary components can be treated as a two sets of IMFs, $\Re\{\gamma_i(k)\}$ and $\Im\{\gamma_i(k)\}$. The instantaneous amplitudes and phases for each set of IMFs can then be determined.

Phase Synchrony. To measure phase synchrony between x_1 and x_2, bivariate EMD is firstly applied to the complex signal $x_1 + jx_2$ [2]. The instantaneous amplitudes for the real and imaginary components of the decomposition, the $i = 1 \dots M$ IMFs at each time instant $k = 1 \dots K$, are denoted by $\Re\{a_i(k)\}$ and $\Im\{a_i(k)\}$ respectively. The instantaneous phase difference between each IMF component is given by $\psi_i(k)$. The degree of phase synchrony between x_1 and x_2 is given by [2]

$$\phi_i(k) = \frac{H_{max} - H}{H_{max}} \tag{4}$$

where $H = -\sum_{n=1}^N p_n \ln p_n$, the Shannon entropy of the distribution of $\psi_i(k - \frac{W}{2} : k + \frac{W}{2})$ defined by a window of length W, N is the number of bins and p_n is the probability of $\psi_i(k - \frac{W}{2} : k + \frac{W}{2})$ within the nth bin [23]. The maximum entropy H_{max} is given by

$$H_{max} = .626 + 0.4\ln(W - 1) \tag{5}$$

The value of ϕ is between 0 and 1, 1 indicating perfect synchrony and 0 a non-synchronous state. An additional step can be incorporated to model simultaneously for component relevance.

$$\phi_i(k) = \begin{cases} 0, \text{ if } \Re\{a_i(k)\} < \epsilon P_r \\ 0, \text{ if } \Im\{a_i(k)\} < \epsilon P_i \end{cases} \tag{6}$$

where P_r is the power of the original real component (similarly for P_i) and ϵ is an appropriate threshold. Once the phase synchrony information has been estimated, it can be conveniently plotted on a synchrony spectrogram, similar to plots produced in [22]. A synchrony spectrogram is essentially a standard time-frequency spectrogram with amplitude information replaced by values for phase coherence. The information can be represented by $\rho(f, k)$, which denotes the phase synchrony at index k and frequency f.

3   Experimental Setup

To illustrate the effectiveness of the proposed approach a set of ultrasound measurements on a composite material (referred to as the standard in the following) has been considered. The material has a Honeycomb structure with Nomex Core and 48 plies thicknesses. Ultrasonic data were obtained by an ultrasonic reflection technique that uses a single transducer serving as transmitter and receiver (5MHz). The standard contains artificial defects introduced during the manufacturing process and composed of the following materials: Brass Foil, Pressure Sensitive Tape, Dry Peel Ply, (in the following [A] stands for Tape, [F] for Peel Ply, [B] for Brass). In particular brass inserts were introduced to represent voids and delamination), dry peel ply to represent inclusions by means of reflection techniques and adhesive tape to represent inclusions by means of transmission techniques. For each standard, defects were positioned as reported in figure 2. The typical insert locations are:

1. Two plies from topside surface for Brass and Pressure Sensitive Tape and five plies for Peel Ply(top)
2. Mid part thickness (mid)
3. One and two plies from backside surface for Brass and Pressure Sensitive Tape and five plies from backside surface for Peel Ply (bottom)

The data set for the considered standard consists of 193x181 spatial samples. Each spatial sample $x_{u,v}(t)$ consists of 77 measurements of the received signal so that the whole data set consists of a matrix sized 193x181x77. In figure 3,

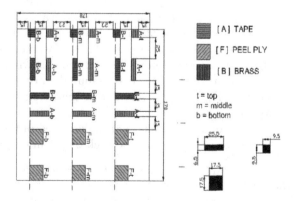

Fig. 2. Defect Scheme

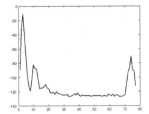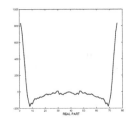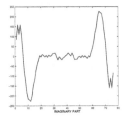

Fig. 3. On the left, an ultrasound signal in A-scan representation is reported. The central and left parts of the figure shows its complex representation after DFT computation.

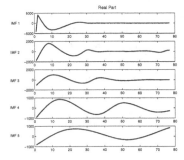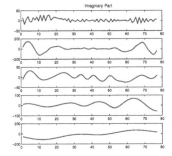

Fig. 4. The EMD decomposition of the DFT of an ultrasound signal: the IMFs relative to the real part of the ultrasound signal are plotted in blue, those relative to the imaginary part are plotted in red

a spatial sample $x_{u,v}(t)$ (i.e. an ultrasound signal in A-scan representation) and its complex representation after DFT computation are shown (real part is plotted in blue, imaginary part in red).

For each input signal, firstly, the DFT was computed and then a complex signal $Z_{u,v}(k) = A_{u,v}(k) + jB_{u,v}(k)$ of length 77 was obtained. The BEMD algorithm was then applied to this complex signal to produce a set of complex IMFs, $\gamma_i(k)$. The real part of this decomposition is shown in blue in figure 4, which denotes the IMFs obtained for the real part $A_{u,v}(k)$, whereas the imaginary component of the IMFs, those representing $B_{u,v}(k)$, are shown in red.

The phase synchrony matrix ρ of the IMFs computed for each complex signal[1] associated with the initial ultrasound signal was then computed obtaining a 250×66 sized matrix for each signal.

Finally the feature vector Y was computed as described in the above section and then a 250 long vector was associated to each ultrasound signal.

---

[1] In many instances, the IMFs were computed by averaging the decomposition of $Z_{u,v}(k) + v$ where v denotes complex white Gaussian noise of variance smaller than that of $Z_{u,v}(k)$. This enables the embedded data modes to be detected more accurately. For more information on noise assisted EMD, the reader is referred to [3].

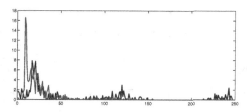

Fig. 5. The feature vector obtained (length 250) for two different signals belonging to a defective area (in blue) and non defective area (in red) respectively

In figure 5 the feature vectors Y for a non defective point (red line) and for a defective point (blue line) are shown.

The feature vectors Y containing phase synchrony patterns were finally supplied as the input to a supervised neural classifier that, for each vector gives a pair of values indicating the class of the considered point in the material. In particular the considered neural network is a 3 layers Back Propagation network with 80 hidden neurons (*tansig* activation functions) and 2 output neurons (*linear* activation functions). Defective points are associated with the output values [1,-1] whereas, for non defective points, [-1,1] is the expected output of the neural network. The net has been preliminarily trained using 407 point (207 belonging to defective areas and 200 belonging to non defective areas). In figure 6 the points selected to train the classifier are indicated (white pixels indicate non defective points, black pixels indicate defective points). As reported, the training points have been chosen from a subset of defective areas in order to evaluate the capability of the proposed approach to generalize knowledge on unseen defects.

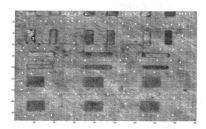

Fig. 6. The points used to train the neural classifier

4 Experimental Results

In this section the capability of the proposed framework to automatically detect defective areas in composite materials will be demonstrated and a comparison with standard techniques will be also provided. In particular three experiments have been carried out:

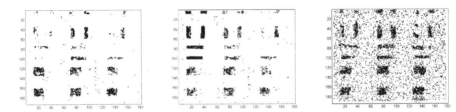

Fig. 7. Defective points detection by supplying as input to the neural network the signal acquired by the receiver (on the left), the wavelet coefficients (in the center) and the synchronization scores (on the right) extracted as described in section 2.

- the ultrasonic signals without any preprocessing have been classified by the neural network (in particular the 77 available samples for each point are given as input to the net after the normalization in the range [-1,1]));
- a classical pre-processing approach based on wavelet decomposition has been applied to the ultrasonic signal and then provided to the classifier (the DB3 family were used and the decomposition was carried out until level 3. The resulting 92 coefficients were then used to represent the ultrasound signal).
- the proposed preprocessing with EMD has been applied as described in the previous section and then the obtained feature vectors have been classified.

It should be noted that, in order to compare the effects that the different preprocessing techniques have on the final results, in all the three experiments, we have used the same training set, the same neural network and we have applied the same post-processing spatial filtering.

In figure 7 the results obtained in the three experiments are shown. In this figure we report the results obtained by the classifier using the same threshold to separate the defect and sound areas.

Figure 7 demonstrates that the wavelet pre-processing increases segmentation performance with respect the case of no pre-processing but it's clear that only the analysis of the synchrony based on complex extension of EMD and performed on DFT coefficients is able to reveal all the defective areas. At the same time, the EMD based approach, erroneously classified as defective a number of non defective points (false positive occurrences) and, moreover, some holes remained in the defective areas (false negative occurrences).

To partially overcome this problem a post-processing filtering has been applied on each of the above segmented images. In particular, dilation and erosion operations have been performed on the images by using the same 3 × 3 square structuring element [15]. After this step the segmentation results in the three considered cases became respectively those shown in figure 8.

The post-processing step both eliminated most of the false positive occurrences and filled most of the holes in the detected defective areas. After this final step it became much more evident the advantage of using EMD based preprocessing: all the defective regions have been detected and only three small regions have been miss-classified as defective ones. On the contrary the wavelet

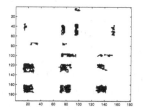 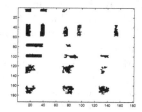 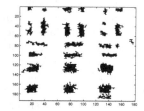

Fig. 8. The regions segmented as defective areas after a spatial filtering based on the connectivity analysis in the case of no preprocessing (on the left), wavelet pre-processing (in the center) and EMD based pre-processing

based pre-processing have not segmented 4 defective regions (practically half of those place on the bottom-side of the standard) whereas in the case of no-preprocessing the miss-detected regions have been much more. On the other side, in the case of wavelet based approach or no-preprocessing, the edge of the detected defective areas are better defined than in case of EMD based pre-processing and this could be very useful in the case where some geometrical description of the detected areas is required. In any case, considering that in the inspection context it is extremely important to detect all the defective areas, even if this could generate some false positive detection, it is possible to conclude that the EMD based pre-processing could be considered the best choice.

References

1. Boller, C., Chang, F.-K., Fujino, Y.: Encyclopedia of Structural Health Monitoring. John Wiley & Sons, Chichester (2009)
2. Looney, D., Park, C., Kidmose, P., Ungstrup, M., Mandic, D.P.: Measuring phase synchrony using complex extensions of EMD. In: IEEE/SP 15th Workshop on Statistical Signal Processing, pp. 49–52 (2009)
3. Wu, Z., Huang, N.E.: Ensemble Empirical Mode Decomposition: A noise-assisted data analysis method. Center for Ocean-Land-Atmosphere Studies, vol. 193 (2004)
4. Shah, A.A., Ribakov, Y., Hirose, S.: Nondestructive evaluation of damaged concrete using nonlinear ultrasonics. Materials and Design 30(3), 775–782 (2009)
5. Ruiz-Reyes, N., Vera-Candeas, P., Curpian-Alonso, J., Cuevas-Martýnez, J.C., Blanco-Claraco, J.L.: High-resolution pursuit for detecting flaw echoes close to the material surface in ultrasonic. NDT&E International 39, 487–492 (2006)
6. Ravanbod, H.: Application of neuro-fuzzy techniques in oil pipeline ultrasonic non-destructive testing. NDT&E International 38, 643–653 (2005)
7. Bettayeb, F., Rachedi, T., Benbartaoui, H.: An improved automated ultrasonic NDE system by wavelet and neuron networks. Proceedings of Ultrasonics International 42(1-9), 853–858 (2003)
8. Gil Pita, R., Vicen, R., Rosa, M., Jarabo, M.P., Vera, P., Curpian, J.: Ultrasonic flaw detection using radial basis function networks (RBFNs). Proceedings of Ultrasonics International 42(1-9), 361–365 (2003)
9. Margrave, F.W., Rigas, K., Bradley, D.A., Barrowcliffe, P.: The use of neural network in ultrasonic flaw detection. Measurement 25, 143–154 (1999)

10. Nath, S.K., Balasubramaniam, K., Krishnamurthy, C.V., Narayana, B.H.: An Ultrasonic Time of Flight Diffraction Technique for Characterization of Surface-Breaking Inclined Cracks. Materials Evaluation 67(2), 141–148 (2009)

11. Rizzo, P., Bartoli, I., Marzani, A., Lanza di Scalea, F.: Defect Classification in Pipes by Neural Networks Using Multiple Guided Ultrasonic Wave Features Extracted After Wavelet Processing. J. Pressure Vessel Technol. 127(3), 294–303 (2005)

12. Wagle, S., Katoa, H.: Ultrasonic wave intensity reflected from fretting fatigue cracks at bolt joints of aluminum alloy plates. NDT&E Int. 42(8), 690–695 (2009)

13. Meksen, T.M., Boudraa, B., Boudraa, M.: Defects Clustering using Kohonen Networks during Ultrasonic Inspection. IAENG International Journal of Computer Science 36(3) (2009) (online publication)

14. D'Orazio, T., Leo, M., Distante, A., Guaragnella, C., Pianese, V., Cavaccini, G.: Automatic Ultrasonic Inspection for Internal Defect Detection in Composite Materials. Independent NonDestructive Testing and Evaluation 41(2), 145–154 (2008)

15. Gonzales, R.C., Woods, R.E.: Digital Image Processing. Prentice-Hall, Englewood Cliffs (2008)

16. Huang, N.E., Shen, Z., Long, S.R., Wu, M.L., Shih, H.H., Quanan, Z., Yen, N.C., Tung, C.C., Liu, H.H.: The empirical mode decomposition and the Hilbert spectrum for nonlinear and non-stationary time series analysis. Proceedings of the Royal Society A 454, 903–995 (1998)

17. Tanaka, T., Mandic, D.P.: Complex Empirical Mode Decomposition. IEEE Signal Processing Letters 14(2), 101–104 (2007)

18. Rilling, G., Flandrin, P., Goncalves, P., Lilly, J.M.: Bivariate Empirical Mode Decomposition. IEEE Signal Processing Letters 14(12), 936–939 (2007)

19. Umair Bin Altaf, M., Gautama, T., Tanaka, T., Mandic, D.P.: Rotation Invariant Complex Empirical Mode Decomposition. In: Proceedings of the Int. Conf. on Acoustics, Speech and Signal Processing (ICASSP), vol. 3, pp. 1009–1012 (2007)

20. Looney, D., Mandic, D.: Multi-Scale Image Fusion using Complex Extensions of EMD. IEEE Transactions on Signal Processing 57(4), 1626–1630 (2009)

21. Cohen, L.: Instantaneous anything. In: Proc. of the IEEE International Conference on Acoustics, Speech and Signal Processing (ICASSP), vol. 5, pp. 105–108 (1993)

22. Sweeny-Reed, C.M., Nasuto, S.J.: A novel approach to the detection of synchronisation in EEG based on empirical mode decomposition. Journal of Computational Neuroscience 23(1), 79–111 (2007)

23. Tass, P., et al.: Detection of n:m Phase Locking from Noisy Data: Application to Magnetoencephalography. Physical Rev. Letters 81(15), 3291–3294 (1998)

Different Regions Identification in Composite Strain-Encoded (C-SENC) Images Using Machine Learning Techniques

Abdallah G. Motaal[1], Neamat El-Gayar[1,2], and Nael F. Osman[1,3]

[1] Center for Informatics Sciences, Nile University, Egypt
[2] Faculty of Computers and Information, Cairo University, 12613 Giza, Egypt
[3] Radiology Department, School of Medicine, Johns Hopkins University, USA
abdallah.motaal@nileu.edu.eg
{nelgayar,nosman}@nileuniversity.edu.eg

Abstract. Different heart tissue identification is important for therapeutic decision-making in patients with myocardial infarction (MI), this provides physicians with a better clinical decision-making tool. Composite Strain Encoding (C-SENC) is an MRI acquisition technique that is used to acquire cardiac tissue viability and contractility images. It combines the use of black-blood delayed-enhancement (DE) imaging to identify the infracted (dead) tissue inside the heart muscle and the ability to image myocardial deformation from the strain-encoding (SENC) imaging technique. In this work, various machine learning techniques are applied to identify the different heart tissues and the background regions in the C-SENC images. The proposed methods are tested using numerical simulations of the heart C-SENC images and real images of patients. The results show that the applied techniques are able to identify the different components of the image with a high accuracy.

1 Introduction

Imaging of the heart anatomy and function using magnetic resonance imaging (MRI) is an important diagnostic tool for heart diseases. Delayed-enhancement (DE) MRI after contrast agent administration is used to differentiate viable from infarcted myocardial tissue. Also, myocardial contractility pattern can be characterized from functional MR images. By combining the information from the viability and functional images, three different tissue types can be distinguished: 1) healthy myocardium; 2) infarcted myocardium; and 3) non-contracting, but viable, tissue, which could represent hibernating tissue [1]. The identification of the hibernating myocardium is very important as it is the tissue that will mostly benefit from revascularization [2]. Inversion-recovery imaging is considered the gold standard technique for acquiring DE MR Images [3]. The obtained image has T1-weighted contrast, and after the administration of the contrast agent, high signal intensity from infarcted myocardium is obtained. Strain-encoding (SENC) MR is used for directly imaging myocardial strain [4] where it is based on applying parallel planes of saturated magnetization to the cardiac tissue with initial tagging frequency, ω_o, which

F. Schwenker and N. El Gayar (Eds.): ANNPR 2010, LNAI 5998, pp. 231–240, 2010.
© Springer-Verlag Berlin Heidelberg 2010

depends on the slice thickness and slice profile in the z-direction. Two images are then acquired with different demodulation frequencies, in the z-direction, from which a functional image of the heart is obtained [4]. Composite SENC have been developed, where it shows both functional and viability information of the heart in single acquisition [1].

Previously, a method is proposed to identify different heart tissues from MRI C-SENC images using an unsupervised multi-stage fuzzy clustering technique. The method was based on sequential application of the fuzzy c-means (FCM) and iterative self-organizing data (ISODATA) clustering algorithms [5]. In a more recent work [6], a bayesian classifier is proposed to identify the background region (air), then the filtered tissue regions are classified into the different tissue types using fuzzy C-means clustering algorithm.

In this work, several classification and clustering techniques are proposed to identify the background and the different tissue types. Numerical simulations and real MR images of patients are used to validate the segmentation techniques, which show excellent results. The paper is organized as follows; section 2 briefly describes the C-SENC. Section 3 describes the proposed system. Section 4 describes the used data sets, explains the details of the experiments conducted, and the results are presented. Finally, the paper is discussed and concluded in sections 5.

2 Theory of C-SENC

The SENC technique is used to measure the local strain of deforming tissues. In SENC MRI, the magnetization of the object under test at point y and time t is modulated in the slice-selection direction with a sinusoidal pattern of a spatial frequency, $\omega_0(y, t)$—which is initially uniform everywhere. Because of the contraction of the LV, myocardial deformation occurs, and the tag pattern moves and undergoes deformation that makes the tissue's new frequency, $\omega(y, t)$, proportionally changing with the degree of deformation at the pixel y. The resulting image intensity at this pixel can be given by [4]:

$$I(y, t; \omega_T) = \int_{-\infty}^{\infty} M(y, z, t)S(z)e^{-j\omega_T z}dz , \qquad (1)$$

where y is the pixel location, ω_T is the tuning value, and $I(y, t; \omega_T)$ is the signal intensity. The equation shows that the signal intensity at a certain point and time is the integral in the slice-selection direction of the longitudinal magnetization multiplied by the encode phase factor over the slice profile $S(z)$. Using some simplifications [4], the resulting image is approximated to

$$I(y, t) = \rho(y, t)S(\omega_T - \omega(y, t)) , \qquad (2)$$

where $\rho(y,t)$ represents the proton density of the voxel, and $S(\omega)$ is the Fourier transform of the slice profile. ω_T is called the tuning frequency, which is determined during the image acquisition by an applied tuning gradient. It is noticed that the function $S(.)$ is shifted in proportion to the change in the tagging frequency, ω, which depends on the tissue deformation. Two images are acquired at two tuning frequencies, ω_a and ω_b, and from these two images we can estimate the local frequency of the slice using the following relation [4]:

$$\omega(y,t) = \frac{\omega_a |I(y, t; \omega_a)| + \omega_b |I(y, t; \omega_b)|}{I(y, t; \omega_a) + I(y, t; \omega_b)}, \tag{3}$$

where $I(y, t; \omega_a)$ and $I(y, t; \omega_b)$ are the images acquired at tuning values ω_a and ω_b respectively. Since the strain is the change in length per unit length,

$$\varepsilon(y,t) = \frac{\Delta L}{L} = \frac{L_t - L_i}{L_i} = \frac{\dfrac{1}{\omega_t} - \dfrac{1}{\omega_o}}{\dfrac{1}{\omega_o}}, \tag{4}$$

thus, the tissue strain at location y and time t can be estimated from [4]:

$$\varepsilon(y, t) = (\frac{\omega_o}{\omega_t} - 1) \times 100, \tag{5}$$

where ω_o is the initial tagging frequency and ω_t is the tags frequency at time t.

In C-SENC [1], the SENC pulse sequence is modified to acquire an image at the beginning of the acquisition with no-tuning to capture the signal from the recovering DC magnetization. Thus the first image differentiates between normal and infarcted myocardium because of their different longitudinal recovery rates after contrast agent injection, the second image, low-tune image, shows the static tissue, and the third image, high-tune image, shows the contracting tissue.

3   Proposed System

Several classifiers are used to classify the different image regions. First, an input vector, $x = [f_1, f_2, f_3, l]$, is constructed, where f_1 is a feature that represents the pixel intensity in the high-tune image, f_2 is a feature that represents the pixel intensity in the low-tune image, f_3 is a feature that represents the pixel intensity in the no-tune image, and l is the label of the current pixel, where $l \in \{$'healthy', 'infarcted, 'hibernating', 'background'$\}$ regions. Fig. 1 shows the input data.

f_1	f_2	f_3	l
H.T intensity	L.T intensity	N.T intensity	Label

Fig. 1. The input data vector

In case of using unsupervised learning clustering algorithm, simple K-means, a labeling stage is added after the clustering stage and the input data becomes 3-D only, as the fourth cell will be omitted. The labeling stage uses information from the average-pixel intensity in each identified cluster. The rules used in the labeling stage are usually verified by medical domain experts. Fig. 2 shows the proposed system.

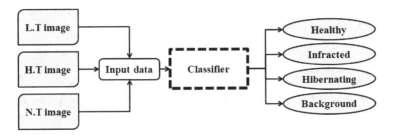

Fig. 2. The block diagram of the proposed system

4 Experimental Results

Simulated images with signal-to-noise ratio 21 dB are generated for C-SENC images. This data represents test data containing regions of healthy, infracted, hibernating tissues and background. The image is divided into four regions, where the upper quarter represents a healthy contracting tissue, the second quarter from above represents hibernating tissue, the third quarter from the above represents an infracted tissue, and the lower quarter represents the background region. This data is used to validate the proposed method. The imaging parameters used in simulation are: The slice selection is 10 mm, matrix size is 100 X 100, low and high tuning frequencies are 0.2 mm^{-1} and 0.3 mm^{-1}, respectively, and the initial tagging frequency is 0.21 mm^{-1}. Human volunteer is scanned on *Philips Acheiva* 3T MR scanner with same imaging parameters of the simulation. C-SENC images are acquired 10–15 min post contrast agent injection [7]. The simulated and the Real C-SENC images are used to construct the input data, that is fed into the classification block to classify the different components in the image. Fig. 3, and Fig. 4 show simulated and real C-SENC images.

4.1 Simulated Data Results

We have chosen 6 popular classifiers, multilayer perceptron (MLP), support vector machine (SVM), radial basis function (RBF), decision trees (DT), bayes classifier, and simple k-means clustering technique. All implementation are carried out using the

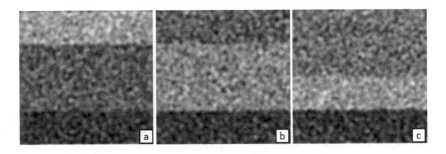

Fig. 3. Simulated (a) high-tune, (b) low-tune, (c) T1-weighted C-SENC image

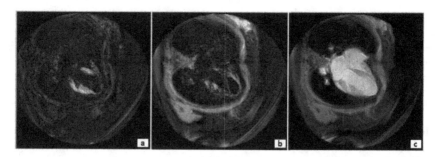

Fig. 4. Real (a) high-tune, (b) low-tune, (c) T1-weighted C-SENC image

WEKA library [8]. For all the supervised learning techniques, we use 10 fold cross validation and the default parameters that are set in WEKA. After constructing the input vector from the simulated images, we feed this data into the classifier block.

In case of MLP, the time taken to build the model is 43.27 seconds. Table 1. shows the confusion matrix. The correctly identified regions are 87.09% while, 12.91% are wrongly classified.

Table 1. The confusion matrix when using multilayer perceptron with 10 folds cross validation

	Healthy	Hibernating	Infarcted	Background
Healthy	2354	96	35	14
Hibernating	83	1953	438	27
Infarcted	39	516	1943	1
Background	4	35	3	2459

Support vector machine (SVM) is also used, where the time taken to build the model is 1.02 seconds. Table. 2 shows the confusion matrix. The correctly identified regions are 87.77% while, 12.23% are wrongly classified.

Table 2. The confusion matrix when using SVM with 10 folds cross validation

	Healthy	Hibernating	Infarcted	Background
Healthy	2379	77	32	11
Hibernating	94	1906	472	29
Infarcted	46	427	2025	1
Background	5	26	3	2467

J48 decision tree is also used, where the time taken to build the model is 0.61 seconds. The constructed tree size is 131 with 66 leaf nodes. Table. 3 shows the confusion matrix. The correctly identified regions are 87.02% while, 13.98% are wrongly classified.

Table 3. The confusion matrix when using J48 DT with 10 folds cross validation

	Healthy	Hibernating	Infarcted	Background
Healthy	**2324**	109	44	22
Hibernating	114	**1988**	384	15
Infarcted	50	497	**1950**	2
Background	25	23	2	**2451**

Radial basis function (RBF) is also used, where the time taken to build the model is 9.41 seconds. Table. 4 shows the confusion matrix. The correctly identified regions are 87.49% while, 12.51% are wrongly classified.

Table 4. The confusion matrix when using RBF with 10 folds cross validation

	Healthy	Hibernating	Infarcted	Background
Healthy	**2375**	80	34	10
Hibernating	92	**1904**	475	30
Infarcted	39	448	**2011**	1
Background	12	27	3	**2459**

Naive Bayes classifier is also used, where the time taken to build the model is 0.09 seconds. Table. 5 shows the confusion matrix. The correctly identified regions are 87.57% while, 12.43% are wrongly classified.

Table 5. The confusion matrix when using Bayes classifier with 10 folds cross validation

	Healthy	Hibernating	Infarcted	Background
Healthy	**2389**	73	28	9
Hibernating	106	**1902**	470	23
Infarcted	53	428	**2018**	0
Background	9	41	23	**2448**

Finally, unsupervised machine learning technique, simple k-means (SKM), is used where the Euclidean distance is used as a distance function. Table. 6 shows the confusion matrix. The correctly clustered regions are 84.44% while, 15.56% are wrongly clustered.

Table 6. The confusion matrix when using SKM

	Healthy	Hibernating	Infarcted	Background
Healthy	**2371**	84	27	17
Hibernating	108	**1876**	480	37
Infarcted	60	412	**2026**	1
Background	4	20	2	**2475**

In case of using the SKM algorithm, no time is needed to build a model. It is the fastest technique in the proposed techniques. But, as stated in section 3, labeling stage is needed when using unsupervised learning technique.

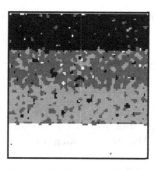

Fig. 5. The result after applying SKM on the simulated C-SENC images

Fig. 5 shows the result after applying SKM on the simulated C-SENC images. As shown in the figure it was able to cluster the 4 regions in the simulated images. Also, we can notice that there are misclustered pixels.

From the previous results, we notice that the majority in the wrongly classified or clustered regions is in the infracted and hibernating regions, and this is due to the near characteristics between those regions, since they are nearly have the same pattern in the low-and high tune images, and slightly differs in the no-tune image as shown in the simulated image.

After testing all the previous techniques, we found that SVM, Naive Bayes classifier, and SKM are the best classifiers in terms of time and accuracy. Although the MLP has a good accuracy but it is time consuming, also J48 DT has a good accuracy but the tree size is large.

4.2 Real MR Data Results

Real C-SENC images are acquired, and samples from different regions are manually selected to test the SVM, Naive Bayes and SKM classification techniques. Fig. 6 shows the different regions that are selected from the image.

For simplicity, 3 regions only are selected, healthy, infracted and background, as in real MR images, if there is a hibernating tissue, it just appear in a very fine region. And it is practically impossible to determine hibernating tissues visually.

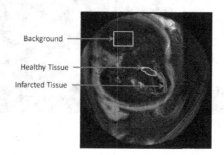

Fig. 6. Different selected regions represent healthy, infracted tissue and background

Table 7. The confusion matrix when using Bayes classifier with 10 folds cross validation

	Background	Healthy	Infarcted
Background	555	0	0
Healthy	0	168	47
Infarcted	0	8	256

Firstly, Naive Bayes algorithm with 10 fold cross validation is used. Table. 7 shows the confusion matrix. The correctly identified regions are 95.04% while, 4.95% are wrongly classified.

SVM with 10 fold cross validation is also used, where the time taken to build the model is 0.66 seconds. Table. 8 shows the confusion matrix. The correctly identified regions are 93.59% while, 6.4% are wrongly classified.

Table 8. The confusion matrix when using SVM with 10 folds cross validation

	Background	Healthy	Infarcted
Background	555	0	0
Healthy	0	163	48
Infarcted	10	8	246

Finally, simple k-means (SKM) is used and table. 9 shows the confusion matrix. The correctly clustered regions are 88.55% while, 11.45% are wrongly clustered.

Table 9. The confusion matrix when using SKM

	Background	Healthy	Infarcted
Background	555	0	0
Healthy	0	141	70
Infarcted	44	4	216

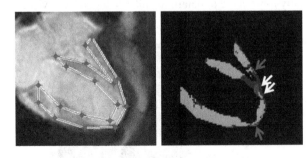

Fig. 7. The myocardium has been clustered to three different types of tissues

Fig. 7 shows a long-axis view of the heart where we define a region of interest in the heart muscle (myocardium). The myocardium is clustered to different groups, healthy, infracted and hibernating tissues. The white arrows point to potentially infracted regions, while the red arrows point to potentially hibernating regions. The healthy regions of the myocardium appear in green color.

5 Summary and Conclusion

In conclusion, in this work, we attempt to investigate machine learning techniques like MLP, SVM, Bayesian, RBF, DTs and SKM to identify the different heart tissues and background in C-SENC cardiac images. The proposed techniques allows for objectively identifying divergent heart tissues, which would be potentially important for clinical decision-making in patients with MI. The proposed method is tested using numerical simulations of the heart C-SENC images of MI and real images of patients. The results show that the proposed techniques are able to identify the different components of the image with a high accuracy. The proposed techniques vary in the computation time and accuracy. In our opinion using SVM, Bayes classifiers gives the best results, as they have high accuracy and low computation time. The misclassified instances is due to the poor SNR, 21 dB, however; recently new imaging techniques are proposed, where better quality images are obtained, which enhance the classification results.

Acknowledgment

This work is supported by PDP grant from ITIDA agency, Ministry of Communication and Information technology, Egypt. We would like to thank Eng. Esraa and Eng. Rana for their valuable discussion.

References

1. Ibrahim, E.-S.H., Stuber, M., Kraitchman, D.L., Weiss, R.G., Osman, N.F.: Combined Functional and Viability Cardiac MR Imaging in a Single Breathhold. Magn. Reson. Med. 58, 843–849 (2007)
2. Watzinger, N., Saeed, M., Wendland, M.F., Akbari, H., Lund, G., Higgins, C.B.: Myocardial viability: magnetic resonance assessment of functional reserve and tissue characterization. J. Cardiovasc. Magn. Reson. 3, 195–208 (2001)
3. Kim, R.J., Wu, E., Rafael, A., Chen, E., Parker, M.A., Simonetti, O., Klocke, F.J., Bonow, R.O., Judd, R.M.: The use of contrast-enhanced magnetic resonance imaging to identify reversible myocardial dysfunction. N. Engl. J. Med. 343, 1445–1453 (2000)
4. Osman, N.F., Sampath, S., Atalar, E., Prince, J.L.: Imaging longitudinal cardiac strain on short-axis images using strain-encoding MRI. Magn. Reson. Med. 46, 324–334 (2001)
5. Ibrahim, E.-S.H., Weiss, R.G., Stuber, M., Spooner, A.E., Osman, N.F.: Identification of Different Heart Tissues from MRI C-SENC Images Using an unsupervised Multi-Stage Fuzzy Clustering Technique. J. Magn. Reson. Imaging 28(2), 519–526 (2008)

6. Motaal, A.G., El Gayar, N., Osman, N.F.: Automated Cardiac-Tissue Identification in Composite Strain-Encoded (C-SENC) Images Using Fuzzy C-means and Bayesian Classifier. Accepted in 4th International Conference on Bioinformatics and Biomedical Engineering (iCBBE 2010), Chengdu, China (2010)
7. Sharma, P., Socolow, J., Patel, S., Pettigrew, R., Oshinski, J.: Effect of Gd-DTPA-BMA on blood and myocardial T1 at 1.5T and 3T in humans. Magn. Reson. Imaging 23, 323–330 (2006)
8. Witten, L.I., Frank, E.: Data Mining: Practical Machine Learning Tools and Techniques with Java Implmentations. Morgan Kaufmann, San Francisco (1999)

Exploiting Neural Networks to Enhance Trend Forecasting for Hotels Reservations

Athanasius Zakhary[1], Neamat El Gayar[1,2], and Sanaa El-Ola. H. Ahmed[1]

[1] Faculty of Computers and Information, Cairo University, 12613 Giza, Egypt
{athanasius.it,sana.ola}@gmail.com
[2] Center for Informatics Science, School of Communication and
Information Technology, Nile University, Giza, Egypt
nelgayar@nileuniversity.edu.eg

Abstract. Hotel revenue management is perceived as a managerial tool for room revenue maximization. A typical revenue management system contains two main components: Forecasting and Optimization. A forecasting component that gives accurate forecasts is a cornerstone in any revenue management system. It simply draws a good picture for the future demand. The output of the forecast component is then used for optimization and allocation in such a way that maximizes revenue. This shows how it is important to have a reliable and precise forecasting system. Neural Networks have been successful in forecasting in many fields. In this paper, we propose the use of NN to enhance the accuracy of a Simulation based Forecasting system, that was developed in an earlier work. In particular a neural network is used for modeling the *trend* component in the simulation based forecasting model. In the original model, *Holt's* technique was used to forecast the trend. In our experiments using real hotel data we demonstrate that the proposed neural network approach outperforms the Holt's technique. The proposed enhancement also resulted in better arrivals and occupancy forecasting when incorporated in the simulation based forecasting system.

1 Introduction

The importance of hotel arrivals and hotel occupancy forecasting stems from its need when designing a hotel revenue management system. Revenue management is the science of managing a limited amount of supply to maximize revenue by dynamically controlling the price/quantity offered [1] [2]. While cutting edge revenue management systems are currently in use by the airline industry, only recently has it been considered for the hotel industry. Because of the large number of existing hotels, the amount of possible total savings, if one implements optimized revenue management systems, is potentially large.

Revenue management forecasting methods fall into one of three types: Historical booking models, advanced booking models and combined models. Historical booking models only consider the final number of rooms or arrivals on a particular stay night. Advanced booking models only include the build up of reservations

F. Schwenker and N. El Gayar (Eds.): ANNPR 2010, LNAI 5998, pp. 241–251, 2010.
© Springer-Verlag Berlin Heidelberg 2010

over time for a particular stay night. Combined models use either regression or a weighted average of historical data and advanced booking models to develop forecasts. A review of forecasting methods for all three types is found in [3] and [6]. In this study, a particular interest is devoted to reservations data as they are very rich and contain very useful information indicating the actual demand to come.

A simple but efficient forecasting technique that is based on reservations data is the *Pickup* method. The *Pickup* forecasting model is a popular advanced booking model which exploits the unique characteristics of reservations data instead of relying only on complete arrival histories to make better forecasts. The main idea of using the pickup method is to estimate the *increments* of bookings (to come) and then aggregate these increments to obtain a forecast of total demand in the future [1]. *Pickup* is defined as the number of reservations picked up from a given point of time to a different point of time over the booking process [3]. A good comparison among the different variations of the pickup forecasting method and applying it on simulated hotel data can be found in [4].

In the theory of forecasting there have been two competing philosophies. The first one is based on developing an empirical formula that relates the value to be forecasted with the recent history (for example ARIMA-type or exponential smoothing models). The other approach focuses on developing a model from first principles that relates the value in question with the available variables/parameters etc, and simulate that model forward to obtain the forecast. Because the majority of real-world systems are either intractable or very complex to model, most forecasting applications follow the first approach. In contrast, Zakhary et al [5] present a forecasting application following the second approach, namely *Forecasting Hotel Arrivals and Occupancy Using Monte Carlo Simulation.*

In this paper, we propose an enhancement to the above mentioned model. In particular, a neural networks model is proposed to provide a multi-step ahead forecast for the *trend* component of the simulation model. In our experiments we demonstrate that the proposed neural network based model provides a more accurate estimate for the trend and results in better arrivals and occupancy forecasting when incorporated in the overall forecasting system. Our case study was based on real reservations data from Plaza hotel [1], Alexandria, Egypt.

The paper is organized as follows: Next section briefly describes the simulation based forecasting system, specifically the estimation of the system components. Section 3 reviews Neural Networks and its usage in forecasting. Section 4 presents the details of the proposed method for forecasting the trend. Section 5 shows the experiments and finally the paper is concluded in Section 6.

2   Forecasting Arrivals and Occupancy Using Simulation

This section briefly describes the simulation based forecasting model. For more details refer to [5]. The forecasting simulator consists of two main modules: *the*

---

[1] Plaza Hotel, Alexandria, Egypt is a 4 stars, Mid-size, business as well as sea-side hotel. http://www.plazaegypt.com/home.htm

analysis module and *the simulation module*. The analysis module takes as an input the historical reservation records. It then analyzes these data and uses it to extract many parameters and components like: trend, seasonality, booking curves, cancellations dynamics, length of stay, etc. Distributions of these components are deduced from the data directly. For every distribution or parameter, a suitable approach is devised to estimate it. For example, the seasonal index is estimated using the multiplicative seasonal decomposition [6]. The length of stay and room type distributions are estimated using a simple frequency based distribution estimator. The effects of the average daily reservations (i days prior to arrival) and the trend and seasonality are estimated by assuming that the net reservation rate $b(i)$ equals a normalized reservation rate $b_n(i)$ times a trend component $t_r(t)$ times a seasonal component $s(t)$, where t is the arrival date, as follows:

$$b(i) = b_n(i) * t_r(t) * s(t) \tag{1}$$

The reservation rates $b_n(i)$ are estimated by grouping the reservation curves to high, medium and low according to their corresponding seasonality indexes, and estimating a template for each seasonal index. The simulation module takes the parameters and components estimated by the analysis module as an input to generate forward reservation records that would take place in the future. Analyzing these generated reservation cases; one can obtain realistic predictions for occupancy, arrivals, and revenue in the future. One advantage of this simulation approach is that one obtains distributions of future key parameters (reservations, arrivals, etc). This is performed by running the simulator many times from the current (deterministic) starting point (determined by the current snapshot of existing reservations), thereby producing many paths, each obtained by the different random components of the future reservations process.

3 NN in Forecasting

A neural network (NN) is a semi-parametric model, inspired by how the brain processes information. It consists of a network of neurons (or nodes) that perform a weighted sum operation, followed by applying a nonlinear squashing function. Some of these nodes are hidden nodes that perform intermediate computations, and feed into the output node (which produces the final output). There are typically many free parameters in the network (called the weights). If the number of parameters or the number of hidden nodes is left unchecked, the network can overfit the data [7].

NNs can be useful for nonlinear processes that have an unknown functional relationship and as a result are difficult to fit [8]. The main idea with NNs is that inputs, or dependent variables, get filtered through one or more hidden layers each of which consist of hidden units, or nodes, before they reach the output variable. Next the intermediate output is related to the final output [9]. One major application area of NNs is forecasting. Refer to Kline et al [10]. for a good survey of the literature. A general problem with nonlinear models is the "curse of model complexity and model over parametrization".

NNs offers a natural alternative to traditional forecasting techniques. They have three great advantages over traditional forecasting methods:

1. They have universal approximation capabilities,
2. They can recognize "on their own" implicit dependencies and relationships in data,
3. They can learn to adapt their behavior (their prediction) to changed conditions quickly and without complication.

4   Proposed Trend Component Forecasting Using NN

As shown in Equation 1, the trend $t_r(t)$ is a key component in forecasting the reservations and then arrivals and occupancy. Enhancing the accuracy of estimating the trend will definitely lead to a more precise forecasting model. For this reason a forecasting model is applied for predicting the trend in the considered forecast horizon. Due to the importance of this component to the whole system, Artificial Neural Networks is used to forecast it.

Investigating the nature of the trend curve shown in Figure 1, it is obvious that it changes very frequently and abruptly. In addition, the training process in neural networks usually envolves intensive computations. For these reasons, it is better to simplify the forecasting problem. This could be done by converting

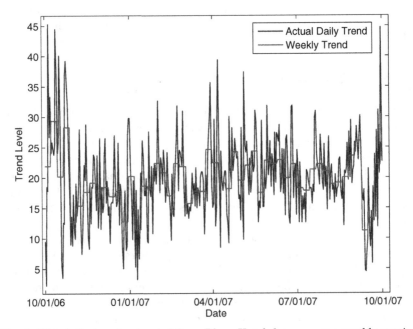

Fig. 1. The daily trend extracted from Plaza Hotel data versus a weekly version

the daily trend into a weekly trend by averaging the trend values of each week. In this case, the points to be forecasted will shrink by a factor of seven. For example, to forecast three months, we need to forecast 14 weekly points instead of 92 daily points. Each forecasted point is then considered a forecast for the corresponding week.

The proposed approach is to train a set of different neural networks for forecasting the future trend levels. These neural networks will vary in forecast horizon, number of inputs, formation of inputs, and the number of hidden nodes. For each forecast horizon, the best three neural networks are used in forecasting. The forecast value is the average of their forecasts. The proposed method can be described in steps as follows:

1. Prepare the inputs for Neural Networks
2. Adjust the NN parameters (Number of hidden units, Learning rate, ...)
3. Train the different NNs
4. Forecast using the best trained NNs

In the following subsections, we describe the sequence in more details.

4.1   Prepare Inputs

In multi step ahead forecasting using NN there are three well-known ways: the *iterative* approach, the *direct* approach, and the *parallel* approach [10]. In the *iterative* approach the model is trained on a one-step ahead basis. The trained model is then used to forecast one step ahead (i.e. one week ahead). The forecasted value is then fed as input to the model to forecast the subsequent point, and continue in this manner until the end of the forecast horizon. In the direct approach, a different network is used for every point in the future to be forecasted. For example, if the forecast horizon is 14 weeks, so 14 networks are built, each one forecasting a specific week in the future. When forecasting, these 14 networks are applied to obtain the forecast of the required forecast horizon. Of course, every network is trained separately. In the parallel approach, only one network is used with a number of outputs equal to the length of the horizon to be forecasted. The network is trained in a way so that output number k produces the kth step ahead forecast. In this work, we mainly focus on the *direct* approach. Neural networks are trained with a number of inputs that varies from two inputs to six inputs. Input/output pairs used in training can be built using different ways. As follows, we describe the different ways we suggest to build the input/output pairs:

− M-I: input lags are consecutive irrespective of the forecast horizon
− M-II: input lags are interleaved in proportion to step-ahead being forecasted
− M-III: consists of sampling in the same way as M-II. Instead of skipping time series values, averages are formed. The target output is also averaged
− M-IV: is same like M-III without averaging output
− M-V: is same like M-I , but one of the inputs is the average of eight consequent inputs. This is to account for the effect of longer term average

Given the time series $x(t-m), ..., x(t-3), x(t-2), x(t-1), x(t), x(t+1), x(t+2), x(t+3), ..., x(t+n)$. The generated input/output pairs for horizon=2 and number of lags =2 for this series using the different aforementioned methods are:

- M-I: $x(t-1), x(t) \Rightarrow x(t+2)$
- M-II: $x(t-2), x(t) \Rightarrow x(t+2)$
- M-III: $average(x(t-3), x(t-2)), average(x(t-1), x(t)) \Rightarrow average(x(t+1), x(t+2))$
- M-IV: $average(x(t-3), x(t-2)), average(x(t-1), x(t)) \Rightarrow x(t+2)$
- M-V: $average(x(t-8), ..., x(t-1)), x(t) \Rightarrow x(t+2)$

4.2   Adjust Parameters and Train Neural Networks

A five-fold validation procedure is used to select the best number of hidden nodes [11]. The following candidate values are considered for the number of hidden nodes: NH $= [0, 1, 3, 5]$. Note there is a possibility of having zero hidden nodes (NH $= 0$), meaning simply a linear model. Balkin and Ord [12] have shown that the possibility of switching to a linear model for some time series improved performance. Concerning the other less key parameters and model details, The sigmoid is used as activation functions for the hidden layer, and a linear for output layer. Training is performed using the Levenberg-Marquardt algorithm for 500 epochs, using a momentum term 0.2, and an adaptive learning rate with initial value 0.01, an increase step of 1.05 and a decrease step of 0.7.

Trend variable $s_{des}(t)$ is converted to weekly averages. Neural Networks are built with variations in forming training data (five variations) and number of lags (five variations). For each forecast horizon, the best three NNs are chosen.

4.3   Forecast Trend Using Trained NNs

For each forecast horizon, formulate the input according to the selected NN structure. Forecast the value of the required horizon in weeks. The output of each forecast horizon is the mean value of the three selected NN of that horizon. The next step is to convert the weekly forecasted points to daily forecasts. This is done by simply repeating each forecast seven times. The last step is to additively tune the level of the forecast. This is done by fitting a straight line on the past data and then getting the forecast of this linear regression for the interval to be forecasted. The mean value of the NN forecast is additively set to be that of the regressed forecast. This is done by subtracting the NN forecast mean and adding the linear regression mean.

Next section describes the results of applying the proposed NN model versus the Holt's technique used primarily in the simulator based forecasting system to forecast the trend obtained from Plaza Hotel for three intervals.

5   Experiments and Results

The proposed model is trained using the trend data of a complete year. The model is tested by forecasting the subsequent 3 months. Table 1 lists the 3

Table 1. The in-sample and the three months ahead forecast periods for the three forecasted snapshots

Snapshot No.	In-Sample Period	Forecast Period
1	1-Oct-2006 - 30-Sep-2007	1-Oct-2007 - 31-Dec-2007
2	1-Oct-2006 - 31-Oct-2007	1-Nov-2007 - 31-Jan-2008
3	1-Oct-2006 - 30-Nov-2007	1-Dec-2007 - 29-Feb-2008

different training and testing periods that were used in our experiment. We compare the performance of the proposed model to Holt's forecasting method. A detailed discussion to this technique can be found in [13]. We have chosen to compare the proposed NN Forecasting technique to Holt's forecasting method for two reasons:

1. The Holt's technique is used for forecasting the trend in our original simulation based forecasting system [5],
2. By definition Holt's is a suitable technique for forecasting data that have trend and no seasonality[13].

We use as error measure the symmetric mean absolute percentage error, defined as follows for both the the arrivals and occupancy time series:

$$SMAPE = \frac{1}{M} \sum_{m=1}^{M} \frac{|\hat{y}_m - y_m|}{(|\hat{y}_m| + |y_m|)/2} \tag{2}$$

where y_m and $\hat{y}_m$ are the actual time series value and the forecasted value respectively. Also, M is the total number of points that are forecasted.

Due to the limited length of the actual data (17 complete months), only three different forecast snapshots could be used. Thus we could not apply test of significance on the results. Figure 2 depicts the trend versus the NN forecast and Holt's forecast for the first interval. Table 2 compares the SMAPE error measures of the proposed model to the Holt's method. Analyzing the results, it is obvious that Neural Network forecasting method has outperformed Holt's methods.

In other cases, trend can go upward or downward. To test the performance of the NN model in such cases, an upward trend is generated by multiplying the original trend of plaza in a straight line with an upward trend. The model is trained using the modified trend. Forecast of the first interval against the modified trend is shown in Figure 3. Table 3 shows the corresponding SMAPE error measures. Similarly, trend has been modified with a downward trend and the results are shown in Figure 4 and Table 4. Investigating the results one can see that NN model has stable performance whether the trend is stationary, upward or downward. Finally in Table 5 we compare the overall forecast SMAPE for the Out-of-Sample periods for the original Monte Carlo Model [5] to the enhanced model that uses the proposed NN trend forecasting.

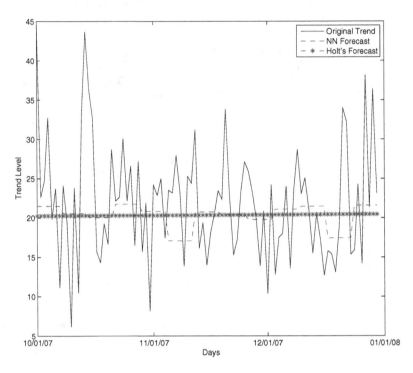

Fig. 2. The original trend component of Plaza Hotel versus its forecasts using the proposed NN model and Holt's in the 1st interval

Table 2. The SMAPE error measures of the proposed NN model versus Holt's for forecasting the trend of Plaza Hotel

	NN model	Holts
1st Interval	0.2574	0.2697
2nd Interval	0.2537	0.2589
3rd Interval	0.2625	0.2808
Average	0.2541	0.270867

Table 3. The SMAPE error measures of the proposed NN model versus and Holt's for forecasting an upward modified version of trend of Plaza Hotel

	NN model	Holts
1st Interval	0.2674	0.2729
2nd Interval	0.2424	0.2589
3rd Interval	0.2625	0.2808
Average	0.25743	0.2709

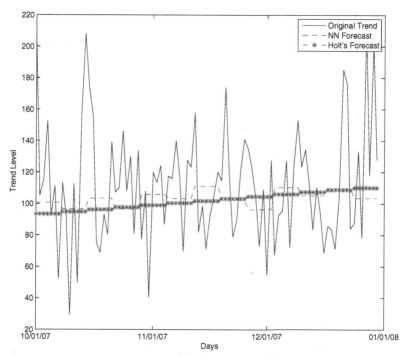

Fig. 3. The modified upward trend component of Plaza Hotel versus its forecasts using the proposed NN model and Holt's in the 1st interval

Table 4. The SMAPE error measures of the proposed NN model versus and Holt's for forecasting an downward modified version of trend of Plaza Hotel

	NN model	Holts
1st Interval	0.2737	0.2846
2nd Interval	0.2364	0.2482
3rd Interval	0.2786	0.2921
Average	0.2629	0.2750

Table 5. The Overall Forecast SMAPE for the Out-of-Sample Periods for the Original Monte Carlo Model versus Modified Model

	Arrivals	Occupancy
Original Monte Carlo Model	43.9	37.7
Modified Monte Carlo Model	42.21	37.1

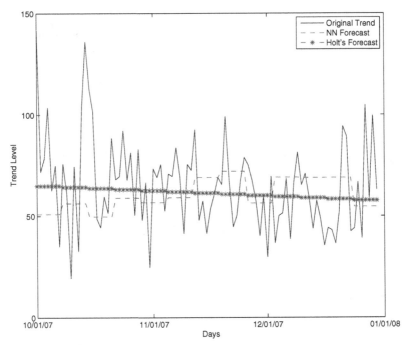

Fig. 4. The modified downward trend component of Plaza Hotel versus its forecasts using the proposed NN model and Holt's in the 1st interval

6   Conclusions

In this paper we propose an enhancement to a simulation based forecasting model previously developed. In particular we suggest using NNs to estimate the trend component, which is one of the key components that the forecasting model is based upon. In the original model, Holt's method is used for trend forecasting. Experimental results on real hotel data show that the use of NN increases the accuracy of trend forecasting by 5%. This improvement is also reflected in the overall arrivals and occupancy results of the overall simulation based forecasting system.

We believe that the gain from using NN for estimating the trend component in the forecasting system can be more significant in hotel environments in which the trend component is the main contributor to the reservation process. This is of particular importance to us since we plan to test and deploy the developed forecasting system in various hotel types (resort, city hotels, airport hotels, ...etc) that exhibit different reservation patterns.

Acknowledgment

This work is part of the "Cross Industry Data Mining" research project within the Egyptian Data Mining and Computer Modeling Center of Excellence. We would like to express our deep gratitude to Dr Hanan Kattara and "Plaza Hotel" family, Alexandria, Egypt for providing us with the reservation data to test and develop our models. We would like to acknowledge the useful discussions with Dr Amir F. Atiya of Dept Computer Engineering, Cairo University and Dr Hisham El-Shishiny of IBM Center for Advanced Studies in Cairo.

References

1. Talluri, K.T., Van Ryzin, G.J.: The Theory and Practice of Revenue Management. Springer Science+Buisness Media, Inc. (2005)
2. Ingold, A., McMahon-Beattie, U., Yeoman, I. (eds.): Yield Management. Continuum, 2nd edn. (2003)
3. Weatherford, L.R., Kimes, S.E.: A comparison of forecasting methods for hotel revenue management. International Journal of Forecasting 99, 401–415 (2003)
4. Zakhary, A., El Gayar, N., Atiya, A.F.: A comparative study of the pickup method and its variations using a simulated hotel reservation data. ICGST International Journal on Artificial Intelligence and Machine Learning 8, 15–21 (2008)
5. Zakhary, A., Atyia, A., El-Shishiny, H., El Gayar, N.: Forecasting hotel arrivals and occupancy using monte carlo simulation. Journal of Revenue and Pricing Management (to appear)
6. Frechtling, D.: Forecasting Tourism Demand: Methods and Strategies. Butterworth Heinemann, Oxford (2001)
7. Haykin, S.: Neural Networks: A Comprehensive Foundation. Prentice-Hall, Englewood Cliffs (1999)
8. Darbellay, G.A., Slama, M.: Forecasting the short-term demand for electricity: Do neural networks stand a better chance? International Journal of Forecasting 16, 71–83 (2000)
9. Gooijer, J.G.D., Hyndman, R.J.: 25 years of IIF time series forecasting: A selective review. Monash Econometrics and Business Statistics Working Papers 12/05, Monash University, Department of Econometrics and Business Statistics (2005)
10. Kline, D.M., Zhang, G.P.: Methods for multi-step time series forecasting with neural networks. Neural Networks for Business Forecasting, 226–250 (2004)
11. Kohavi, R.: A study of cross-validation and bootstrap for accuracy estimation and model selection. In: Proceedings International Joint Conference on Artificial Intelligence, IJCAI (1995)
12. Sandy, J.K.O., Balkin, D.: Automatic neural network modeling for univariate time series. International Journal of Forecasting 16(4), 509–515 (2000)
13. Gardner, E.S.: Exponential smoothing: The state of the art Part II. International Journal of Forecasting 22, 637–666 (2006)

VLSI Architecture of the Fuzzy Fingerprint Vault System

Sung Jin Lim[1], Seung-Hoon Chae[1], and Sung Bum Pan[1,2,*]

[1] Dept. of Information and Communication Engineering, Chosun Univ.,
375, Seosuk-dong Dong-gu, Gwangju, 501-759, Korea
[2] Dept. of Control, Instrumentation, and Robot Engineering, Chosun Univ.,
375, Seosuk-dong Dong-gu, Gwangju, 501-759, Korea
{gigasj83,ssuguly}@gmail.com, sbpan@chosun.ac.kr

Abstract. User authentication using fingerprint information provides convenience as well as strong security at the same time. However, serious problems
may cause if fingerprint information stored for user authentication is used
illegally by a different person since it cannot be changed freely as a password
due to a limited number of fingers. Recently, research in fuzzy fingerprint vault
system has been carried out actively to safely protect fingerprint information in
a fingerprint authentication system. In this paper, we propose hardware
architecture for a geometric hashing based fuzzy fingerprint vault system. The
proposed architecture consists of the software module and hardware module.
The hardware module performs the matching for the transformed minutiae in
the enrollment and verification hash table. We also propose a hardware
architecture which parallel processing technique is applied for high speed
processing.

Keywords: Fingerprint authentication, fuzzy vault, geometric hashing, fuzzy
fingerprint vault.

1 Introduction

The authentication system based on biometric information offers greater security and
convenience than the traditional methods of personal verification. The biometrics
such as fingerprint, iris, and voice has been received considerable attentions, which
refers the personal biological or behavioral characteristics used for verification or
identification. Since biometrics cannot be lost or forgotten like passwords, biometrics
has the potential to offer higher security and more convenience for the users. The
fingerprint is chosen as the biometrics for verification in this paper. Owing to their
uniqueness and immutability, fingerprints are today the most widely used biometric
features. If the biometric data are compromised, the user may quickly run out of the
biometric data to be used for authentication and cannot re-enroll[1-2]. Recently, study
on fuzzy fingerprint vault which fuzzy vault theory is applied to fingerprint
authentication has been carried out after Juels and Sudan proposed the fuzzy vault

---

* Corresponding author.

F. Schwenker and N. El Gayar (Eds.): ANNPR 2010, LNAI 5998, pp. 252–258, 2010.
© Springer-Verlag Berlin Heidelberg 2010

theory. Fuzzy fingerprint vault is a cryptology method which secret key and fingerprint information of the user are combined to obtain a secret key for only the right user. Real minutiae of the user is protected by creating a polynomial using the user's secret key and organizing the fingerprint template of the user with the user's real minutiae after creating the chaff minutiae randomly[3].

While study to simultaneously protect user fingerprint information and secret key of that user using fuzzy fingerprint vault is being reported, they cannot be realized because the alignment process was omitted due to absence of the fingerprint. To solve this problem, a method which geometric hashing technique is applied to the fuzzy fingerprint vault system was proposed[4-7]. The geometric hashing technique is an object authentication algorithm which object information is extracted and stored in a database and searched after being geometrically transformed[8].

In this paper, we propose hardware architecture for a fuzzy fingerprint vault system based on geometric hashing. The proposed architecture is performed by combining the software and hardware modules. The software module consists of modules for fingerprint minutiae information extraction, fingerprint template generation, fingerprint hash table generation and database storage. The hardware module consists of the matching module, verification module and memory to store the enrollment hash table and verification hash table. The matching module compares the transformed minutiae of the enrollment hash table and transformed minutiae of the verification hash table. The verification module takes the role of the calculation according to the result of the matching module. In addition, we propose a hardware architecture which parallel processing technique is applied for high speed processing of the fuzzy fingerprint vault system. Software module is identical to the previously proposed hardware architecture. The hardware module consists of the memory for storing the enrollment hash table and verification hash table, matching modules and the verification module.

The organization of the paper is as follows. Section 2 introduces fuzzy fingerprint vault based on the geometric hashing. Section 3 explains the hardware architecture for fuzzy fingerprint vault based on the geometric hashing, Section 4 shows the experimental results and Section 5 concludes.

2 Fuzzy Fingerprint Vault

Juels and Sudan proposed a scheme for crypto-biometric system called fuzzy vault. This is method which can protect the user's important secret key and biometric information using fuzzy concept. Clancy et al. proposed a fuzzy fingerprint vault based on the fuzzy vault of Juels and Sudan[4-5]. Using multiple minutiae location sets per finger, they first find the canonical positions of minutia, and use these as the elements of the set A. They added the maximum number of chaff minutiae to find R that locks. However, their system inherently assumes that fingerprints are pre-aligned. This is not a realistic assumption for fingerprint-based authentication schemes.

The architecture of the fuzzy fingerprint vault system of Chung et al. [6] consists of two processes: enrollment and verification processes as shown in Fig. 1. Enrollment process consists of minutiae information acquisition stage, enrollment hash table generation stage again. In minutiae information acquisition stage, minutiae information includes real minutiae of a user and chaff minutiae generated randomly.

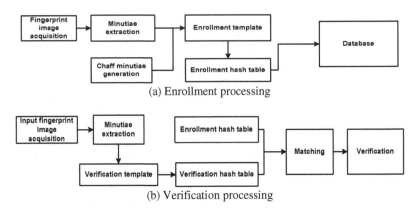

(a) Enrollment processing

(b) Verification processing

Fig. 1. Fuzzy fingerprint vault system

It is challenging to perform fingerprint verification with the protected template added by chaff minutiae. And then, Chung et al. applies modified geometric hashing. According to the geometric characteristics of the minutiae information, a table, called an enrollment hash table, is generated. Let $m_i = (x_i, y_i, \theta_i, t_i)$ represent a minutia and $L = \{m_i \mid 1 \leq i \leq r\}$ be a locking set including the real and chaff minutiae. In L, the real and chaff minutiae can be represented by $G = \{m_i \mid 1 \leq i \leq n\}$ and $C = \{m_i \mid n+1 \leq i \leq r\}$, respectively. Note that, the enrollment hash table is generated from L. In the enrollment hash table generation stage, an enrollment table is generated in such a way that no alignment is needed in the verification process for unlocking vault by using the geometric hashing technique. That is, alignment is pre-performed in the enrollment table generation stage.

After the enrollment process, the verification process to separate the chaff minutiae(C) from the real minutiae(G) in the enrollment minutiae table should be performed. In the verification process, minutiae information(unlocking set U) of a verification user is obtained and a table, called verification table, is generated according to the geometric characteristic of the minutiae. Then, the verification table is compared with the enrollment minutiae table, and the subset of real minutiae is finally selected. Note that, the verification table generation stage is performed in the same way as in the enrollment process. In comparing the enrollment and verification minutiae tables, the transformed minutiae pairs with the same coordinates, the same angle, and the same type are determined. The minutiae pairs having the maximum number and the same basis are selected as the subset of real minutiae(G). Also, any additional alignment process is not needed because pre-alignment with each minutia is executed in the enrollment and verification minutiae table generation stages.

3   Hardware Architecture of the Fuzzy Fingerprint Vault

To implement the hardware system of the fuzzy fingerprint vault, the proposed architecture was performed by integrating software and hardware modules as shown in Fig. 2. The enrollment processing for the fuzzy fingerprint vault system consists of steps for real minutiae extraction, chaff minutiae generation, fingerprint template generation,

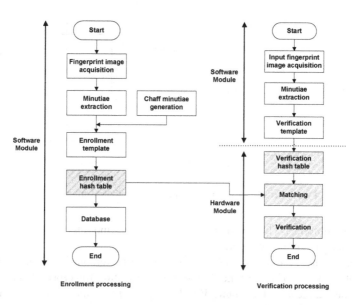

Fig. 2. Flow diagram of the fuzzy fingerprint vault system

fingerprint hash table generation and fingerprint database storage. The verification processing consists of steps for input fingerprint minutiae extraction, fingerprint hash table generation, matching and verification (Candidate list generation step).

The matching step and verification step of the verification processing is performed in hardware. Since the amount of computation of the matching step for the verification processing increases as the number of chaff minutiae increases in the enrollment processing, it is advisable for the matching step and verification step to be performed in hardware. On the other hand, the enrollment processing and fingerprint hash table generation step in the verification processing are performed in software. The proposed hardware module consists of two memories(enrollment and verification hash table), matching module and verification module as shown in Fig. 3. Hash table for each is organized of transformed minutiae. The enrollment fingerprint transformed minutiae is created by geometric transformation of user's real minutiae and chaff minutiae that was inserted to protect this in the enrollment processing. The verification fingerprint transformed minutiae is created by geometric transformation of minutiae of the fingerprint for verification.

The enrollment hash table can be expressed as $E = \{tr_i \mid 0 \leq i \leq r\text{-}1\}$ where r is the number of enrollment fingerprint templates. tr_i is the hash table which is created through geometric hashing after selecting m_i among the fingerprint template as the reference point and consists of r-1 transformed minutiae. The verification fingerprint transformed minutiae can be expressed as $V = \{tr_j \mid 0 \leq j \leq s\text{-}1\}$ where s is the number of verification fingerprint templates. The matching module consists of the Compare module and Count module. The Compare module is the module that compares the enrollment fingerprint transformed minutiae and verification fingerprint transformed minutiae and the Count module calculates the number of corresponding

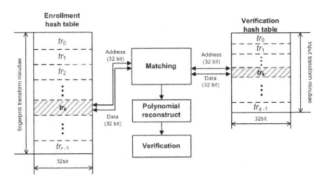

Fig. 3. Structural diagram of the proposed hardware module

transformed minutiae. The verification module is the module that aligns calculated similarity in high similarity order.

It assumes the number of enrollment fingerprint templates to be r, tr of the enrollment hash table E to be r, number of the verification fingerprint template to be s and tr of the verification hash table V to be s. First, tr_0 of the verification hash table and tr_0 of the enrollment hash table is input to the matching module. In the Compare module of the matching module, tr_0 of the verification hash table and tr_0 of the enrollment hash table are compared. After comparing the coordinate, angle and type of the transformed minutiae for the two tr_0 that were input, whether or not they match is sent to the Count module. The Count module calculates the number of transformed minutiae that match. Then, tr_1 of the enrollment hash table is input to the matching module and compared with the tr_0 of the verification hash table. After comparison of all tr in the enrollment hash table with tr_0 of the verification hash table are completed using the same method, tr_1 of the verification hash table is input to the matching module. Up to tr_{s-1} of the verification hash table is compared by executing this repeatedly. The verification module performed alignment of the calculated similarity. By using the number of corresponding transformed minutiae, similarity is measured and candidate list is generated in high similarity order.

In this paper, we also propose a hardware architecture which parallel processing technique is applied to reduce the matching time. Separation of the software and hardware modules is identical to the previously proposed hardware architecture. The proposed parallel processing hardware module consists of two memories storing the enrollment hash table and verification hash table, matching module and verification module as shown in Fig. 4. For parallel processing, the number of matching modules used in the hardware module is two. While the architecture of the matching module is similar to the previously proposed hardware architecture, it is different because the input enrollment fingerprint transformed minutiae is output after the comparison. The architecture of the verification module is identical to the previously proposed hardware architecture. When the number of the enrollment fingerprint templates is r and the number of the verification fingerprint templates is s, the enrollment hash table E consists of r and verification hash table V consists of s.

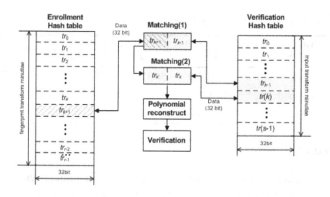

Fig. 4. Structural diagram of the proposed parallel processing technique hardware module

4 Experimental Results

The number of real minutiae that was used in the hardware architecture experiment for the fuzzy fingerprint vault system based geometric hashing proposed in this paper is between a maximum of 90 and minimum of 16. The experiment was performed by adding 100, 200, 300 and 400 chaff minutiae. The software module was realized by using C language in Visual C++ 6.0.

For hardware module implementation, a Spartan 3E starter board was used. The development board contains a Xilinx XC3S500E FPGA. The hardware module was designed by using VHDL in Xilinx ISE 9.2. Hardware simulation was performed in Moledmsim XE 6.0.

Table 1 shows the major resources that were used when the hardware architecture for the proposed fuzzy fingerprint vault and hardware architecture which the parallel processing technique was applied were implemented in the development board. Since the parallel processing hardware architecture has matching modules whose number equals that of the verification fingerprint transformed minutiae when there is one matching module in the proposed hardware architecture, the amount of resources that are used is large. It can be seen that the amount of hardware is about 56% higher for the number of slices, about 28% higher for the number of slice flip flops and 58% higher for the total number of 4 input LUTs in the parallel processing hardware architecture.

Execution time of the hardware module for the fuzzy fingerprint vault system was measured with number of chaff minutiae equal to 100, 200, 300 and 400. As shown in Table 2, real time processing is possible even when the number of chaff minutiae is increased to improve security in the proposed hardware architecture.

Table 1. Major resources when matching module is implemented in a Xilinx Spartan 3E FPGA

Number of matching module	Number of Slices	Number of Slices Flip Flop	Total Number of 4 input LUTs
1	419 out of 4,656(9%)	161 out of 9,312(2%)	668 out of 9,312(7%)
2	496 out of 4,656(10%)	246 out of 9,312(2%)	762 out of 9,312(8%)

Table 2. Required number of cycles according to the number of chaff minutiae

Chaff minutiae / Module	Number of chaff minutiae			
	100	200	300	400
software	30,128,840	56,376,516	75,087,761	110,009,410
1 matching module	10,023,288	12,027,946	18,763,596	26,461,481
2 matching module	8,173,713	9,808,456	15,301,191	21,578,603

5 Conclusion

While a user authentication system using fingerprint information provides convenience and strong security at the same time, serious problems may cause if the fingerprint information is used illegally or leaked. In this paper, we proposed hardware architecture for a geometric hashing based fuzzy fingerprint vault system. The matching module of the proposed hardware architecture was performed so that all transformed minutiae in the enrollment fingerprint hash table are matched with each transformed minutiae in the verification fingerprint hash table. In addition, the hardware architecture of the matching system which parallel processing technique was applied for high speed processing of the system organizes matching modules in number equal to the number of transformed minutiae in the input fingerprint hash table and matches them simultaneously. Execution time of the proposed system was 0.24 second for 36 real minutiae and 200 chaff minutiae and 0.53 second for 400 chaff minutiae. In addition, execution time for hardware architecture with 2 matching modules which parallel processing technique was applied was 0.18 second and 0.47 second respectively for the same condition. Based on the experimental result, it was verified that real-time fingerprint authentication is possible by using the hardware architecture of the proposed fuzzy fingerprint vault system and high speed processing is possible by applying parallel processing technique.

References

1. Maltoni, D., Maio, D., Jain, A.K., Prabhakar, S.: Handbook of Fingerprint Recognition. Springer, Heidelberg (2003)
2. Uludag, U., Jain, A.: Securing fingerprint template: Fuzzy vault with helper data. In: Conf. on Computer Vision and Pattern Recognition Workshop, pp. 163–172 (2006)
3. Juels, A., Sudan, M.: A fuzzy vault scheme. In: IEEE International Symposium on Information Theory, p. 408 (2002)
4. Clancy, T., Kiyavash, N., Lin, D.: Secure smartcard-based fingerprint authentication. ACM SIGMM Multim., Biom. Met. & App., 45–52 (2003)
5. Uludag, U., Pankanti, S., Jain, A.K.: Fuzzy vault for fingerprints. In: Kanade, T., Jain, A., Ratha, N.K. (eds.) AVBPA 2005. LNCS, vol. 3546, pp. 310–319. Springer, Heidelberg (2005)
6. Chung, Y., Moon, D., Lee, S., Jung, S., Kim, T., Ahn, D.: Automatic alignment of fingerprint features for fuzzy fingerprint vault. In: Feng, D., Lin, D., Yung, M. (eds.) CISC 2005. LNCS, vol. 3822, pp. 358–369. Springer, Heidelberg (2005)
7. Lee, S., Moon, D., Jung, S., Chung, Y.: Protecting secret keys with fuzzy fingerprint vault based on a 3D geometric hash table. In: Beliczynski, B., Dzielinski, A., Iwanowski, M., Ribeiro, B. (eds.) ICANNGA 2007. LNCS, vol. 4432, pp. 432–439. Springer, Heidelberg (2007)
8. Wolfson, H., Rigoutsos, I.: Geometric hashing: an overview. IEEE Computational Science and Engineering 4, 10–21 (1997)

Clustering Very Large Dissimilarity Data Sets

Barbara Hammer[1] and Alexander Hasenfuss[2]

[1] University of Bielefeld, CITEC, Germany
[2] Clausthal University of Technology,
Department of Computer Science, Germany

Abstract. Clustering and visualization constitute key issues in computer-supported data inspection, and a variety of promising tools exist for such tasks such as the self-organizing map (SOM) and variations thereof. Real life data, however, pose severe problems to standard data inspection: on the one hand, data are often represented by complex non-vectorial objects and standard methods for finite dimensional vectors in Euclidean space cannot be applied. On the other hand, very large data sets have to be dealt with, such that data do neither fit into main memory, nor more than one pass over the data is still affordable, i.e. standard methods can simply not be applied due to the sheer amount of data. We present two recent extensions of topographic mappings: relational clustering, which can deal with general proximity data given by pairwise distances, and patch processing, which can process streaming data of arbitrary size in patches. Together, an efficient linear time data inspection method for general dissimilarity data structures results. We present the theoretical background as well as applications to the areas of text and multimedia processing based on the generalized compression distance.

1   Introduction

The availability of electronic data increases dramatically in nearly every aspect of daily life, and it is estimated that the amount of electronic data doubles roughly every twenty months. At the same time, the quality and information content of modern data is constantly improving, and, often, data are represented in dedicated structures such as XML files, sequences, or graph structures. While both aspects allow the access to detailed and specialized information according to the current need of the user, this access is far from trivial and often buried in the sheer amount of information. Because of these problems automatic data mining tools play a major role to visualize and customize data such that humans can more easily access the relevant information at hand.

A multitude of different techniques has been developed in the context of interactive information visualization and data mining [23,33]. One very popular technique is offered by the self-organizing map (SOM) [18]: a lattice of neurons is arranged according to the data similarity such that clustering, browsing, and visualization become possible. This way, SOM can serve as data inspection tool or as preprocessing step for further information processing. In consequence, SOM

F. Schwenker and N. El Gayar (Eds.): ANNPR 2010, LNAI 5998, pp. 259–273, 2010.
© Springer-Verlag Berlin Heidelberg 2010

has successfully been applied to diverse areas such as robotics, telecommunication, web and text mining, bioinformatics, etc. [18].

With increasing quality and availability of electronic data, new challenges arise: data sets become larger and larger and, often, data are not available a priori, rather, data are stored in a distributed way and can be accessed only in sequential order. This property makes it infeasible to keep all data items in main memory at once or to sift through the data more than once. In consequence, many standard data mining methods cannot be applied in such situations, such as e.g. standard SOMs with online or batch training. Due to this fact, data mining for very large or streaming data has become one central issue of research. Several approaches have been proposed including extensions of k-means clustering and variants with approximation guarantees, heuristics which rely on sampling and according statistical guarantees or grid based methods, such as e.g. CURE, STING, and BIRCH, or iterative compression approaches which process only a fixed subset of the given data at a time [6,10,1,2,20,9,12,32,35].

Another challenge is the fact that more and more electronic data are stored together with dedicated structures such as XML files, temporal or spatial sequences, trees, or graph structures. These data are typically not represented as standard Euclidean vectors and a comparison of data by means of the Euclidean metric is often not appropriate. Rather, problem-adapted specific choices should be used such as alignment distances, tree or graph kernels, etc. Most classical statistical data mining tools such as SOM have been proposed for vectorial data and, hence, they are no longer applicable in such situations. A variety of methods which extend SOM to more general data structures has been proposed: Statistical interpretations of SOM as considered in [11,17,30,31] allow to change the generative model to more general data. The approaches are very flexible but computationally quite demanding. For specific data structures such as time series, recursive models have been proposed [3,14]. Online variants of SOM have been extended to general kernels [34]. However, these versions have been derived for (slow) online adaptation only. The approach [19] provides a fairly general method for large scale application of SOM to non-vectorial data: batch optimization is extended to general proximities by means of the generalized median. Thereby, prototype locations are restricted to data points which restricts the flexibility of the approach compared to the original setting.

Note that large data sets constitute a particular problem for SOMs for general data structures: if data are characterized by pairwise dissimilarities instead of Euclidean vectors, squared complexity is already necessary to store the relevant information in a dissimilarity matrix. In consequence, training algorithms display at least quadratic complexity which is infeasible for large data sets. Only few proposals address this problem such as [4,5]. Here, information is approximated by a small representative subset of the matrix and extended to the full data afterwards. The drawback of this approach is that a representative subset of the dissimilarity matrix has to be available prior to training, which is commonly not the case for streaming data without direct random access.

In this contribution, we present two ideas which extend SOMs to large dissimilarity data sets such that a method with linear time and constant memory results. On the one hand, we extend SOM to data given by pairwise dissimilarities in a way which is similar to relational fuzzy clustering as derived in [15,16]. We transfer these settings to SOM together with an investigation of the underlying mathematics for general data sets. On the other hand, we rely on extensions of clustering to large data sets by means of patch clustering as proposed in [1,10,6]. We extend this scheme to the relational approach resulting in a constant memory and linear time method. We demonstrate two applications to text files and symbolic musical data.

2   Relational Self-Organizing Maps

2.1   Standard Batch SOM and Corresponding Cost Function

Topographic maps constitute effective methods for data clustering, inspection, and preprocessing. Classical variants deal with vectorial data $x \in \mathbb{R}^n$ which are distributed according to an underlying distribution P in the Euclidean space. The goal of prototype-based clustering algorithms is to distribute prototypes $w^i \in \mathbb{R}^n$, $i = 1, \ldots, k$ among the data such that they represent the data as accurately as possible. A new data point x is assigned to the winner $I(x)$ which refers to the prototype with smallest averaged distance $\sum_{l=1}^{k} h_\lambda(\mathrm{nd}(i,l)) \cdot \|x - w^l\|^2$. Here, $h_\lambda(t) = \exp(-t/\lambda^2)$ denotes an exponential function with neighborhood λ and $\mathrm{nd}(i,l)$ denotes a priorly chosen neighborhood structure of neurons, often induced by a low dimensional lattice in a low-dimensional Euclidean or hyperbolic space, to achieve greater flexibility of the lattice structure [24]. Prototypes are adapted such that SOM optimizes the cost function [17]

$$E_{\mathrm{SOM}}(w) = \frac{1}{2} \sum_{i=1}^{k} \int \delta_{i,I(x)} \cdot \sum_{l=1}^{k} h_\lambda(\mathrm{nd}(i,l)) \cdot \|x - w^l\|^2 \, P(dx)$$

Often, this cost function is optimized by means of an online stochastic gradient descent. Alternatively, if data $x^1, \ldots, x^m$ are available priorly, batch optimization can be done. The corresponding discrete cost function is given by

$$E_{\mathrm{SOM}}(w, x) = \frac{1}{2} \sum_{i=1}^{k} \sum_{j=1}^{m} \delta_{i,I(x^j)} \cdot \sum_{l=1}^{k} h_\lambda(\mathrm{nd}(i,l)) \cdot \|x^j - w^l\|^2 .$$

This is optimized by an iterative optimization of assignments and prototypes until convergence in Batch SOM, see Algorithm 1. The neighborhood cooperation is usually annealed to $\lambda \to 0$ during training such that the quantization error is optimized in the limit of small neighborhood size. It has been shown in [8] that this procedure converges after a finite number of steps to a local optimum of the SOM cost function for fixed λ. In the following theoretical considerations, we will always assume a fixed and small neighborhood parameter λ. This consideration approximately corresponds to final stages of training.

Algorithm 1. Batch SOM

---

input

    data $\{x^1, \ldots, x^m\} \subset \mathbb{R}^n$;

begin

   init w^i randomly;

   repeat

      set $I(j) = \mathrm{argmin}_i \sum_{l=1}^{k} h_\lambda(\mathrm{nd}(i,l)) \cdot \|x^j - w^l\|^2$;

      set $w^i = \sum_j h_\lambda(\mathrm{nd}(I(j),i))x^j / \sum_j h_\lambda(\mathrm{nd}(I(j),i))$;

   until convergence;

   return w^i;

end.

2.2 Pseudo-Euclidean Embedding of Dissimilarity Data

In the following, we restrict the presentation to an overview of the key ingredients, omitting a few details and all proofs, which can be retrieved from [13]. Here, we deal with the application of SOM to settings where no Euclidean embedding of the data points is known. We will more generally deal with data points x^i which are characterized by pairwise dissimilarities $d_{ij} = d(x^i, x^j)$ which are symmetric and 0 if $x^i = x^j$. D denotes the corresponding matrix of dissimilarities. It can hold that no isometric Euclidean embedding can be found, i.e. SOM for Euclidean data is not applicable in this case. Discrete data such as DNA sequences or strings and alignment of those constitute one example.

For every matrix D, the so-called pseudo-Euclidean embedding exists, i.e. a vector space with symmetric bilinear form $\langle \cdot, \cdot \rangle$ (which need not be positive definite) and embeddings x^i of the points x^i can be found such that $d(x^i, x^j) = \langle x^i - x^j, x^i - x^j \rangle$ [26]. More precisely, define the centering matrix $J = I - 1/m \mathbf{1}\mathbf{1}^t$ with identity matrix I and the vector $\mathbf{1} = (1, \ldots, 1) \in \mathbb{R}^m$. Define the generalized Gram matrix associated to D as $G = -\frac{1}{2} \cdot JDJ$. Obviously, this matrix is symmetric and, thus, it can uniquely be decomposed into the form $G = Q\Lambda Q^t$ with orthonormal matrix Q and diagonal matrix of eigenvalues Λ with p positive and q negative entries. Taking the square root of Λ allows the alternative representation in the form

$$G = XI_{pq}X^t = Q|\Lambda|^{1/2} \begin{pmatrix} I_{pq} & 0 \\ 0 & 0 \end{pmatrix} |\Lambda|^{1/2}Q^t$$

where I_{pq} constitutes a diagonal matrix with p entries 1 and q entries -1, i.e. $X = Q_{p+q}|\Lambda_{p+q}|^{1/2}$ where only $p + q$ nonzero eigenvalues of Λ are taken into account. We can define the symmetric bilinear form in $\mathbb{R}^{p+q}$

$$\langle x, y \rangle_{pq} = \sum_{i=1}^{p} x_i y_i - \sum_{i=p+1}^{p+q} x_i y_i \, .$$

Then, the columns of X constitute vectors x^i with pairwise dissimilarities $d_{ij} = \langle x^i - x^j, x^i - x^j \rangle_{pq}$. This embedding is referred to as pseudo-Euclidean embedding of the data points. The values $(p, q, m - p - q)$ are referred to as signature.

Since the algorithm of Batch SOM (Algorithm 1) relies on vector operations and the computation of dissimilarities only, this algorithm can directly be used for arbitrary dissimilarity data embedded in pseudo-Euclidean space: given D, we can compute corresponding vectors x such that $\langle x^i, x^j \rangle_{pq} = d_{ij}$ holds for all data points. For these vectors and the corresponding dissimilarities, Batch SOM can directly be applied whereby the bilinear form $\langle x - w, x - w \rangle_{p,q}$ takes the role of the squared Euclidean norm $\|x - w\|^2$ in the algorithm.

However, there are two drawbacks of this naive procedure: on the one hand, we rely on the pseudo-Euclidean embedding the computation of which is costly (cubic complexity) and the interpretation of which is not clear. The prototypes are points in this pseudo-Euclidean space and probably depend on the chosen embedding, i.e. another vectorial embedding which shares the properties of the pseudo-Euclidean one could probably lead to fundamentally different prototypes. On the other hand, a connection of the SOM algorithm in pseudo-Euclidean space to the SOM cost function or, more desirable, a cost function which is independent of the concrete embedding is not clear as well as connected properties such as convergence of the algorithm.

Now we first reformulate batch SOM in pseudo-Euclidean space such that it becomes independent of the concrete pseudo-Euclidean embedding and it is based on the dissimilarity matrix D only. Afterwards, we discuss a connection to a cost function and questions concerning the convergence of the algorithm.

2.3   Relational SOM

The key observation to reformulate batch SOM independent of a concrete embedding of points consists in the fact that prototypes are located on special positions of the vector space only. Prototypes have the form

$$w^i = \frac{\sum_j h_\lambda(\mathrm{nd}(I(j), i)) x^j}{\sum_j h_\lambda(\mathrm{nd}(I(j), i))} = \sum_j \alpha_{ij} x^j$$

where $\sum_j \alpha_{ij} = 1$. Further, one can compute that for every linear combinations $\sum_j \alpha_{ij} x^j$ and $\sum_j \alpha'_{ij} x^j$ with $\sum_j \alpha_{ij} = 1$ and $\sum_j \alpha'_{ij} = 1$ and symmetric bilinear form $\langle \cdot, \cdot \rangle$ with $d_{ij} = \langle x^i - x^j, x^i - x^j \rangle$ the following is valid:

$$\left\langle \sum_i \alpha_i x^i - \sum_i \alpha'_i x^i, \sum_i \alpha_i x^i - \sum_i \alpha'_i x^i \right\rangle = (\alpha')^t D\alpha - \frac{1}{2} \cdot \alpha^t D\alpha - \frac{1}{2} \cdot (\alpha')^t D\alpha'$$

These observations immediately allow us to reformulate batch SOM in pseudo-Euclidean space independent of the concrete vector embedding of points. The key issues are to substitute prototype locations $w^i = \sum_{ij} \alpha_{ij} x^j$ by coefficient vectors $\alpha_i = (\alpha_{i1}, \ldots, \alpha_{im})$ and to substitute the computation of dissimilarities

---

Algorithm 2. Relational SOM

---

input

    symmetric dissimilarity matrix with zero diagonal $D \in \mathbb{R}^{m \times m}$;

begin

    init $\alpha_{ij} \geq 0$ such that $\sum_j \alpha_{ij} = 1$;

    repeat

        compute $\text{dist}_{ij} = [D\alpha_i]_j - \frac{1}{2} \cdot \alpha_i^t D\alpha_i$;

        set $I(j) = \operatorname{argmin}_i \sum_{l=1}^{k} h_\lambda(\text{nd}(i, l)) \cdot \text{dist}_{jl}$;

        set $\alpha_{ij} = h_\lambda(\text{nd}(I(j), i))) / \sum_j h_\lambda(\text{nd}(I(j), i)))$;

    until convergence;

    return α_{ij};

end.

---

of prototypes and data points by means of the above formula. The resulting algorithm, Relational SOM, is displayed as (Algorithm 2).

This algorithm is equivalent to vectorial SOM under the identity $\boldsymbol{w}^i = \sum_{ij} \alpha_{ij} \boldsymbol{x}^j$ whereby it does not rely on an explicit embedding but on the pairwise dissimilarities only. This shows that the winner assignments of SOM in pseudo-Euclidean space are independent of the concrete embedding.

2.4 Convergence

Since relational SOM is equivalent to batch SOM in pseudo-Euclidean space, it follows immediately that the algorithm converges towards a fixed point of the algorithm whenever the embedding is in fact Euclidean: in this situation, standard SOM is implicitly performed in the (Euclidean) space and the corresponding guarantees of the standard batch SOM as e.g. revisited in [8] hold. If the pseudo-Euclidean embedding is a non-Euclidean embedding, i.e. negative eigenvalues occur, the guarantees of standard batch SOM do not hold and one can indeed find situations where divergence of the algorithm is observed, i.e. the algorithm ends up in cycles. Batch SOM in Euclidean space relies on a subsequent optimization of the underlying cost function with respect to prototype locations and winner computations. Since an optimum is picked at both steps, the cost function decreases in every epoch in the Euclidean setting and because of the finite number of different possible winner assignments, convergence of the algorithm can be observed after a finite number of epochs.

For the non-Euclidean setting, it turns out that the subsequent computation of prototypes and winner assignments does not necessarily compute optima in the two steps. While it can easily be seen that the winner assignment as given in SOM is optimal, the choice of prototypes as generalized mean of the data points according to the winner assignment is not: it can happen that a saddle point

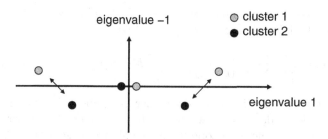

Fig. 1. Example of points in pseudo-Euclidean space for which relational clustering does not converge to a fixed point. It is indicated by arrows which points cyclically change their cluster assignments.

with worse cost function value instead of an optimum is picked if the negative directions of the bilinear form are dominant in this situation. Then an optimum would lie at the borders of the solution space since it would be beneficial to push prototypes away from the data with respect to the negative directions of the pseudo-Euclidean space. An efficient determination of the global optimum, however, is not possible because non-convex quadratic programming is NP hard [29,25]. Further, even if optima at the borders of the solution space could be computed easily in this setting, their usefulness can be doubted and numerical problems are likely to occur. Thus, the choice of the prototypes as generalized mean, which optimizes the problem only with respect to the positive directions of the pseudo-Euclidean space, seems a reasonable compromise in this situation.

Due to this fact, divergence can occur. Consider, for example, pseudo-Euclidean space with signature $(1,1,0)$ and the points as displayed in Fig. 1. For small enough neighborhood range λ, the cluster assignment of four points changes in every epoch as indicated by the arrows. Note, however, that this setting heavily relies on a large contribution of the negative axes to the bilinear form. In practice, the positive directions usually by far outweigh the negative directions such that this problem seems unlikely in practice: we did not observe divergence of the algorithm for real life data sets in a single experiment so far.

2.5   Dual Cost Function

The SOM cost function as introduced above constitutes a reasonable evaluation measure of batch SOM in pseudo-Euclidean space, albeit cases exist at least in theory where no local optimum of this cost function is actually found by batch SOM in this setting as discussed above. This cost function relies on the pseudo-Euclidean embedding and has no meaning if only the dissimilarity matrix D is available. Fortunately, one can substitute the SOM cost function by a so-called dual function for every fixed point of relational SOM which does not depend on the concrete pseudo-Euclidean embedding but which relies on D and the winner assignments only. More precisely, the following equality holds: assume a fixed point of relational SOM and batch SOM in pseudo Euclidean space, respectively, is found, characterized by winner assignments $I(j)$ and corresponding prototypes $\boldsymbol{w}^i$. Then, we find $E_{\text{SOM}}(\boldsymbol{w}, \boldsymbol{x}) = E_{\text{SOM}}^{\vee}(I, \boldsymbol{x})$ where

$$E_{\text{SOM}}^{\vee}(I, \boldsymbol{x}) := \sum_i \frac{1}{4\sum_j h_\lambda(\text{nd}(I(j), i))} \sum_{jj'} h_\lambda(\text{nd}(I(j), i))) h_\lambda(\text{nd}(I(j'), i))) d_{jj'}$$

The dual SOM cost function relies on winner assignments only and it measures the dissimilarities of data points within clusters as defined by the winner assignments, averaged over the prior topological neighborhood structure. As such, it is independent of a vectorial embedding of data but constitutes a valid cost function for clusterings based on dissimilarity data only. Since it coincides with the standard SOM cost function for fixed points of the algorithm, we can conclude that also the value of the standard batch SOM cost function is independent of the concrete vectorial embedding of data for fixed points of the algorithm. Further, for the euclidean case, also the overall structure of these two cost functions coincides in the sense that a one-one correspondence of global optima of the SOM cost function and its dual holds. For non-Euclidean settings, winner assignments can correspond to saddle points of the SOM cost function.

We would like to mention that the identification of the dual SOM cost function gives rise to another (at least theoretical) possibility to deal with non-Euclidean dissimilarities D. In analogy to relational k-means clustering as discussed e.g. in [28,21], the shape of the dual SOM cost function is approximately independent of constant shifts of off-diagonal elements of D for small neighborhood λ and common lattice structures nd. More precisely, if $\tilde{d}_{ij} = d_{ij} + d_0 \cdot (1 - \delta_{ij})$, d_0 denoting a positive constant and δ_{ij} the Kronecker delta, one can find for any fixed winner assignment I

$$E^{\vee}(I, \tilde{d}_{ij}) - E^{\vee}(I, d_{ij}) = \frac{1}{4}\left(\sum_{ij} h(I(j), i) - k\right)$$

where k denotes the number of prototypes. If we assume that the neighborhood function $\text{nd}(i, j)$ is nonnegative and it equals 0 if and only if $i = j$, then the term $\sum_{ij} h(I(j), i)$ converges to m for small λ, i.e. the difference of the two cost functions becomes a constant, and hence local optima can be observed at exactly the same places. Since I is discrete, this already holds for approximations of the limit setting.

Since every symmetric dissimilarity measure becomes Euclidean for large shifts of the off-diagonal elements, one could in principle first use the minimum shift which is necessary to make data Euclidean, and apply standard SOM in Euclidean space or its relational counterpart, afterwards, whereby convergence is guaranteed. Unfortunately, as demonstrated in [13], the numerics can become more difficult for shifted data, since differences in the dissimilarities are more and more annihilated such that relational SOM for non-Euclidean data can lead to better results than relational SOM for the corresponding shifted Euclidean data. Hence, the direct application of relational SOM might be advisable albeit convergence is not guaranteed in general.

3   Patch Processing

The time and space complexity of relational SOM constitutes one major drawback of the method: space complexity is linear, and time complexity is quadratic with respect to the number of data points (assuming a constant number of prototypes and epochs of the method). Note that the information which represents the data, the dissimilarity matrix D, itself is quadratic with respect to the number of data, i.e. every method which is linear necessarily has to pick only subsets of the full matrix D, neglecting possible relevant information of D.

Obviously, its complexity makes relational SOM unsuitable for large data sets. Therefore, we propose an approximative method which yields to an intuitive linear time and constant space algorithm, using patch clustering as introduced for Euclidean SOM in [1]. Our assumption is that single data points are available in consecutive order. Further, we assume the existence of a method to compute the pairwise dissimilarity of data on demand. These assumptions are quite reasonable if a very large (possibly distributed) data set is dealt with for which some problem specific dissimilarity measure is used. It is in general infeasible to precalculate the full dissimilarity matrix D in this case, such that methods which reduce the necessary information on demand to only a linear part of matrix D are favourable.

The main idea of the proposed method is to process data in patches which fit into main memory. Every patch is compressed by means of the resulting prototypes and their multiplicities resp. a limited-space approximation thereof. This information serves as additional input for the next patch computation such that all already seen data are implicitly taken into account in every step.

More precisely, we process m data points in n_p patches of priorly fixed size $p = m/n_p$. We assume divisibility of m by n_p for simplicity. A patch P_t is represented by the corresponding portion of the dissimilarity matrix D:

$$P_t = (d_{sl})_{s,l=(t-1)\cdot p+1,\ldots,t\cdot p} \in \mathbb{R}^{p \times p}$$

which represents the dissimilarities of points $(t-1)\cdot p + 1,\ldots,t\cdot p$. For every patch, the prototypes of the former patch are added with multiplicities according to the size of their receptive fields, and relational SOM is applied to this extended patch. Thereby, it is no longer possible to represent prototypes by means of the full coefficient vectors α_i since this would require linear space for storage and, in addition, computing the dissimilarity of prototypes and data points would eventually require the full (quadratic) dissimilarity matrix D - both demands are infeasible for large data sets. Therefore, we use an approximation of the prototype representation by a finite number K of coefficients with maximum contribution and (since these can be more than K) which are closest to the respective prototype. Formally, a K-approximation refers to the K indices j_1, $\ldots, j_K$ corresponding to points $x^{j_1}, \ldots, x^{j_K}$ with smallest dissimilarity to w^i.

This definition gives rise to a formalization of extended patches which are processed in turn in patch processing and which specify the parts of the matrix D necessary for the algorithm. Assume the current patch P_t is considered. Assume

N_{t-1} refers to the index set of the K-approximation of all prototypes obtained in the previous step. When considering k prototypes, the size of this set is $|N_{t-1}| = k \cdot K$. For the next round of patch processing, dissimilarity clustering is applied to the points corresponding to the indices in N_t and the data from the current patch, i.e. we need the following part of the dissimilarity matrix

$$P_i^* = \begin{pmatrix} d(N_{t-1}) & d(N_{t-1}, P_t) \\ d(N_{t-1}, P_t)^t & P_t \end{pmatrix}$$

where $d(N_{t-1}) = (d_{uv})_{u,v \in N_{t-1}}$ denotes the inter-dissimilarities of points from the K-approximation, and $d(N_{t-1}, P_t) = (D_{uv})_{u \in N_{t-1}, v=(t-1) \cdot p+1, \ldots, t \cdot p}$ denotes the dissimilarities of points in the K-approximation and the current patch. We refer to P_i^* as extended patches.

Based on these data handling techniques, patch relational SOM can be defined as iterative processing of patches enriched by the K-approximation of prototypes from the previous patch. The prototypes contribute to the new clustering task according to the sizes of their receptive fields R_i, i.e. a prototype w^i is counted with multiplicity $|R_i|$. Correspondingly, every point x^j in N_{t-1} contributes with multiplicity $m_j = |R_i|/K$. It is straightforward to extend relational SOM to deal with multiplicities m_j of point x^j. The only change concerns the update of the coefficients α_{ij}, which is enriched by the multiplicities. Taking this as subroutine, patch relational SOM can be performed as shown in Algorithm 3.

After processing, a set of prototypes together with a K-approximation is obtained which compresses the full data set. An inspection of prototypes is easily possible by looking at the points which are closest to these prototypes.

Note that the algorithm runs in constant space if the size p of the patches is chosen independently of the data set size m. Similarly, under this assumption, the fraction of the distance matrix which has to be computed for the procedure is of linear size $\mathcal{O}(m/p \cdot p) = \mathcal{O}(m)$ and the overall time complexity of patch clustering is of size $\mathcal{O}(m/p \cdot p^2) = \mathcal{O}(mp) = \mathcal{O}(m)$, assuming constant p. Hence, a linear time and constant space algorithm for general dissimilarity data results which is suited for large data sets, if constant patch size is taken according to the available space. Note that this algorithm, unlike alternatives such as [4,5], does not require a representative subpart of the matrix D prior to training, rather, data can be processed in the order as they are accessed, i.i.d. and non i.i.d. cases leading to virtually the same final result of the algorithm due to the incorporation of the statistics of all already seen data in every patch.

4   Experiments

We present two applications of relational SOM with underlying hyperbolic lattice structure [24]. In both cases, data are treated as symbolic strings, for which general dissimilarity measures such as alignment or n-gram kernels are available. In this contribution we rely on a dissimilarity measure which has been derived from information theoretic principles and which can be considered as an efficient

---

Algorithm 3. Patch Relational SOM

---

begin

    cut the first patch P_1;

    apply relational NG to $P_1 \rightarrow$ prototypes W_1;

    compute the K-approximation N_1 of W_1;

    update multiplicities m_i of N_1;

    set $t = 2$;

    repeat

        cut the next patch P_t;

        construct extended patch P_t^* using P_t and N_{t-1};

        set multiplicities of points in P_t to $m_i = 1$;

        apply relational NG with multiplicities to $P_t^* \rightarrow$ prototypes W_t;

        compute K-approximation N_t of W_t;

        update multiplicities m_i of N_t;

        $t \rightarrow t + 1$;

    until $t = n_P$

    return prototypes W_{n_P};

end.

---

approximation of a universal dissimilarity measure which minorizes every reasonable dissimilarity. The normalized compression distance (NCD) [7] is defined as

$$\mathrm{NCD}(x, y) = \frac{C(xy) - \min\{C(x), C(y)\}}{\max\{C(x), C(y)\}}$$

where x and y are strings, xy its concatenation, and $C(x)$ denotes the size of a compression of the string x. For our experiments, bzip2 was used.

4.1   Mapping MIDI Files

In this application, classical music represented in the MIDI format should be mapped. The data set consists of a selected subset of 2061 classical pieces from the *Kunst der Fuge* archive. To extract relevant strings which represent the MIDI files, a directed graph was extracted based on pitch differences and relative timings as intermediate step. As discussed in [22], the representation is widely invariant to shifts of the overall pitch and scalings of the overall tempo, thereby emphasizing on the melody underlying the musical piece and abstracting from the specific orchestration. The paths of the graph are then concatenated and represented in byte code, leading to a string representation suited for NCD. A relational hyperbolic SOM with 85 neurons was trained. Figure 2 shows the projection of the hyperbolic grid into the Euclidean plane. Obviously, the composers sharing a common style or epoch are mostly situated close to each other.

4.2   Mapping Newsgroups

The goal of this applications is to map 183,546 articles from 13 different news-groups in a semantically meaningful way by means of a relational HSOM. The texts were preprocessed by removing stop words and applying word stemming [27]. Patch processing was applied using relational HSOM with 85 neurons and a 3 approximation of the prototypes. Since the full dissimilarity matrix would occupy approx. 250 GB space, it is no option to process the data at once or precalculate the full dissimilarity at once. Instead, 183 patches of around 1000 texts were taken and the distances of these patches were precalculated, resulting in only around 12 MB space. In addition, since patch processing requires the dissimilarities of the extended patches, around $1000 \cdot 85 \cdot 3$ dissimilarities had

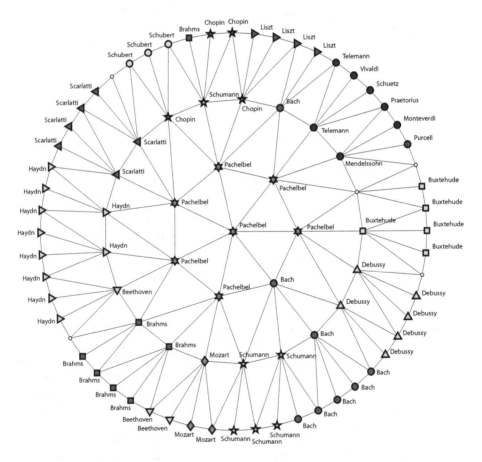

Fig. 2. Relational Hyperbolic Self-Organizing Map for 2061 classical pieces which are compared by symbolic preprocessing and NCD. Labeling of the neurons is based on majority vote.

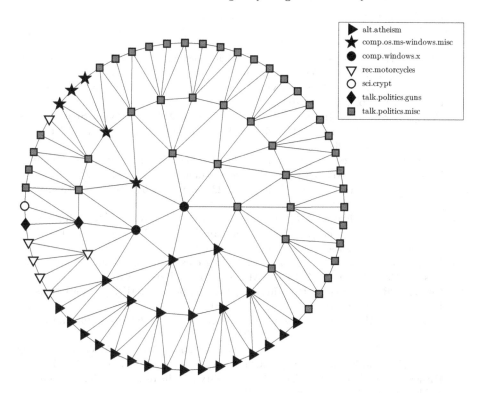

Fig. 3. Visualization of 183,546 newsgroup articles using patch relational HSOM, displaying the topographic arrangement of the biggest newsgroups on the map. In this case, patch processing reduces the required space of the full dissimilarity matrix from approx. 251 GB to only around 13 MB.

to be computed on demand for every patch. This way, the whole computation could be performed on a common desktop computer in reasonable time.

The outcome is depicted in Fig. 3. Clearly, the data arrange on the lattice according to their semantic meaning as mirrored by the corresponding newsgroup, such that visualization and browsing become possible.

5   Conclusions

We have presented an extension of SOM which is capable of mapping huge dissimilarity data sets in constant space and linear time such that visualization and data inspection become possible for the user. The main ingredients are an extension of SOM to a relational setting which corresponds to the implicit application of SOM in pseudo-Euclidean space, and a decomposition of the full clustering procedure into separate pieces by means of patch processing. Interestingly, the method is quite intuitive and can be linked to the standard SOM cost function or its dual, such that extensions such as supervision [13] are easily possible.

References

1. Alex, N., Hasenfuss, A., Hammer, B.: Patch clustering for massive data sets. Neurocomputing 72(7-9), 1455–1469 (2009)
2. Badoiu, M., Har-Peled, S., Indyk, P.: Approximate clustering via core-sets. In: Proc. STOC, pp. 250–257 (2002)
3. De, G., Barreto, A., Araujo, A.F.R., Kremer, S.C.: A Taxonomy for Spatiotemporal Connectionist Networks Revisited: The Unsupervised Case. Neural Computation 15(6), 1255–1320 (2003)
4. Belongie, S., Fowlkes, C., Chung, F., Malik, J.: Spectral partitioning with indefinite kernels using the Nyström extension. In: Heyden, A., Sparr, G., Nielsen, M., Johansen, P. (eds.) ECCV 2002. LNCS, vol. 2352, pp. 531–542. Springer, Heidelberg (2002)
5. Bezdek, J.C., Hathaway, R.J., Huband, J.M., Leckie, C., Kotagiri, R.: Approximate data mining in very large relational data. In: Dobbie, G., Bailey, J. (eds.) Proc. Australasian Database Conference, pp. 3–13 (2006)
6. Bradley, P.S., Fayyad, U., Reina, C.: Scaling clustering algorithms to large data sets. In: Proc. KDD, pp. 9–15. AAAI Press, Menlo Park (1998)
7. Cilibrasi, R., Vitanyi, M.B.: Clustering by compression. IEEE Transactions on Information Theory 51(4), 1523–1545 (2005)
8. Cottrell, M., Hammer, B., Hasenfuss, A., Villmann, T.: Batch and median neural gas. Neural Networks 19, 762–771 (2006)
9. Domingos, P., Hulten, G.: A General Method for Scaling Up Machine Learning Algorithms and its Application to Clustering. In: Proc. ICML, pp. 106–113 (2001)
10. Farnstrom, F., Lewis, J., Elkan, C.: Scalability for clustering algorithms revisited. SIGKDD Explorations 2(1), 51–57 (2000)
11. Graepel, T., Obermayer, K.: A stochastic self-organizing map for proximity data. Neural Computation 11, 139–155 (1999)
12. Guha, S., Rastogi, R., Shim, K.: CURE: an efficient clustering algorithm for large datasets. In: Proc. ACM SIGMOD Int. Conf. on Management of Data, pp. 73–84 (1998)
13. Hammer, B., Hasenfuss, A.: Topographic mapping of large dissimilarity data sets, Technical Report IFI-01-2010, Clausthal University of Technology (2010)
14. Hammer, B., Micheli, A., Sperduti, A., Strickert, M.: Recursive self-organizing network models. Neural Networks 17(8-9), 1061–1086 (2004)
15. Hathaway, R.J., Bezdek, J.C.: Nerf c-means: Non-Euclidean relational fuzzy clustering. Pattern Recognition 27(3), 429–437 (1994)
16. Hathaway, R.J., Davenport, J.W., Bezdek, J.C.: Relational duals of the c-means algorithms. Pattern Recognition 22, 205–212 (1989)
17. Heskes, T.: Self-organizing maps, vector quantization, and mixture modeling. IEEE TNN 12, 1299–1305 (2001)
18. Kohonen, T.: Self-Organizing Maps. Springer, Heidelberg (1995)
19. Kohonen, T., Somervuo, P.: How to make large self-organizing maps for non-vectorial data. Neural Networks 15, 945–952 (2002)
20. Kumar, A., Sabharwal, Y., Sen, S.: A simple linear time (1+epsilon)- approximation algorithm for k-means clustering in any dimensions. In: Proc. IEEE FOCS, pp. 454–462 (2004)
21. Laub, J., Roth, V., Buhmann, J.M., Müller, K.-R.: On the information and representation of non-Euclidean pairwise data. Pattern Recognition 39, 1815–1826 (2006)

22. Mokbel, B., Hasenfuss, A., Hammer, B.: Graph-based Representation of Symbolic Musical Data. In: Torsello, A., Escolano, F., Brun, L. (eds.) GbRPR 2009. LNCS, vol. 5534, pp. 42–51. Springer, Heidelberg (2009)

23. Nisbet, R., Elder, J., Miner, G.: Handbook of Statistical Analysis and Data Mining Applications. Academic Press/Elsevier (2009)

24. Ontrup, J., Ritter, H.: Hyperbolic self-organizing maps for semantic navigation. In: Dietterich, T., Becker, S., Ghahramani, Z. (eds.) Advances in Neural Information Processing Systems, vol. 14, pp. 1417–1424. MIT Press, Cambridge (2001)

25. Pardalos, P.M., Vavasis, S.A.: Quadratic programming with one negative eigenvalue is NP hard. Journal of Global Optimization 1, 15–22 (1991)

26. Pekalska, E., Duin, R.P.W.: The Dissimilarity Representation for Pattern Recognition – Foundations and Applications. World scientific, Singapore (2005)

27. Porter, M.F.: An algorithm for suffix stripping. Program 14(3), 130–137 (1980)

28. Roth, V., Laub, J., Kawanabe, M., Buhmann, J.M.: Optimal cluster preserving embedding of nonmetric proximity data. IEEE TPAMI 25(12), 1540–1551 (2003)

29. Sahni, S.: Computationally related problems. SIAM Journal on Computing 3(4), 262–279 (1974)

30. Seo, S., Obermayer, K.: Self-organizing maps and clustering methods for matrix data. Neural Networks 17, 1211–1230 (2004)

31. Tino, P., Kaban, A., Sun, Y.: A generative probabilistic approach to visualizing sets of symbolic sequences. In: Kohavi, R., Gehrke, J., DuMouchel, W., Ghosh, J. (eds.) Proc. KDD 2004, pp. 701–706. ACM Press, New York (2004)

32. Wang, W., Yang, J., Muntz, R.R.: STING: a statistical information grid approach to spatial data mining. In: Proc. VLDB, pp. 186–195 (1997)

33. Wong, P.C., Thomas, J.: Visual Analytics. IEEE Computer Graphics and Applications 24(5), 20–21 (2004)

34. Yin, H.: On the equivalence between kernel self-organising maps and self-organising mixture density network. Neural Networks 19(6), 780–784 (2006)

35. Zhang, T., Ramakrishnan, R., Livny, M.: BIRCH: an efficient data clustering method for very large databases. In: Proc. ACM SIGACT-SIGMOD-SIGART Symp. on Principles of Database Systems, pp. 103–114 (1996)

Author Index